Cyberpragmatics

Pragmatics & Beyond New Series (P&BNS)

Pragmatics & Beyond New Series is a continuation of Pragmatics & Beyond and its Companion Series. The New Series offers a selection of high quality work covering the full richness of Pragmatics as an interdisciplinary field, within language sciences.

For an overview of all books published in this series, please see *http://benjamins.com/catalog/pbns*

Volume 213

Cyberpragmatics. Internet-mediated communication in context
by Francisco Yus

Cyberpragmatics

Internet-mediated communication in context

Francisco Yus
University of Alicante

John Benjamins Publishing Company
Amsterdam / Philadelphia

∞™ The paper used in this publication meets the minimum requirements of American National Standard for Information Sciences – Permanence of Paper for Printed Library Materials, ANSI Z39.48-1984.

Library of Congress Cataloging-in-Publication Data

Yus, Francisco.

Cyberpragmatics : internet-mediated communication in context / Francisco Yus.

p. cm. (Pragmatics & Beyond New Series, ISSN 0922-842X ; v. 213)

Includes bibliographical references and index.

1. Pragmatics. 2. Digital communication. 3. Communication--Data processing. 4. Cognitive psychology. I. Title.

P99.4.P72Y87 2011

401'.450285--dc23 2011025494

ISBN 978 90 272 5619 5 (Hb ; alk. paper)

ISBN 978 90 272 8466 2 (Eb)

John Benjamins Publishing Co. · P.O. Box 36224 · 1020 ME Amsterdam · The Netherlands

John Benjamins North America · P.O. Box 27519 · Philadelphia PA 19118-0519 · USA

*This book is dedicated
to the memory of Enrique Alcaraz Varó,
an invaluable researcher, teacher and friend
who left us in 2008*

Acknowledgements

I would like to thank Luz Gil, Bryn Moody and Carmen Soler for their stylistic suggestions on a draft of this book.

Table of contents

CHAPTER 5

The virtual conversation 151

CHAPTER 6

You've got mail 219

Introduction

This book is the last stage in my proposal of a cognitive pragmatics analysis of Internet-mediated communication and interaction. For this specific approach to communication on the Net, I coined the term *cyberpragmatics* (Yus 2001a, 2010b). In short, *cyberpragmatics* aims at applying pragmatics to Internet users' interactions, specifically cognitive pragmatics and, within that, relevance theory, which has proved to be useful for the explanation of face-to-face communication and also of asynchronous communication, as happens with literature (Sperber & Wilson 1986, 1995). Today's Internet-mediated communication typically involves massive exchanges of messages of a written, audio-visual and multimodal quality, and most of them with an oral connotation. This is why typed texts often appear to be hybrids between the stability and rigidity of the written (i.e. typed) text, on the one hand, and the spontaneity and ephemeral quality of speech, on the other.

Although this book adopts an explicitly pragmatic approach, it also mentions other studies on Internet communication insofar as they shed light on the quality of interactions on the Net. But the central theoretical framework will be cognitive pragmatics and specifically relevance theory, as I have already pointed out. Hence, throughout the book there is an underlying certainty that although the Internet might exhibit attributes and strategies that are inherent in this medium, all of them can ultimately be explained within cognitive pragmatics. Communication is a human ability, a human resource and there is no difference between interpreting the messages that we are sent in physical contexts and doing so in virtual scenarios. The only thing that varies is the way communication is achieved, together with the means that human beings have devised to engage in interactions. With the Internet, geographical limitations and lack of physical co-presence are overcome.

The Internet is also a broad field of research with a wide range of research options. In this book several of them are addressed, specifically those that, one way or another, have to do with user-to-user communication on the Net. The book starts with a chapter in which several pragmatic assumptions are commented upon and specifically the theoretical foundations of relevance theory.

The second chapter analyses social gatherings on the Internet and how users *present* themselves, in Goffman's sense, in virtual settings. Terms such as *virtual community* will be analysed and related to their physical counterparts, but this apparently dichotomous approach to community (physical or virtual) is then revised and a more realistic proposal of hybrid personal networks of interactions is proposed (Yus 2007b), with a mixture of physical and virtual properties. This mixture fits today's increasing tendency to spend an important part of our lives in virtual scenarios. It also affects the notion of identity, which is shaped in different kinds of physical-virtual groupings and interactive environments.

Other forms of self-presentation on the Internet will also be studied in this chapter, for example the traditional personal web page and the nickname (*nick*), the *alter ego* of the real person actually typing on the keyboard while interacting on the Net.

The third chapter addresses the quality of the processing of information on web pages or websites. Ever since Tim Berners-Lee devised the "html language" (hypertext markup language)[1] and its implementation on the web page, most of the communication that takes place on the Internet is "html-based." Initially, when the Internet was not so popular, there were different applications for retrieving information and surfing the Net. Nowadays, on the contrary, most Internet resources and interactive applications are located on websites, including social networking sites, library catalogues, chat rooms, online games, *cybernewspapers*, etc.

The chapter starts with the application of terms such as intentionality, cognitive environments and mutual manifestness (typical in relevance-theoretic research) to the apparently uni-directional flow of communication on web pages between authors and readers. Next, I will review relevance-related studies that focus on users' satisfaction obtained from the information that they retrieve from search engines such as *Google*. Since this is a "software-user" type of communication, it is not an area that should be covered by a pragmatics of human communication (user-to-user), within which *cyberpragmatics* is included. However, several studies fitting relevance theory have addressed users' inferences on the output of these search engines, and these can lead to interesting conclusions on how users obtain positive rewards from the output of the search.[2] This fact justifies the inclusion of a heading in this book.

Later in the chapter a more genuine aspect of *cyberpragmatic* research will be analysed: the (ir)relevance of users' communication using websites, with headings such as the role of "sender users" and "addressee users," the levels or patterns of interactivity, the management of the vast amount of information available on the Net and the parallel danger of *infoxication* (mental *intoxication* due to an excess of

1. Together with his colleagues, he also developed the HTTP protocol (HyperText Transfer Protocol) and the Uniform Resource Locator for websites.

2. See Yus (2000a), thematic section 12, for a list of these relevance-theoretic studies of search engines and information retrieval systems in general.

information to be processed), which often reduces relevance due to a loss of interest provoked by increased mental effort. In this sense, it will be shown how the web page normally demands some kind of reinterpretation of the two variables on which the estimation of relevance depends, because of this massive availability of information that the Internet offers.

Lastly, in this third chapter there is a section devoted to the term *usability* and its conceptualization from cognitive pragmatics, and two additional sections on the transference of offline (i.e. printed) discourses to the Internet and the pragmatic consequences of the adaptation of these discourses to the specificity of the Net (screen size, link-mediated structure, scrolling of text, etc.). Specifically, I comment on printed newspapers turned into *cybernewspapers* and printed advertisements turned into *banners* and *pop-up ads* (among other discursive formats).

Chapter four examines forms of interaction that are relatively new in the history of Internet communication but have had tremendous impact on the way people interact on the Internet, that is, the ones under the label of *Web 2.0*: blogs, social networking sites (such as *MySpace*, *Facebook* or *Tuenti*) and the short-messaging microblog *Twitter*, which combines typical features of SMS (short messages of up to 140 characters) and a social networking orientation.

The fifth chapter is devoted to virtual synchronous conversations. The chapter starts with an analysis of chat rooms and instant messaging (i.e. *Messenger*) and their place in the oral-written and visual-verbal dichotomies. Their most interesting feature is what I call *oralized written text*, that is, the use of several strategies to turn typed text into a more expressive and speech-connoted kind of discourse that allows for the communication, to a certain extent, of the nonverbal behaviour (vocal and visual) that typically accompanies human interactions in situations of physical co-presence and makes it possible to convey not only thoughts, but also the feelings and emotions attached or associated with them in a more effective way.

Chapter five also covers the extension of virtual conversations to virtual worlds where users interact with one another by means of 3D avatars that exhibit a certain ability to convey information nonverbally. Among all the virtual worlds available nowadays, I focus on *Second Life*. This heading is preceded by an analysis of several proposals to enrich the interactions in chat rooms and instant messaging contextually. Finally, a short heading is devoted to videoconferencing, probably one of the most contextualized means of communication on the Internet nowadays.

Chapter six deals with one of the oldest kinds of Internet communication: electronic mail (e-mail). The chapter is divided into two main sections, one devoted to describing the main features of e-mail, sometimes difficult to place in the oral/written dichotomy; and a second section in which the main elements of an electronic message (sender, addressee, subject line…) are studied with special emphasis on their role in the eventual (un)successful interpretation.

Chapter seven is devoted to politeness on the Internet. A review of different theories that deal with this topic is proposed and then applied, where possible, to a medium typically prone to exaltation and unrestricted expression of emotions (due to the lack of physical co-presence of the interlocutors) that often lead to rudeness and impoliteness.

Finally, chapter eight suggests future research within *cyberpragmatics*. It takes into account today's constant technological advances that facilitate human communication and interaction, as happens, for instance, with the mobile phone. In fact, many of these future research issues include the increasingly important role that the mobile phone plays today for people accessing the Net and the different ways in which Net discourses are produced and interpreted on the computer screen and on the small screen of the mobile phone, with interesting pragmatic consequences.

CHAPTER 1

Pragmatics, context and relevance

1. Pragmatics and the use of language

Since prehistoric times, we humans have been fascinated by our ability to use words and transfer our thoughts to other people. In an attempt to understand the special qualities of this gift of language, compared to the sounds produced by animals,[1] we have always reflected on language, how it is learned, which part of the brain is in charge of producing and interpreting language, and so on.

But it was not until the 20th century when this interest in language, now called linguistics, acquired a truly scientific status. Saussure's pioneering research that gave birth to *Structuralism* and Chomsky's *Generative Grammar* placed linguistics on the right track towards the maturity that it exhibits nowadays.

Regardless of this label of *science* that linguistics deserves, it should be stressed that human language is such a complex phenomenon that in the development of linguistics a number of branches, schools, or perspectives have appeared, which deal with different aspects of language, and often overlap to some extent. Hence, utterance (1) would arouse the interest of linguists according to their different linguistic perspectives, who would draw different conclusions:

(1) The cat is on the mat.

Among many other approaches inside linguistics, a LEXICOLOGIST would analyse the semantic fields of *cat* and *mat* and their intersections or overlappings with similar terms like lynx, rug, feline or carpet. By contrast, a linguist specializing in SEMANTICS would be more interested in sentence organization, the referents of the words and how they provide a context-free sense to the whole sentence

---

1. Indeed, the linguist Charles Hockett proposed in 1960 to restrict the term "human language" only to vocal signs with an arbitrary relationship with their referents (i.e. words), leaving aside nonverbal behaviour such as paralanguage of kinesics. Nowadays, by contrast, many studies have stressed the importance of these apparently *marginal* aspects of human communication. In Chapter 5 of this book, for example, I will comment on the problems that Internet users face when they try to compensate, in text-based conversations, all the nonverbal connotations that people normally communicate in face-to-face situations and are absent in this kind of interaction on the Net (see Yus 2005a).

(cat – member of feline family, domestic, common, has claws, purrs ...). Finally, a linguist interested in GENERATIVE GRAMMAR would analyse the rules that human beings abide by in the production and interpretation of this grammatically correct sentence.

However, until pragmatics appeared as an alternative approach to linguistic analysis, few linguists had been interested in the meaning of utterance (1), now as (2b), in situation (2a), and with the intended interpretation (2c):

(2) a. [*Peter and Mary own a cat which has the habit of sitting on its mat so that its owner knows that it is hungry*].
b. Peter (to Mary): "The cat is on the mat."
c. The cat is hungry (go feed the cat, please).

The main contribution of pragmatics is, precisely, the certainty that it is impossible to analyse language outside the context in which it is produced and interpreted. Actually, human beings are rarely (if ever) truly literal when we speak; quite on the contrary, we tend to leave implicit all the information that we guess our interlocutors should be able to obtain by themselves. In other words, speakers base the interpretation that they intend to produce with the utterance (the so-called *speaker meaning*, opposed to the zero-context *sentence meaning*) on the interlocutors' ability to access the necessary information that will enable them to interpret the utterance adequately. An utterance such as (2b) will be misunderstood if Mary is unable to access the contextual information (2a) and derive the interpretation (2c) (she might think that she is merely being informed of the location of the cat). Therefore, pragmatics aims at explaining the role of context in interpretations like (2c) and also, for example, in the incorrect interpretation of utterance (3b) in situation (3a) by a child who, wrongly, chooses the literal interpretation of her mother's utterance. Pragmatics is also interested in why the answers (4b–e) to the question (4a) are perfectly valid and acceptable although they do not answer the question explicitly:

(3) a. [*A child entering a house*].
b. Mother: "Wipe your feet, please."
c. [*The boy takes off his shoes, full of mud, and carefully rubs his feet against the rug*]. (Peccei 1999: 1)

(4) a. Where is my box of chocolates?
b. I was hungry.
c. I've got to catch a train.
d. Where is your slimming plan?
e. The children were in this room this morning.
(Smith & Wilson 1983: 163)

In short, pragmatics is undoubtedly the branch of linguistics that has proved to be most suitable for the analysis of everyday communication, partly because of the lack of interest in context of previous paradigms[2] such as *Structuralism* or *Generative Grammar* (see Yus 1997a: 17–18, 2003a). Specifically, in the study of Internet-mediated communication cognitive pragmatics and, more particularly, Sperber & Wilson's relevance theory, can be useful when explaining the role that context plays in the eventual quality of interpretations, which is also the main objective of *cyberpragmatics*.

However, although pragmatics underlines the importance of context in human communication, this initial movement has evolved into a more diversified approach to language and, inevitably, to an array of pragmatic branches or schools that somehow gives the impression of a certain lack of homogeneity within this *paradigm* (Nuyts 1987), to the extent that it is even difficult to define pragmatics. Levinson (1983), for example, devoted an entire chapter to this task without arriving at a wholly satisfactory definition. Some possible definitions include "the study of factors that govern our choice of language in social interaction and the effects of our choice on others" (Crystal 1987: 120), "the study of how utterances have meanings in situations" (Leech 1983: x), and "the study of how more gets communicated than is said" (Yule 1996: 3).

2. Sperber and Wilson's relevance theory

In this book several pragmatic aspects of Internet-mediated communication are analysed. As the main theoretical framework, I will use Sperber & Wilson's (henceforth S&W) relevance theory (1986, 1995), a cognitive theory that has made a major contribution to our understanding of how we produce and interpret language. As its name indicates, the underlying hypothesis in this theory is that interpreting *stimuli* (verbal or nonverbal communicative acts) is subject to an inherently human search for relevance in the information that we process, an aspect that is rooted in human psychology. Under subsequent headings some important tenets of the theory will be commented upon. These will be used later in the description of users' communicative behaviour on the Internet.[3]

2. I am using *paradigm* in the sense proposed by Thomas Kuhn, that is, as theoretical models that provide scientific foundations and strategies that researchers within the paradigm invariably follow (Kuhn 1975, see Alcaraz Varó 1990).

3. For general comments on this theory see Blakemore (1992), Carston & Powell (2005), Escandell Vidal (1996a), S&W (1987), Vicente (1999), W&S (2002a, 2002b) and Yus (1996a, 1997a: 79–136, 1998a, 2003a, 2006a, 2010a). Bibliographical references of studies that use this theoretical model can be found in Yus (1998b, 2000a).

2.1 The code model versus the inferential model

Traditionally, communication has been regarded as an almost automatic coding-decoding activity, the so-called *code model* in which the words uttered by the speaker are interpreted in a machine-like way by the interlocutor.[4] Needless to say, communication is a much more complicated task and is subject to inferential hypotheses on whose validity there is no guarantee or certainty, what S&W call the *inferential model*. S&W do not deny the existence of coding in communication (we do use words to communicate our thoughts), but they locate it only at the initial stages of interpretation, and argue that mere coding is insufficient for understanding utterances and, consequently, must be combined with inference. In Wilson's (1994: 47) words, an inferential approach is needed because

> utterance interpretation is not a simple matter of decoding, but a fallible process of hypothesis formation and evaluation [...] Because of mismatches in memory and perceptual systems, the hearer may overlook a hypothesis that the speaker thought would be highly salient, or notice a hypothesis that the speaker had overlooked. Misunderstandings occur.[5] The aim of a theory of communication is to identify the principles underlying the hearer's (fallible) choices.

In short, any type of utterance interpretation, either in physical settings (offline communication) or on the Internet, involves an initial stage of decoding of the words (said, typed) and a second stage of turning the schematic identification of these words into fully contextualized propositions that match the sender's intended interpretation.

2.2 Ostension and intention

S&W assume the conceptualization of communication as grounded in intentionality, an approach that can be traced back to researchers such as Grice or Strawson, according to whom it is important to identify the speaker's intention to communicate for a correct interpretation of the utterance. In this sense, S&W propose two kinds of intention, the *informative intention* (the intention to inform the interlocutor of something) and the *communicative intention* (the intention to

4. In Sperber's (1994: 181) words, according to the *code model* "failures of communication occur when encoding or decoding isn't done properly, or when noise damages the sound signal, or, more significantly, when the codes of the interlocutors are not properly matched [...] If this explanation is correct, then the ability to communicate linguistically shouldn't be described as intelligent at all."

5. See Yus (1998c, 1998d, 1999a, 1999b) for an application of this theory to misunderstandings.

alert the interlocutor to this informative intention). With the latter, the speaker can effectively draw the interlocutor's attention and direct it to his/her intentions. That is, not only the intention to inform is *manifest*, but *mutually manifest* to both interlocutors (see next heading). In this case, the stimulus acquires an *ostensive* quality. In this sense, linguistic communication is very useful because whenever a person talks to us we immediately identify at least his/her communicative intention.

A consequence of this picture of communication is that only the interactions that satisfy both types of intention are worth analysing by pragmatics, while accidental transmissions of information (the ones in which the speaker simply gives off or *exudes* information without an intention[6] to communicate) are outside its scope of analysis. For example, in the similar situations (5) and (6), only the latter would deserve pragmatic analysis, even though in both of them inferential operations are performed in order to make sense of the other person's behaviour:

(5) [*Tom is walking along a street that leads to the railway station and sees his friend Peter, who is carrying a suitcase. While walking fast, he looks at his watch, worried. Tom infers correctly that Peter is hurrying to catch a train*].

(6) [*Tom is walking along a street that leads to the railway station and sees his friend Peter, who is carrying a suitcase. Tom crosses the street and waves at him. Peter waves back and, while looking at him, he points at his watch, with a worried expression. Tom infers correctly that Peter is trying to communicate that he is hurrying to catch a train and has no time to talk to him*].

2.3 Manifestness. Cognitive environments

People construct different concepts and representations of the world, just as their personal experiences are different. S&W call this array of information *cognitive environments*, which are formed in the following way: the facts about the world are manifest to a person only if this person is capable of representing them mentally and accepting their representations as valid (S&W 1987:699). The sum of all facts that are manifest to a person makes up his/her *cognitive environment*. The *total cognitive environment* of individuals consists not only of the facts that they know, but also of all the facts that they are capable of knowing at a specific time and place (ibid.).

6. In Yus (1997b, 1998e) a pragmatic *verbal-visual model of communication* was proposed, made up of sixteen cases, and half of them were devoted to the analysis of (non)verbal behaviours that transfer information with no underlying communicative intention.

S&W propose an extension of the term cognitive environment to assumptions (weaker than facts that are known) and introduce the notion of *degrees of manifestness*, since every time we identify a stimulus, some assumptions about it are more accessible than others. Needless to say, information that is simply manifest to individuals is weaker than the information that they know and of which they already have a mental representation (S&W 1986: 40). Besides, since manifest information is weaker than known information, they propose the notion of *mutual manifestness*, which replaces the more traditional term *mutual knowledge*. According to S&W, *mutual manifestness* does not have the same limitations as *mutual knowledge*.

In a nutshell, during conversations the interlocutors are exposed to a great deal of contextual information of a physical and conceptual quality. In this sense, S&W propose the term *mutual cognitive environment* for the amount of information that is manifest to both interlocutors in a specific situation. Inside this cognitive environment, the information that both interlocutors are aware that they share is called *mutually manifest assumptions*. Therefore, communication is basically an attempt to make certain information ("a set of assumptions" in relevance-theoretic terminology) mutually manifest to both interlocutors. This applies to any kind of communication including Internet-mediated communication.

2.4 (Non-demonstrative) inference and deduction

In the course of interpreting an utterance, hearers (and Internet users) identify its logical form, that is, the sequence of words that make up the utterance, regardless of context. Then, they construct its propositional form inferentially, together with a hypothesis as to the intended explicit and/or implicated interpretation (*explicature* and *implicature*, respectively), and in parallel to the retrieval of the necessary contextual information to obtain the interpretation(s). S&W claim that addressees engage in a *mutual parallel adjustment* of explicit information, implicated conclusions and context.

Specifically, inference is a mental operation that individuals perform in their assessment of other people's communicative and informative intentions and on which they base their own utterances. Inference is affected by a number of contextual factors, including the socio-cultural specificity of the speech community to which interlocutors belong:

> In order to become a competent speaker of a natural language it is not enough to know a number of rules for grammatical construction, semantic and phonological reference, etc., but it is also necessary to use a wide range of "commonsense" knowledge and inferences and principles about other people's inner and intentional worlds. (Belinchón et al. 1992: 184)

Inference, then, would fill the gap between the semantic representation of the utterance (context-free, decoded) and what the utterance actually communicates (in context, inferred) (S&W 1987: 697). Inference is defined as the process by which an assumption is accepted as true o probably true depending on the truth or probable truth of other assumptions.

S&W defend a model which stresses the role of the addressee when inferring the speaker's communicative and informative intentions, that is, when generating hypotheses as to the purpose of an utterance. Specifically, they defend a notion of *nondemonstrative inference*, since there is no way in which we can ensure which cognitive operations lead to a correct inference, nor are there any ways to measure, a priori, the eventual success of the production of inferences: "in *demonstrative inference* [...] the truth of the premises guarantees the truth of the conclusions. In *nondemonstrative inference*, the truth of the premises merely makes the truth of the conclusions probable" (S&W ibid.: 701).

However, although human inference is not, in a strict sense, logical, it does use deductive rules that individuals apply spontaneously before confirming interpretive hypotheses. Inferential processes work with the individual's previous mental representations and factual assumptions about the world. Each new inferred stimulus is combined with pre-existing ones in order to modify and improve the general picture of the world that everybody possesses. Deduction would be the result of combining new information *P* to old information *C* stored in the person's mind. S&W call this cognitive operation *contextualization*. This operation may lead to *cognitive effects*,[7] which are produced when *C* is altered in some relevant way, leading to the strengthening, suppression or combination of previous assumptions.

2.5 Sources of information in a context

According to relevance theory, context is a subset of one's assumptions about the world that is used in the interpretation of stimuli. This entails a dynamic and mental view of context opposed to the traditional and static view of context as "given beforehand" or "taken for granted" that we can find in certain pragmatic research.

7. Initially called *contextual effects*. In later publications, S&W distinguished between positive and negative cognitive effects, the former being the only ones that are beneficial to the person. Obviously, human beings search for relevance in terms of "positive" cognitive effects. From now on, whenever the term *cognitive effects* is mentioned in this book, it will mean "positive cognitive effects."

Context can be divided into a number of *informative sources* from which the individual can gather information when inferring the speaker's (or Internet user's) intended interpretation. For example, the hearer can gather information from previous utterances in the same conversation (the processing of which is still vivid in the hearer's short-term memory store), from the physical setting, from encyclopaedic knowledge, etc.[8] Imagine, for instance, that in situation (7a) the speaker intends to make mutually manifest his intention to be ironic with the utterance (7b):

(7) a. [*A cold, windy and rainy evening in London; two people trapped in a traffic jam*].

b. [*Smiling, with a noticeable ironic tone of voice*] When a man is tired of London, he is tired of life.

The hearer of (7b) can gather information from a number of contextual sources[9] that invalidate the possibility that the speaker is willing to communicate the literal meaning of his words: (1) the fact that the utterance is a famous quote attributed to Dr. Johnson; (2) the hearer's encyclopaedic knowledge of the weather in London; (3) information from the physical context (rain, wind); (4) the speaker's smile and tone of voice; and (5) the hearer's information of the speaker's overall opinion about London. All of this contextual saturation helps the hearer to identify the intended irony almost immediately.

Therefore, for every hearer there is an initial context of comprehension, made up of a previous utterance in the conversation, but this context can (and often *has to be*) varied and widened in the hearer's search for the relevance of the utterance. This operation may be beneficial or detrimental to that search for relevance: "since variations in context may increase or decrease the relevance of the proposition that is being processed, the goal of reaching an optimal level of relevance may constrain the choice of context" (W&S 1986: 593).

8. In subsequent chapters I will analyse the discursive strategies to which Internet users resort in order to compensate for the loss of contextual information that is provoked by the *cybermedia* that are still used as text-based communication, compared to other "richer" media.

9. In Yus (1998f, 2000b, 2000c, 2006b, 2007a, 2009a) I argue that the human being is capable of processing information from multiple contextual sources, either simultaneously or in sequence, as part of the inherent search for the relevance of the utterance. I also argue that the redundancy produced by the multiple activation of these sources (or the contextual saturation provided by a sequential activation) may reduce the mental effort required for interpreting irony effectively. This multiple or sequential activation is the basis of my proposal of a *criterion of optimal accessibility to irony*.

2.6 Relevance: Interest (cognitive effects) vs. processing effort

Cognitive effects are generated when the addressee processes new assumptions (e.g. assumptions from an utterance which has just been said) against information already stored or highly accessible. Hence, it is not only a matter of identifying assumptions, but also a task of evaluating the outcome of adding these assumptions to already stored or accessible information. In general, the higher the number of cognitive effects, the more relevance will be obtained. Especially interesting for relevance theory is the cognitive combination of new information and contextual information (stored or accessible assumptions) that generates *contextual implications*. These can only be obtained from the union of both types of information. For example, only by combining the new information provided by Ann's answer in (8a) and contextual information (8b) (accessible from Tom's encyclopaedic knowledge), can a contextual implication such as (8c) be derived:

(8) a. Tom: Hey Ann! Are you going to Mike's party on Saturday?
Ann: My parents are away this weekend.

b. [*Ann has a grandmother who needs constant care. Normally it is her parents that look after her, but if they are away, Ann has to do it*].

c. Ann can't go to the party on Saturday.

An assumption that generates no cognitive effect is irrelevant: "an assumption is relevant in a context if and only if it has some contextual effect in that context" (S&W 1986: 22). However, this is only one side of the coin. There also has to be some way of measuring degrees of relevance, of explaining how a certain context is selected and new information is processed against this context for the derivation of effects. This idea of degrees of relevance is covered by the claim that all information-processing demands mental effort and that the greater this effort is, the lesser the relevance will be. In conclusion, the definition of relevance is formulated as two conditions (S&W 1995: 265–266):

Condition a. An assumption is relevant to an individual to the extent that the positive cognitive effects achieved when it is optimally processed are large.

Condition b. An assumption is relevant to an individual to the extent that the effort required to achieve these positive cognitive effects is small.

2.7 Presumption of relevance, principle of relevance

The degree of relevance of stimuli is variable and context-dependent. An utterance may be very relevant in one context and utterly irrelevant in another context. Thus, there is no guarantee that assumptions will invariably be regarded as relevant, let alone that they will end up being processed at a conceptual level. However, ostensive stimuli, those which comply with the communicative and informative intentions (with which the speaker calls the interlocutor's attention towards an intention to communicate information) do carry a presumption or expectation of their eventual relevance. Consequently they are initially worth processing. Of course, the addressee plays an important part in the eventual (ir)relevance of the stimulus (e.g. an utterance), which will depend on the effort demanded for its processing and the number of cognitive effects that can be derived. In this cognitive model there is a kind of "division of labour" between the speaker (devising an utterance that communicates his/her thoughts efficiently) and the hearer (assuming the effort required to process the utterance, accessing contextual information and selecting one interpretation among the choice of interpretations of the same utterance in a specific situation). Interpretation, in short, is a trade-off between the search for interest in utterances (i.e. cognitive effects) and the mental effort demanded for the derivation of these effects. The *presumption of optimal relevance* is, logically, divided into two premises (S&W 1995: 267 and 270):

Premise a. The ostensive stimulus is relevant enough to be worth the audience's processing effort.

Premise b. The ostensive stimulus is the most relevant one compatible with a communicator's abilities and preferences.

Processing the information that an utterance makes manifest is subject to risks and effort: the risk of not knowing exactly which assumptions, among the array of assumptions that every communicative act makes manifest, are the ones that the speaker intends the hearer to select (Blakemore 1992: 21) and the effort to select a proposition and to process it in context. This is why every ostensive act of communication carries a presumption of its eventual relevance; the speaker is aware of the mental effort that he/she is demanding from the interlocutor and makes manifest the presumption that processing the utterance is going to be worth the effort. Besides, human beings are biologically geared to paying attention to potentially relevant stimuli and, since every act of ostensive behaviour carries a presumption of relevance (they call this presumption *principle of relevance*), ostensive communication is bound to be successful.

The hearer, in short, has to identify which assumptions the hearer is trying to make mutually manifest and satisfy this presumption of relevance, and also has to make hypotheses as to the content of these assumptions and access the intended contextual information that aids in selecting the right interpretation. All of these cognitive operations could be too effort-demanding if the hearer engaged in the task of assessing every possible interpretation and then decided which of them was the intended one (i.e. the most relevant). S&W (1986: 167) propose, instead, that the first interpretation that provides the highest number of cognitive effects in exchange for the least effort is the one that the hearer is bound to select, dismissing, at the same time, any other competing interpretations whose balance (of effects/effort) is not so optimal.

In later publications the *principle of relevance* has been divided into two parallel principles, one of a purely cognitive quality (*cognitive principle of relevance*), that claims that human beings are geared to the maximization of relevance, and a more communication-connoted one (*communicative principle of relevance*) and the inherent object of research within a cognitive pragmatics of human communication.

However, the cognitive principle of relevance is also important, since it governs all the cognitive behaviour of the individual and this principle is deeply rooted in human psychology and hence unavoidable. We resort to this principle when, in our daily lives, we filter out all the potentially irrelevant information from the world around us (for instance when we are unable to remember most of the people that pass by us in the street). A search for relevance guided by this *cognitive principle* has played a major evolutionary role in human survival. Additionally, we tend to select from context only the amount and quality of information that might be useful in order to draw interesting conclusions (the availability of contextual information is vast, but we have developed an ability to access only the potentially relevant information, the one that really aids in our interpretation of the world and of other people's utterances). Finally, we are also geared to combining new information with already stored information in order to draw relevant conclusions. All this is covered by this *cognitive principle of relevance.* An example of how old and new information are combined in a search for relevance and allows for interesting conclusions is provided in (9)–(11) below (Yus 2010a):

(9) New information (visual input):
A yellow Mercedes is parked near our Department.

(10) Information already available (from encyclopaedic knowledge):
a. Professor Smith, who supervises my thesis, owns a yellow Mercedes.
b. Professor Smith usually takes the bus to the university.
c. Only when he intends to stay at university till late in the evening does he drive his car to university (since there are no late buses returning to where he lives).

(11) (Relevant) conclusion (inferred by combining (9) and (10)):
This evening I will be able to discuss with him at length how my thesis is progressing.

S&W claim that in a situation where (9) is processed, (11) would be relevant since it can only result from the combination of (9) and (10). A similar procedure also applies to linguistic communication and, as I claim in this book, also to Internet-mediated communication.

This cognitive principle is at work in any kind of processing of stimuli, both verbal and visual (and also applies to one's own thoughts in a specific situation, some of which are more likely to be entertained than others), but S&W narrow this broad application of the principle to the analysis of verbal communication, as we can deduce from the main objective of relevance theory: "to identify underlying mechanisms, rooted in human psychology, which explain how humans communicate with one another" (S&W 1986: 32).

Of special interest is the fact that the application of these cognitive and communicative principles is influenced by the attributes of the medium used. In the case of Internet-mediated communication, certain procedures for obtaining and filtering information are frequent and these differ from the ones used in other channels of communication. For example, it is very common, especially among adolescents who have mastered the use of the Net, to engage in so-called *multi-tasking*. Adolescents usually chat with other users while answering messages from *Messenger* and searching for information in a portal, and in all of these simultaneous activities these adolescents are also searching for relevance and dismissing potentially irrelevant information, even though some supplementary mental effort is required in order to engage in this multi-tasking successfully (see Baron 2008a, 2008b). In fact, it could well be the case that multi-tasking might be altering the way information is processed and stored and, eventually, the very organization and functioning of the human brain (see Yus 2007c, 2008b; Salvucci & Taatgen 2011). As a consequence of paying attention to multiple sources of potential relevance and trying to process all of them in parallel, Internet users might develop a reluctance to devote cognitive resources to stimuli that do not offer immediate reward or involve deferred relevance, as happens when we read a long novel (see Carr 2010).[10]

Within pragmatics, the *communicative principle of relevance* ("every act of ostensive communication conveys the presumption of its own optimal relevance") is

10. "A novel has a message, but you have to work a lot to understand it, it takes effort, the author will not give it to you" (Umberto Eco, in a conversation with Javier Marías, *El País*, 22-1-2011, Babelia, p. 16).

the one specifically applied to verbal communication. When this principle is satisfied (normally, whenever anybody addresses us, but also when processing documents such as novels, news items, web pages, blogs, etc.), addressees undertake an interpretive task which aims at selecting the most appropriate interpretation from the range of interpretations that the utterance (or text[11]) might have in the current context (W&S 2002a: 256). In principle, hearers will follow the following general inferential procedure:

a. Follow a path of least effort in constructing an interpretation of the utterance (and in particular in resolving ambiguities and referential indeterminacies, in going beyond linguistic meaning, in supplying contextual assumptions, computing implicatures, etc.).
b. Stop when their expectations of relevance are satisfied.

3. Cyberpragmatics

Over the last few years, the development of technologies that help humans to communicate with one another beyond physical barriers has been enormous. Their impact on how language is produced and interpreted is also an important aspect and an inherent object of analysis for pragmatics.

We are now, more than ever, aided in our eagerness to communicate with one another by technologies that keep us in permanent connection with other people. And the Internet, together with the mobile phone, are impressive technologies that make it possible to share information and engage in conversations and interactions with other people regardless of where they are physically located. Nowadays, the Net is used for fostering communities and shaping one's identity (see Chapter 2), searching and retrieving information, reading documents on web pages (see Chapter 3), creating and developing social networking sites and interactive blogs (see Chapter 4), engaging in synchronous virtual conversations (see Chapter 5), and exchanging e-mails (see Chapter 6), among other possible forms of interaction.

Cyberpragmatics was coined in 2001 for a cognitive pragmatics study of Internet-mediated communication (see Yus 2001a, 2001b, 2010b, forthcoming a). Its main interest is the analysis of how information is produced and interpreted within the Internet environment. It is also interested in how users access contextual information (often limited if compared to other context-saturated situations

11. From now on, the term "utterance" will cover all kinds of coded information communicated on the Net, oral or written (typed).

such as face-to-face conversations) in order to fill in the informative gaps between what users type on the keyboard and what they really intend to communicate. There are important pragmatic consequences (related to variations in accessibility to context) that the different forms of communication on the Net exhibit, and therefore there are also consequences on the amount and quality of information communicated and interpreted therein. Specifically, a number of hypotheses on Internet-mediated communication make up the foundations of *cyberpragmatics*:

1. On the Internet, the "addresser users" have communicative intentions and devise their utterances with the expectation that these intentions will end up being relevant to the other users and that their utterances will be interpreted correctly. Since users are aware that, in principle, there are multiple ways in which their utterances can be coded, they type (or talk in voice-enabled Internet communication) with the expectation that these coded utterances will be adequate evidence to lead "addressee users" effectively towards the intended interpretation.
2. Internet users use inferential strategies when they interpret messages on the Net, and these do not differ from the ones used for the comprehension of utterances in oral conversations shaped by physical co-presence. We are equipped with a biologically evolved tendency to maximize the relevance of the utterances that we process, but we do not apply different inferential procedures for our interpretation of stimuli (verbal and nonverbal) in physical or virtual contexts.
3. Internet users expect their interlocutors to be able to access the necessary amount of contextual information that will allow them to arrive at a correct interpretation of their utterances. In the same way, their interlocutors will invariably access contextual information as a necessary stage in a relevance-oriented interpretation of these utterances.
4. The attributes of the different *cyber-media* (chat rooms, e-mail, *messenger*, web pages, social networking sites…) influence the quality of the user's access to contextual information, the amount of information obtained, the interpretation selected, the cognitive effects derived and the mental effort involved in obtaining these effects.

Figure 1.1 summarizes these claims. *Cyberpragmatics* analyses communicative exchanges that take place among Internet users using the different *cyber-media* available. "Sender users" predict that their interlocutors will draw relevant conclusions by accessing the necessary contextual information. In the same way, "addressee users" will search for relevance in the utterances (or pictures, videos,

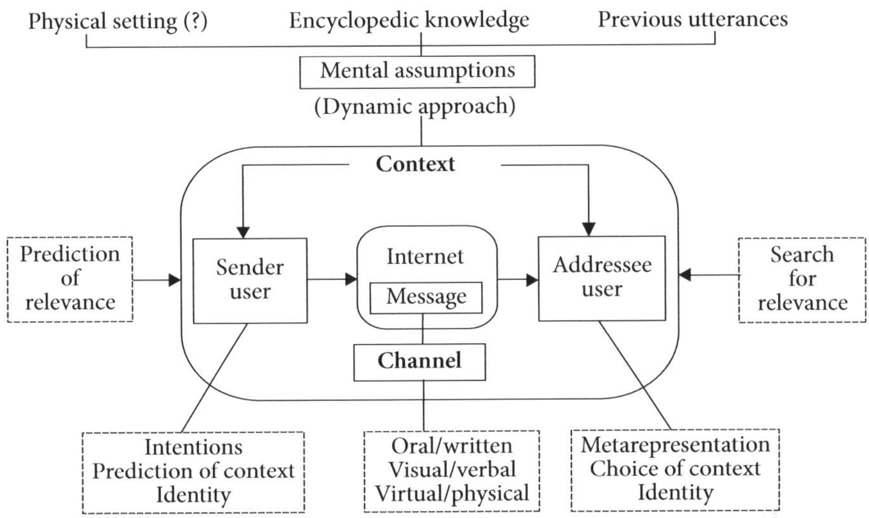

Figure 1.1 Internet-mediated communication according to *cyberpragmatics*

podcasts…) that they process. Consequently, context plays a major role both in the production and interpretation of information on the Net, in the same way as in face-to-face interactions.

It should be stressed that S&W propose a much more dynamic view of context, specifically as a subset of the person's mental store of information, than the one proposed in the early decades of pragmatic research, as I have already pointed out above. The actual access to contextual information is constrained by the interpretive requirements of the stimulus (utterance, picture, nonverbal behaviour…) being processed. On the Internet, the accessibility to contextual information is also constrained by the quality of the different channels of communication and their place in the oral/written, visual/verbal and synchronous/asynchronous dichotomies, which alters, to a greater or lesser extent, the way in which users predict and obtain relevance. Besides, these *cyber-media* are in constant evolution, and new qualities are constantly added that alter the position of the medium in the aforementioned dichotomies. For instance, traditional text-based chat rooms have incorporated voice and video (*web cam*) in the last few years, and this development has had consequences regarding the quality and quantity of contextual information that can be accessed during interpretation and on the overall assessment of relevance by the users of chat rooms.

4. Cyberpragmatics, cognition and the Internet

Information and communication technologies (ICT) and their users influence one another. On the one hand, people often make innovative (and often surprising) uses of these technologies, which affect the subsequent designs of new models and also the quality of the interactions that occur between people when they use these technologies. An example would be SMS communication between mobile phones (*texting*). This was initially a marginal feature of the design of mobile phones, but the growing (and surprising) popularity of this service, especially among adolescents, led to redesigning the device, the introduction of new effort-saving typing options, etc.[12]

Besides, the position that a *cyber-medium* occupies in the oral/written, visual/verbal and synchronous/asynchronous dichotomies influences, as I have already stated, the quality and quantity of interactions on the Net.[13] A central objective of analysis within *cyberpragmatics* is to determine to what extent these qualities of *cyber-media* affect the estimation of relevance, that is, how they affect the assessment of the cognitive effects that may be derived and the mental effort demanded in return. In this sense, a number of "alterations" of relevance may be produced by the different qualities of these media for Internet-mediated communication (see Yus 2007c, 2008b, 2010c, forthcoming a).

1. There is a close link between the estimation of relevance and the general use of technologies and interfaces for Internet communication. Indeed, users who have problems getting connected to the Net, who do not master the protocols and

12. In this book there is no chapter devoted to the mobile phone and SMS *texting*. However, it is now evident that mobile phones and the Internet are undergoing an increasing process of mutual hybridity. The new smart phones allow users to surf the Net, send e-mails, engage in virtual conversations, access and update their social networking sites, etc. As I will argue in Chapter 8, *cyberpragmatic* research will have to widen its scope in the near future and cover Internet-mediated interactions maintained through mobile phones and *texting* (another type of virtualized communication in which the physical location of the interlocutors is unimportant). Besides, as has been reported by the *Pew Internet Project* (Horrigan 2009), the use of the mobile phone has an impact on the way Internet is used. Something similar happens with modern televisions, typically outside Internet research, but which are currently being "invaded" by the Internet as yet another type of terminal for accessing the Net (see Alandete 2009a; Grau 2010).

13. Kwasnik & Crowston (2005:79–80) wonder "whether digital genres emerge from what people do on the web, or whether the technology itself affords ways of doing things that people can then discover and exploit." There is no easy answer, since users tend to *repurpose* technologies, as in the SMS *texting* mentioned above. Even more problematic, according to these authors, is the fact that "many technologies are converging – voice, image, text, databases, computing – creating opportunities for combining and recombining genres of many different kinds in inventive ways and for unexpected purposes," that is, creating alterations in the aforementioned dichotomies.

software for virtual interactions, etc. will face increased effort even before engaging in interactions. Expert users, by contrast, will make the best use of the technology and of the options for interaction that it offers at a low cost in mental effort. Even the type of discourse exhibited indicates which users are experts in using technologies for Internet-mediated communication (Yus 2003b). For example, in chat rooms novice users tend to type a lot of text, as in (12), while users that master the techniques of virtual conversations will leave implicit (i.e. not typed) all the information that they guess that their interlocutors will be able to supply (infer) by themselves, as in (13a) which, in the context of the topic of this chat room (the official website of *Operación Triunfo*, a TV contest for would-be singers, similar to *Face Academy* in Britain), would really communicate the information in italics in (13b):

(12) <mariabisb> rosa tiene una voz bonita pero le falta mucha autoridad en el escenario. En eso le dan 100 vueltas chenoa y bisbal, y manu.
[*Rosa has a nice voice but she lacks authority on the stage. Chenoa and Bisbal and Manu beat her hands down in that*].[14]

(13) a. <Bisbaal> y creo q n tienen dntro d la academia.
[*I think they don't have in the academy*].

b. <Bisbaal> y[*o*] creo q[*ue*] [*los concursantes de Operación Triunfo*] n[*o*] tienen [*un ordenador conectado a Internet*] d[*e*]ntro d[*e*] la academia [*de Operación Triunfo*] [*en la que están concursando para convertirse en cantantes profesionales*].
[*I think that the contestants of Fame Academy don't have a computer connected to the Internet in the Academy where they are trying to become professional singers*].

2. One of the objectives of *cyberpragmatics* is to analyse the challenge that Internet users face when they attempt to connote their messages so that they can be closer to speech, that is, when they try to compensate for the lack of oral attributes of typed text. The cues-filtered quality of text typed on the keyboard may demand, on some occasions, supplementary mental effort and it may be more difficult to identify other users' underlying intentions, attitudes, feelings and emotions, thus also altering the eventual relevance of text-based communication (Yus 2005a).

Therefore, all *cyber-media* can be placed on a *scale of contextualization* ranging from highly context-saturated media (videoconferencing, Internet-enabled phone calls, chat rooms with web cam, etc.) to highly cues-filtered text-based

14. The samples collected for this book were mainly taken from Spanish *cyber-media*. Wherever possible, I will provide a translation in italics following similar linguistic strategies as the ones found in the Spanish samples.

media (traditional chat rooms, e-mail, instant messaging, etc.).[15] Although in all the cases there is an information-gap that has to be filled inferentially, in media that provide few options for contextualization an increased level of mental effort might be produced during this inferential activity. In other words, according to the so-called *underdeterminacy thesis*, there are always information-gaps to be filled inferentially, both between what the speaker says and what the speaker intends to communicate, and between what the speaker says and what the hearer picks up as an interpretation, as summarized in (14) below:

(14) a. What the speaker intends to communicate.
[*only resembles…*].
b. What the speaker literally says.
[*only resembles…*].
c. What the hearer interprets.

Crucially, the *cyber-media* that are unable to communicate effectively the vocal and visual information that is available in face-to-face conversations, for example the interfaces of text-based chat rooms, might generate supplementary information-gaps to be filled inferentially by the addressee, as summarized in (15):

(15) a. What the "sender user" intends to communicate.
[*only resembles…*].
b. What the "sender user" could have said (in a context-saturated face-to-face conversation).
[*only resembles…*].
c. What the "sender user" actually types.
[*only resembles…*].
d. What the "addressee user" could have listened to (in a context-saturated face-to-face conversation).
[*only resembles…*].

15. Several authors have proposed similar scales. Baltes et al. (2002) propose a scale around two axes: one of synchrony and the other of presence/absence of oral and visual cues. The scale predicts one extreme of high contextual saturation, as in face-to-face conversations, and another extreme of minimal contextual support, as in the traditional written letter, and in between we would place media such as teleconferencing, chat rooms and e-mail. Hard af Segerstad & Ljungstrand (2002) make a similar proposal of a scale, this time on a single axis: on the left we would place asynchronous media with minimal feeling of "social presence"; on the right, we would place synchronous media with greater feeling of co-presence. On this single axis the following media would be listed, from left to right: traditional letter, e-mail, chat room, instant messaging (e.g. *Messenger*) and face-to-face interaction.

e. What the "addressee user" actually reads.
[*only resembles…*].
f. What the "addressee user" interprets.

Of course, users resort to a number of techniques in order to connote texts with oral qualities, giving rise to what can be labelled *oralized written text* (Yus 2001a) or *textual deformation* (Yus 2005a), basically consisting of creative spelling, repetitions of letters and punctuation marks, together with the use of *emoticons* (:-)), all of which will be studied in Chapter 5. This deformation is not only intended to enrich typed text with oral connotations, but also to communicate feelings and emotions more accurately. An example would be (16a), sent to a chat room. From this message we can deduce that (16b) would be the resulting proposition after being contextualized by the user, while (16c) would be the explicit interpretation of the utterance (its *explicature* in relevance-theoretic terminology) plus the attitudinal schema favoured by the presence of the question mark. Finally, (16d) would be the resulting proposition after the interpretation of the repeated question mark as a signal of the user's feeling of impatience:

(16) a. <nenita69> alguien de torrejon de ardoz???????
[*Anybody from Torrejón de Ardoz???????*].
b. [*Is there*] anybody from Torrejón de Ardoz [*connected to this channel*] [*who wants to chat with me*]?
c. <nenita69> is asking if anybody from Torrejón de Ardoz [*who is connected to this channel*] [*wants to chat with her*].
d. <nenita69> is asking *with insistence* if anybody from Torrejón de Ardoz [*who is connected to this channel*] [*wants to chat with her*].

3. However, the improved qualities of the design of *cyber-media* for Internet-mediated interactions do not invariably lead to parallel improvements in the quality and quantity of information produced and processed, and also to improvements in the eventual balances of cognitive effects and mental effort. For example, avatar-centred communication is a huge evolution, in terms of contextualization, from traditional chat rooms. But managing the nonverbal behaviour of avatars in virtual worlds such as *Second Life* and, at the same time, typing words for verbal communication with other avatars may be really difficult without a clear reward (see Chapter 5). Besides, many users distrust advances in contextualization because they convey a more faithful or realistic image of themselves (instead of the limited inferences that can be drawn from plain text), which might be counterproductive. Surprisingly, in spite of the availability of rich Internet media for interactions, many users still resort to the "secure" environment of text-based communication that masks or neutralizes personal information (Yus 2001b).

4. As I will argue in Chapter 3, the transference of offline discourses to the specific properties of the Net and the computer screen requires an adaptation of these discourses to the new environment, specifically in terms of how much information from the original text is preserved and organized in a new link-structuring pattern, how the interface is designed for usability, and how the initial balance of effects and effort can be maintained or improved in the new discourse of the Net. Two examples will be analysed in that chapter: the *cybernewspaper* versus the printed paper (see Yus 2003d) and the Internet advertisement (e.g. the banner) versus the printed advertisement (see Yus 2005c).

5. The *social* quality of many texts or discourses created on the Internet may provide a benefit that somehow compensates for the mental effort demanded for their production and interpretation, for example providing a strengthening of group- or community-related assumptions. We are immersed now in the so-called *Web 2.0*, the age of cooperative networks of users, and it comes as no surprise that *Time Magazine* elected the ordinary Internet user as their "person of the year" in 2006, acknowledging users' desire to share information collectively (as in *YouTube* or *Wikipedia*). In these cases, the effort required to produce the vast amount of information made available on the Net is compensated for by a feeling of community, of "being there," that allows for an eventual positive balance in relevance terms.

6. At the same time, this social orientation of user-generated content may be detrimental to the estimation of relevance. For example, an excess of information may reduce the credibility of the source and the user's willingness to process it. To make matters worse, much information on the Internet is not structured under an organizing hierarchy of documents; there is often no "authority" that filters out the useless information and directs the user towards potentially relevant documents, and in this case the user has to take full responsibility for this filtering, adding mental effort to the task of processing information on the Net.

7. Finally, some illegal uses of the Internet may increase processing effort enormously and reduce the eventual relevance of the page accessed. Among them, the following instances can be listed: (a) *spam* e-mail messages, which make it really difficult to browse incoming e-mail (see Chapter 6); (b) some *nicks* in chat rooms are fake ones, and when we click on them with the purpose of having a private conversation with the user who has this *nick*, we are taken to a web page instead, where some product is advertised; (c) the *pop-up* advertisements that annoy users who are trying to read the information on a web page; and (d) hacker assaults on pages such as *Wikipedia*, altering the content of entries and lowering their credibility.

In subsequent chapters I propose to study, with the aid of the cognitive pragmatics framework of relevance theory, the specificity of Internet-mediated communication. But ideas from other theories and models will also be incorporated into each chapter insofar as they contribute to a better understanding of this increasingly popular form of human communication and interaction.

CHAPTER 2

The presentation of self in everyday web use

1. Introduction

On December 18th 1998 the film *You've got mail* was first shown in the USA. In this film, the main characters Kathleen and Joe (played by Meg Ryan and Tom Hanks, respectively) have an intense romance through e-mail communication, with both of them masked behind the *nicks NY152* (Joe) and *Shopgirl* (Kathleen). However, in their physical lives they are enemies that hate each other intensely. This is just one example of how the Internet modifies or moulds the public presentation of people's identities and the challenges that Internet-mediated communication poses for the study of human interactions, not only as just another medium of communication (Belson 1994, Vidal Jiménez 2000), but also as a powerful tool for the definition and development of identities and personalities, together with the creation and consolidation of virtual groups and communities.

The title of this chapter is adapted from the famous micro-sociological analysis by Goffman (1987 [1959]). Goffman's differentiation between the roles that we play in society and the real identity that is hidden behind the "social facade" is undoubtedly applicable to Internet-mediated communication, where users' identities often remain backstage in intimacy, while other electronic identities play their parts in the visible area of the social stage. In this chapter I will show how the individual's identity is influenced, in both cases, by interactions, by the social use of language and by the feeling of community, group or network membership.

2. Discourse and sources of identity

Throughout their lives, people assume a number of discursive features and interactive behaviours that eventually shape them in their growth as human beings. These features arise from a general tendency of humans to gather together and establish social ties, a tendency which Allott (1998) labels as *groupism*. This is why many pragmatic studies have underlined the importance of the social context in human communication (see Akman 2000). Initially, we can represent the links between discourse and identity as an inverted triangle (see Yus 2002a). At the wide top area of the triangle we can place the discursive features of macro-social quality assumed (and often inherited) by the individual such as race, sex, nationality or specific speech community membership. In the middle part of the triangle we can place social groups whose membership the individual chooses and which are often linked to inherent jargons that mark frontiers of discursive specificity. Finally, at the narrow bottom part of the triangle we can place the individual as a unique holder of personal identity (the self) whose discursive features, shaped as a unique idiolect, differentiate this individual from the others.

One of the main sources of identity is the speech community (Gumperz 1971:114, 1989). Sometimes, as happens in Quebec or Catalonia, the language of a community may even be the subject of heated political debate, which reinforces the ties that bind people to their shared language and hence stresses their group identity. But individuals may also choose to belong to specific social groups related to specific jargons. This belonging enhances their *intra*-group identity, complemented with their *inter*-group identity of not belonging to other speech groups or communities. Very often, as in well-known *urban tribes*, specific jargons are linked to strong submissions to certain codes or patterns of nonverbal behaviour, including *artifactual communication*, that is, communication through objects such as clothing, complements (e.g. *piercings*) and other visual symbols of strong group identification.[1] Finally, the bottom vertex of the triangle would be occupied by the person's individual identity (self), which is shaped and moulded through conversational interactions with others in daily life. In fact, human beings are constantly negotiating their discursive identity with other people, a process which Boxer & Cortés-Conde (1997:282) call *relational identity.* In this sense, Goffman (1987) describes human beings as *interactive constructions*, in which individuals negotiate their the personal images (*faces*) with other people or in which they *position* themselves against others (Davies & Harre 1990).

This three-fold representation of discursive identity as layers in an inverted triangle is *re-inverted*, as it were, on the Internet (Figure 2.1). Indeed, the initial wide area at the top of the triangle, made up of macro-social aspects of discursive identity, undergoes a process of minimization or fragmentation due to the users' ability to interact with other users who belong to speech communities that are geographically and culturally distant. This world-wide interaction may dilute the markers of macro-social discursive identity, while other important macro-social

1. An important term, in this sense, is *social network* (Milroy 1978, 1992; Milroy & Milroy 1992), which represents the intensity of discursive social exchanges within a community and which is to acquire special relevance for research on Internet-mediated communication (because of the growing popularity of social networking sites on the Net and the rise of personal networks of a hybrid physical-virtual quality; see Chapter 4). Le Page's (1986) *acts of identity* are also worth mentioning here, a phrase that describes human beings' tendency to reflect upon the linguistic attributes of those social groups they want to belong to or identify themselves with.

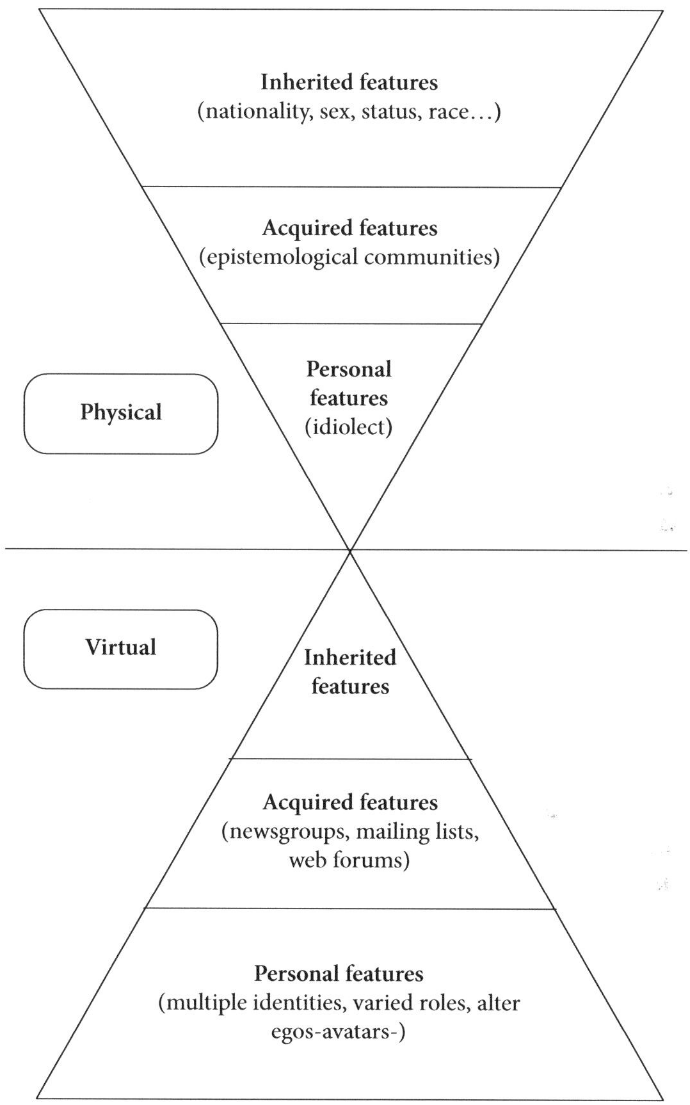

Figure 2.1 Discursive sources of physical-virtual identity

(and inherited) attributes such as the user's sex or race simply disappear in text-based virtual interactions. This is one more aspect of today's globalization, which has led to a physical network society (Castells 1997; Echeverría 1994, 1999), a virtual network society in cyberspace (Garton et al. 1997, Beamish 1995, Reid 1991, Warschauer 2000) and, nowadays, a society of personal networks with a hybrid physical-virtual mixture of interactions (Yus 2005b, 2007b).

The former middle layer of the inverted triangle in offline sources of identity would be similar to that of online groups, but on the Internet these would be replaced with virtual alternatives such as newsgroups, online forums or e-mail distribution lists. But these virtual groups shape, in a similar way, the user's identity by means of a set of tight intra-group markers of discursive identity. Indeed, these social groups on the Net also exhibit jargons and assumed shared information that are only available to those users who belong to the group (Watson 1997: 106, Cutler 1995: 20) and become one more source of group cohesion (Donath 1999, Maldonado 1998, Meyrowitz 1985: 143–144). For example, it is typical of certain newsgroups to use specific abbreviations and acronyms that create discursive barriers of comprehensibility for non-members of the group (Thomsen et al. 1998).

Finally, the former bottom vertex of the inverted triangle that represents the person's identity (self) shaped as idiolect would suffer a process of multiplication and/or fragmentation on the Internet due to the possibility of forming multiple virtual identities that are added to the physical identity, overlap with it or even replace it in extreme cases.[2]

3. The (speech) community

People store a number of commonsense assumptions that emanate from the human environment and our trust in these assumptions is not easily altered by other in-coming stimuli. The fact that we belong to a specific speech community entails the creation and storage of certain archetypical assumptions that we accept as "normal" in the ordinary life of the community. For those belonging to a community it is interesting to observe the extent of the *mutual cognitive environment* that exists among them, that is, to assess which area of the individual cognitive environments of the people of a community is shared by all of them and of which they are all aware (the *mutually manifest* area). Conversations are a good means to determine this area of mutuality. Besides, the reiterative determination of this area generates *community stereotypes*, made up of highly accessible stereotypical

2. Many studies differentiate between *real* and *virtual* identities (also in Yus 2001a). However, this dichotomy is biassed, as if only offline interactions could be *real*. In fact, for many people communication on the Net and their identities therein may be even more important and real than communication and identity formation in traditional physical scenarios. This is why I prefer to use the alternative *physical* versus *virtual* dichotomy for interactions and sources of identity formation. Weinreich (1997) proposes, as a form of compensation for this bias of the real/virtual dichotomy, a differentiation between *sensory world* and *virtual world* (see also Wynn & Katz 1997, Poster 1995: Chapter 2).

schemas.[3] This is part of the tendency of human cognition to form and maintain ties, to weigh one's social prestige against other people's, to assess the effect of our actions on other people's opinions and to predict their plausible replies (Nicolle 2000: 239).[4] Similarly, Jary (1998a: 166) stresses the fact that the stimuli which make assumptions about the social environment manifest tend to be very prominent. The information related to the individual-in-society is very relevant and, at the same time, highly accessible and easy to process due to its archetypical quality (S&W 1986: 88; on the applicability of relevance theory to social issues see S&W 1997 and Coupland & Jaworski 1997, among others). Finally, Gumperz (1977) points out that there are *expectations of co-occurrence*, specific to a particular culture, which people use in their daily interactions, often spontaneously. Frequently, these expectations become prominent in inter- or cross-cultural interactions, in which each participant brings along his/her own cultural specificity, as happens, for example, in inter-cultural business negotiations (Mateo & Yus 2009).

Besides, it should be underlined that in a virtual environment many social attributes are absent due to the lack of physical co-presence of the interlocutors. This absence entails a loss in the amount of mutuality between the users' cognitive environments and a parallel absence of archetypical social conventions to which people tend to resort in their daily interactions (Donath 1996). As Belson (1994) comments, the norms that are habitual in face-to-face communication are no longer conventionalized on the Internet, nor are there many norms for structuring (in)formal messages or for the assessment of politeness. But this statement does not imply that Internet-mediated communication is necessarily doomed to communicative failure, or devoid of effective protocols for interactive behaviour. On the contrary, it will be shown in this book that virtual interlocutors manage to create strategies that make up for the loss of socially connoted conversational cues and of the essential contextual information found in face-to-face conversations

---

3. Žegarac (2007) specifies that this kind of information fits what he calls *central cultural representations*, in the sense that they are valid in different contexts of our daily lives without the danger of misunderstandings.

4. This stereotypical information has been labelled by authors differently. Among others, we can list *script* as a prototypical succession of events for a shared activity (Lindsay & Norman 1983: 704), *frame* as a structure of data for representing an archetypical situation (Minsky 1975: 355) or definition of a situation that is constructed in accordance to organizational principles that govern the events and our subjective involvement in them (Goffman 1974: 10), *schema* as a structure of memory that comprises a number of active structures capable of assessing and transferring information (Bobrow & Norman 1975, quoted in Tannen 1979), and *theme* as a conceptual structure that contains a number of inter-related scripts (Abelson 1975, quoted in Tannen ibid.).

(Matthews 2000: 80). At the same time, *cyber-media* for Internet communication have evolved enormously in the last few years, thus opening new options for contextualization and communicative richness (web cam, sound, 3D environments, videoconferencing…). This evolution has increased the ability to convey and process contextual information of a social or personal quality.

4. The virtual community

Several analysts have underlined the difficulty that the definition of "community" entails (see Fernback 1997: 39). A possible solution is to propose the attributes that a community should possess in order to be given this label. This is what Erickson (1996a) did when proposing the following qualities of communities: belonging, relationships, commitment, values, goods, and perdurability. To these, the following attributes can be added: a shared location, reciprocity, norms and goals (see Yus 2010b: 44–45, de Cindio & Ripamonti 2010, Baym 2010: 72–98).

Are these qualities applicable to virtual communities? Yes, they are, in theory, as can be deduced from the bibliography available on this topic.[5] In general, it can be stated that worries about an excessive dependency on computers (and parallel isolation) of some Internet users is more likely to be found in sociological or philosophical studies on the Net than in the linguistic and pragmatic approach of this book. For example, it has been argued that the virtual community is an effect of the progressive adaptation of human beings to different environments or habitats: natural, urban, and now telematic (Echeverría 1999). Turkle (1996a) also stresses how American life, typically in middle-class suburbs where people hardly know their neighbours, has encouraged people to meet in cinemas, malls and, eventually, electronically in their own homes irrespective of their physical location. She points out how the Internet prevents fruitful interactions among people (Turkle 2011). And in Yus (2007b) a growing tendency towards hybrid (physical-virtual) personal networks of interaction is foreseen (see 5 below). For London (1997), communal life, which he calls *the public sphere*,[6] has fragmented due to an obsession with security and protection, not only from crime or violence, but also from having to talk with people, and hence people take refuge in

5. See, for instance, Jones (1995a, 1997a, 1998a), Smith & Kollock (1999) and Porter (1997).

6. Also called *public space* (Habermas), *civic nuclei* (Mumford), *talk shops* (Barber), or *third place*, together with the house and the workplace (Oldenburg, Schurer). Bibliographical references in London (ibid.).

suburbs that isolate them from other individuals (see Galindo Cáceres 1998). By contrast, if we study both types of community from a discursive-pragmatic point of view, we will realize that people resort to similar strategies of contextualization and intention recognition both in *physical* and in *virtual* communities. But this assertion does not mean that the outcome of interpretations will be equivalent in all cases and situations. The search for relevance of the stimuli that reach us is a universal cognitive activity of human beings and is rooted in the biological architecture of the mind. Therefore, the strategies of production and comprehension of messages, guided by relevance, will not differ essentially in physical and virtual environments, but it is nevertheless undeniable that there are different options for contextualization in either case.

In the bibliography available, virtual communities are often defined according to the tie that bind users together: their desire to share a certain type of information, belief or interest (and the subsequent satisfaction obtained). In other words, the tie of being aware of sharing a certain cognitive environment, for instance:

> Groups of people who congregate electronically to discuss specific topics which range from academic research to hobbies. They are linked by a common interest or profession. There are no geographic boundaries to on-line communities and participants anywhere in the world can participate. (Del'Aquila 1999)

> Social aggregations that emerge from the Net when enough people carry on those public discussions long enough, with sufficient human feeling, to form webs of personal relationships in cyberspace. (Rheingold 1993)

> Incontrovertibly social spaces in which people still meet face-to-face, but under new definitions of both 'meet' and 'face' [...] Virtual communities are passage points for collections of common beliefs and practices that united people who were physically separated. (Stone 1991, quoted in Jones 1998b: 15)

Given the peculiarities of virtual communities, it is understandable that researchers could not avoid the temptation to compare them with their traditional physical counterparts.[7] In this comparison there is often an underlying premise: that both types of community, physical and virtual, are mutually exclusive and that it is necessary to "log on" to virtual communities as a complement to "real" com munities. But, in fact, there is a high level of inter-connectedness between them

7. See Weston (1994), Agren (1997), Baym (1995), Kollock & Smith (1999), Wellman & Gulia (1999), Patterson (1996: Chapter 6), Kling (1996a), Giménez (1997), Q. Jones (1997), van Alstyne & Brynjolfsson (1997), Hamman (1999), Croon 1997), Valtersson (1996), Weinreich (1997), McIlvenny (1999), Cherny (1999), Etzioni (2000) and Yus (2001a: 53–57), among others.

in today's society and the qualities of virtual communities are usually related to similar qualities of the physical counterparts (Baym 1998: 37–38).[8] And nowadays few people log onto their social networks. Rather, it is taken for granted that these people are constantly connected to them. We are now experiencing what William Gibson, who coined the term *cyberspace*, predicted many years ago: that in the future (that is, nowadays) people would no longer pay to get connected to the Internet; quite the opposite: they would pay to get disconnected.

4.1 The linguistic essence of the virtual community

In the past, Internet-mediated communication was basically text-based, and even nowadays the text typed by users is essential in virtual interactions. Analysts such as Cicognani (1998) or Danet (1998), among others, make a general differentiation between types of text-based communities. On the one hand, *synchronous virtual communities* (for example chat rooms), where interlocutors are connected simultaneously to the Net, build up a sort of textual interactive dialogue that disappears as soon as the users stop the connection and switch off the computer. In synchronous communities there are no traces of our presence, nor are there options for a long-lasting form of community. On the other hand, *asynchronous virtual communities* (for example newsgroups) build up an archive of interactions and hence an increasingly complex form of community where stronger communal ties can be fostered (see Lombard & Ditton 1997, Sotillo 2000). The possibility to build up an archive of interactions on the Net turns these communities into rhetorical entities (Bormann, quoted in Thomsen et al. 1998), whose collective meaning arises from an experience and history constructed from the users' contributions.

In both types of community, the interactive key (and, eventually, one of the main sources of virtual identity construction) lies in the text typed by the users (Stuart 1999, Simich-Dudgeon 1999). According to Mitra (1997: 59), the texts exchanged on the Internet are artifacts that keep virtual communities bound

8. An example of the parallelism between virtual and physical communities was the project *Infoville* in Villena (Alicante, Spain). Unlike the virtual community, *Infoville* was not a space separated from the physical community, but an inter-connection between people that shared a physical community and were even neighbours that came across each other in the streets, but who also shared a virtual community as a supplement to their face-to-face encounters (see McInnes 1997). In fact, conversations in offline and online scenarios frequently overlapped without discontinuity. This is a kind of mixture that will be more and more frequent during this century. See also the term *communal computing* (*informática comunitaria*) in Finquelievich (2000).

together, as well as indicators of which direction they are taking. The identities inside the community are mainly created via the ways users present themselves in their discourses. As a consequence, the textual quality of virtual communities is their most outstanding attribute.

On the other hand, the text is useful to link virtual experiences that, on most occasions, suffer from a spatial-temporal fragmentation (or at least restructuring).[9] The text on the Net may remain archived beyond the synchronous connection of the members of the community.[10] This is why Maldonado (1998: 25) qualifies these communities as *transit communities* (*comunidades de paso*). And the classic label of *global village* by McLuhan would also fit this redefinition of the traditional idea of space and time under the new trans-spatial and trans-temporal possibilities that the Internet opens up (see Stille 2000). In other words, "with infinite space and around-the-clock availability, the Internet has made building relationships and community easier than ever before by defying time and space limitations" (Mitra 2010: 51).

The text is also useful as a holder of the user's features of identity when typing and transmitting it on the Internet. In an interactive medium that has removed the user from the body and the body from its spatial-temporal location, only the textual identity remains (see 6 below), although the loss of information is compensated for by technological advances that in the last few years have brought Internet-mediated communication closer to the richness of oral conversations. But this identity tends to a certain idealization of the virtual self, caused by the absence of the contextual clues that normally frame the extent of our impressions of other people's identities. As Stallabrass (1998: 79–80) points out,

> when we can only count on partial information, we tend to fill the gaps with idealized elements. Here there is no danger of infection, pregnancy or violence, but neither is there danger of physical intimacy. The mask that computer-mediated communication provides, unlike the clothes that one wears for a fancy dress party, hides us completely. The gender, sexual orientation, colour, or even the species, everything can change instantly and at one's will.

9. On this issue, see Cicognani (1998: 18), Greenhill & Fletcher (1996: 182), Bruns (1998a, 1998b), Jones (1997a), Boudourides (1997), Reid (1991), and Sandbothe (1998), among others.

10. Nowadays, the perdurability of text is complemented with authentic "repositories" of visual, audio or multimodal information on the Net, which also play an important communal role, as happens with photographs in *Flickr* or videos in *YouTube*.

4.2 Virtual cognitive environments

When two people interact, a number of assumptions about their cognitive environments are manifest to each of them and some may become mutually manifest, and hence part of their *mutual cognitive environment*. This may be essential to guarantee an efficient flow of conversation and interlocutors normally make hypotheses about which assumptions are *mutually manifest* in the course of a conversation. Thus each speaker will predict that certain assumptions are mutually manifest and each hearer will use these assumptions when selecting the speaker's intended interpretation (S&W 1986: 44).

In Internet-mediated communication the conversational tasks of addresser and addressee do not differ from the ones mentioned for face-to-face interactions. Virtual interlocutors on the Net also make hypotheses about the existence and extent of the mutuality in their cognitive environments, as an essential step towards effective communication. However, very often these virtual interlocutors are faced with limited, partial or even inexistent information concerning other users' cognitive environments. For instance, users frequently log onto the Net with a *nick* and their personal features may be constructed only textually with the keyboard. Their bodies and nonverbal behaviour are absent in text-based interactions, as are gestures or paralinguistic contours of the voice, and it is difficult to apprehend essential aspects of the users such as their race, sex, social origin, physical shape or status. The personal representation inside the Internet is not an inevitable consequence of biology, birth or social circumstances but, rather, an easy-to-manipulate incorporeal fabrication.[11] In this sense, there are *information richness theories*, as they are generically labelled in Yus (2007b), such as *Social Presence Theory* (see Byrne 1994, Jaffe et al. 1995), which suggest the need for interlocutors to be aware that they are mutually involved in the conversation, a feeling that decreases – leading even to a total lack of interest in the conversation – when the contextual information available to both interlocutors is reduced due to the qualities of the channel. On this basis, Kiesler et al. (1984) define computer-mediated communication as a channel that de-personalizes. These authors argue that there is social anonymity that is a direct consequence of having to imagine our interlocutors or, in relevance theory terms, of having to make hypotheses on the assumptions that belong to the mutual cognitive environment of interlocutors that are not co-present. Of course, the informative richness of current *cyber-media* increases the overall options for self-disclosure

11. See, for instance, Mitchell (1995), Trott (1996), Cherny (1995a) and Davis (1997).

and hence the options for more intense interpersonal relationships on the Net (see Mesch & Beker 2010).[12]

In general, which assumptions tend to be manifest – or probably mutually manifest – in Internet-mediated communication? Traditionally, those manifested through typed text, but there are continuous, rapid advances in the richness of *cyber-media* that are generating more and more options for contextualization. But on the Internet it seems that the general norms of behaviour in physical communities are inverted: in physical scenarios, people usually identify other people that share an interest with them. When, in the course of conversational interactions, we reveal and identify aspects of mutuality, we tend to gather and form groups tied by these mutual interests. In virtual scenarios, by contrast, we can go straight to the newsgroup or forum where the topic that interests us is treated and, after that, we can discover new areas of mutuality (Kollock & Smith 1996: 116). Similarly, the more users gather together in a newsgroup, the more difficult it turns out to delimit the area that belongs to all the users' mutual cognitive environment or, in Jones' (1995b, 1997a: 17) words, the more difficult it is to establish the *symbolic space* constructed by interactions in the forum, which is the most essential element of cohesion in any community (see also Mitra 1997: 57–60, Erickson 1996b).

An example of a feature whose mutuality is checked by Internet users (and which eventually serves as a marker of community membership) is the use of abbreviations, the repetition of characters and acronyms in newsgroups, chat rooms and instant messaging. As is the case with any specialized jargon that sets up discursive barriers for those outside the group, in these environments for Internet communication the users make hypotheses on the degree of mutuality with other users that allows for correct understanding of these innovative uses of the text typed thorough the keyboard, in a similar way as happens with jargons in specialized communication (see Posteguillo 1997, 2003; Alcaraz Varó et al. 2007).

12. *Information richness theories* is a label that covers theories that, one way or another, address how (or whether) the loss of contextual information produced by the channel generates a loss of interest in the information being processed, with an extreme outcome in the interruption of communication. Among others, these theories would fit this label: (1) *Reduced Social Context Cues Theory* (see Sproull & Kiessler 1986), (2) *Social Information Processing Theory* (see Walther 1992), (3) *Social Identity Theory of Deindividuation Effects* or SIDE (see Spears, Lea & Lee 1990, Spears & Lea 1992, Reicher, Spears & Postmes 1995); (4) *Media Richness Theory* (see Daft & Lengel 1984, Rice 1992), and (5) *Uncertainty Reduction Theory* (see Berger & Calabrese 1975).

5. Towards personal networks of physical-virtual interactions

In the last few years it has become evident that the initial attempt at a differentiation between physical and virtual communities no longer makes sense in a technology-filled society like ours, in which the role that both types of community play in this twenty-first century is getting increasingly blurred (see Yus 2003c, 2005b, 2007b, 2008a). Rather than connecting to virtual communities, nowadays people enjoy multiple physical-virtual possibilities of interaction and social gathering shaped as personal networks that form an intersection and in which the user is a node in a dense inter-relation of friends, relatives, colleagues and acquaintances.

Today's evolution of social interactions is leading to interwoven and hybrid interactions of a physical and virtual quality, and the importance of the former as a solid foundation of community bonding is decreasing enormously. Indeed, at the beginning of the 90s, when Internet started to become popular, traditional physical communities were already undergoing a process of disconnection from their physical foundations, and people were already searching for ties and interactions in places (such as bars, squares, etc.) that were not part of their neighbourhoods. In that decade, the Internet was playing no major part in the formation and development of identities and communities as alternatives to the ones fostered in physical contexts. The Internet was something that one had to log onto, with a poor virtual scenario compared to the physical materiality of classic spaces for social interactions.[13]

By contrast, in this decade of the twenty-first century the changes in both physical and virtual interactions have been enormous. It can be stated that nowadays the communities in physical spaces are suffering from a process of *virtualization*, that is, they are becoming *virtual realities*, since they have definitely lost the physical anchorage that tied them to a delimited space and the prominent role that they used to play in the past. Physical communities have fragmented, extended, disintegrated, losing the boundaries that made it possible to identify them. Now, more than ever, people search for their physical social networks in scattered places. And they massively use technologies such as the mobile phone, which removes the person from the physical anchorage and stresses, instead, the importance of the person regardless of his/her location. Traditional community-

13. This view of "physical better than virtual" can still be found in contemporary research on communities and social networks. For example, Galindo Cáceres (2010) argues that social networks on the Internet are only a configuration of options for individual contact, not for communal relationships. The centre is the individual, the satisfaction at finding someone who fits our interests, which indicates a poor or inexistent social network.

fostering spaces such as the local bar, the main square, the neighbourhood, etc. are no longer important for the communal or interactive needs of the citizens.

At the same time, Internet-mediated interactions are immersed in a process of *materialization* or *physicalization*, since they are no longer spaces which one has to log onto but are, instead, essential options for interactions with other people and they even compete in intensity with face-to-face interactions in physical settings. All the range of options for Internet-mediated interactions that are available to the user in this decade (among others, the 3G services for the mobile phone, chat rooms, videoconferencing, virtual worlds – such as *Second Life* or *World of Warcraft* –, blogs, SMS texting, Twitter, instant messaging, social networking sites, interactive websites and e-mail, among others) are now massively used by people who cannot often differentiate them from physical interactions in terms of communicative satisfaction. Besides, many ties and gatherings on the Internet reach levels of communal intensity that are difficult to find in physical communities. In short, we are heading towards a gradual hybridization between traditional physical spaces for communities, which tend to be more and more virtual, and Internet-supported communities, that are increasingly "physical" and important in today's interactions.

Table 2.1 Media for communication with friends (survey, 2008)

	Men	Women	Total
Instant messaging	19 (90,4%)	56 (86,1%)	75 (87,2%)
Telephone	11 (52,3%)	25 (29%)	36 (41,8%)
SMS	9 (42,8%)	41 (63%)	50 (58,1%)
Skype	3 (14,3%)	6 (9,2%)	9 (10,4%)
E-mail	6 (28,6%)	20 (30,7%)	26 (30,2%)
Mobile phone	17 (81%)	49 (75,3%)	66 (76,7%)
Chat room	--------------	1 (1,5%)	1 (1,1%)
Social networking site	--------------	12 (18,4%)	12 (13,9%)

In December 2008, a survey form was given out to university students from the University of Alicante (Spain).[14] It confirmed this tendency to hybridization, since young people today massively use technologies in parallel to their physical interactions and do not consider them deficient means for keeping in touch with their friends (see Table 2.1). They systematically use instant messaging (87.2%), SMS (58.1%) and the mobile phone (76.7%), and these are not supplements or complements to their physical social networks, but primary sources for managing them and their daily interactions. Indeed, for young people today,

14. 21 male, 65 female, aged 17 (16.2%), 18 (40.6%), 19 (16.2%) and 20 or more (26.7%).

> online and offline lives are connected to each other. Digital worlds are very real to youth – and within their subjective experiences, the "real" and "virtual" may even blend with each other. Therefore, we refrain from using the term "real world" to contrast with "online" or "digital worlds." Instead we will use the terms physical/digital and offline/online to capture both ends of the continuum representing online and offline worlds. (Subrahmanyam & Šmahel 2011:35)

Besides, there is an increasing number of people with friendships (sometimes very intense ones) that are only sustained virtually on the Internet, without ever meeting face-to-face. As can be seen in Table 2.2, more than 80% of informants hold and sustain social relationships exclusively on the Net. For that purpose, the most typical *cyber-medium* was instant messaging (specifically *Messenger*) (68.6%) and social networking sites (*Facebook, Tuenti...*), although nowadays the percentage of the latter is surely much higher. What is surprising, though, is the low percentage of e-mail use for maintaining friendships (23.2%), a medium that is usually considered to be "too cold" or "too serious" by today's youngsters (see Chapter 6).

Table 2.2 Contact only through the Internet

Contact only through the internet?			
	Men	**Women**	**Total**
YES	17 (81%)	53 (81,6%)	70 (81,3%)
NO	4 (19%)	12 (18,4%)	16 (18,6%)
Which medium do you use for communication?			
Messenger	15 (71,4%)	44 (67,6%)	59 (68,6%)
Social networking site	9 (42,8%)	11 (16,9%)	20 (23,2%)
Skype	1 (4,76%)	4 (6,1%)	5 (5,8%)
SMS	-------------	1 (1,5%)	1 (1,1%)
Chat room	-------------	1 (1,5%)	1 (1,1%)
E-mail	6 (28,5%)	14 (21,5%)	20 (23,2%)
Avatars	1 (4,7%)	-------------	1 (1,1%)

The consequences of the current state of hybridization of physical and virtual networks, of the *materialization* of Internet-mediated interactions and the *virtualization* of physical interactions are multiple and, to a certain extent, contradictory. Just as there are still nowadays highly homogeneous neighbourhoods in terms of race, religion or country of origin, with a parallel homogeneous use of language, we can also find interactions with a diffuse, multiple, virtual or physical, but especially hybrid quality. However, the prospects for the future indicate a tendency, in Western technified societies, towards a full mixture of physical-only interactions, Internet-mediated ones and hybrid ones (the

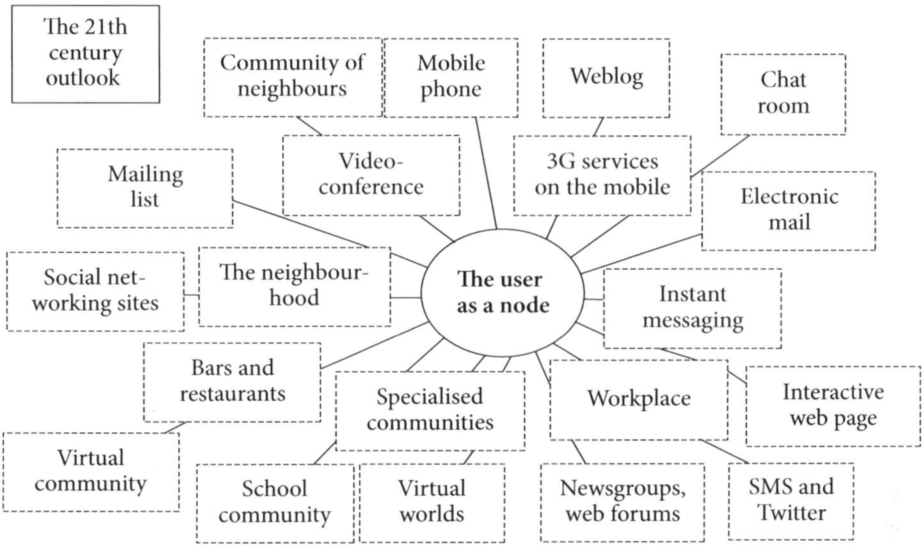

Figure 2.2 Hybridized physical-virtual interactions in the future

latter being increasingly frequent).[15] The image of the user of the future is that of "the person as a node," through whom these types of interactions form intersections, as represented in Figure 2.2. And this picture overlaps with other physical-virtual options for interaction and community bonding. For example, with the aid of *smart phones*, so-called *situated wireless communities* can be created, where the mobile phone aids people in getting "more closely bound with each other through a sense of sharing common physical and/or social contexts" (Sun & Poole 2010: 122). And in these phone-sustained communities we can see clearly the aforementioned physical-virtual hybridization, since in these gatherings "sharing a common physical context leads to stronger joint attention, and sharing a common social context leads to stronger social linkage. As a result,

15. This hybridization does not mean that the user's identity invariably remains identical in physical and virtual scenarios even if the users experience genuine forms of bonding in both types of community, especially if interlocutors do not know each other offline. As Mitra (2010: 60) correctly points out, "the crossover from the cyber community to real life poses a significant concern. There is no guarantee that cyber community identities are completely truthful, so it would be unwise to assume that the online *persona* is the same person in real life. The lack of face-to-face contact, other than through video cameras, removes the best way of judging the true identity of the other person. Relationships and identities are based completely on digital representations, suggesting that sufficient caution is needed before cyber community relationships move into real life."

the sense of physical and social coexistence helps to bind people more closely in wireless communities, leading to 'contextual communality' (ibid.: 123)." Similar ideas will have to be developed for the role of public wi-fi connections in community building (see Hampton et al. 2010).

Of course, being the node at an intersection of hybrid networks entails the non-stop assessment of one's identity and status inside these networks. In this sense, the language and the discursive roles that users adopt in interactions on the Net (for example by assuming or controlling the conversational floor, exhibiting strategies of textual oralization, etc.) are useful ways to undertake this assessment. Examples would be the role of language in instant messaging, what in Chapter 5 will be labelled *ambient awareness*. And the same applies to social networking sites, where the number of posts, the times a post is commented upon, the impact of one's photos and texts on other users in the network of friends, etc. shape users' identities and their prestige in their networks. This obsession with determining one's position in the networks explains why (especially) adolescents engage in the time-consuming and absorbing routine of checking people's profiles and revising their own (Livingstone 2010: 476).

6. Virtual identity

In general, it can be stated that a virtual identity is shaped by using and exchanging texts, pictures or multimodal discourses with other users.[16] This entails a challenge for these users, who have to pay special attention to group demands for an optimal exchange of information, often beyond personal identity construction (Foster 1997).[17] For analysts such as Gheorghiu (2008: 60–61), the social

16. In previous research (for example Yus 2001a), the exchange of texts between users was emphasized as a main source of identity formation. But it is obvious that the evolution in the different *cyber-media* for Internet-mediated communication has favoured the increasing role that other discourse types (e.g. pictures, videos or any multimodal combination) play in today's identity on the Net. See, for instance, Davies (2007) for a study of the role of exchanged pictures (through *Flickr*) in the formation and assessment of identity, both in its social and individual application.

17. This group/individual dichotomy is related to the two most basic forms of characterization that humans use for labelling others. According to Goffman (1983: 176), the characterization that an individual can make of other people thanks to the ability to see and hear them directly is organized around two basic forms of identification: one of a *categoric* quality (which implies placing them in one or several social categories), and the other of an *individual* attribute, which assigns a unique identity to those people based on physical appearance, tone of voice, proper name or any other source of personal differentiation. This double source – categoric and individual – is essential for interactive life.

or cultural component of identity is essential, since it provides individuals with a feeling of belonging and a number of patterns for behaviour. The users can, in this sense, understand each other according to specific rituals, interpersonal interactions and social prestige. Gheorghiu concludes that collective identity surpasses personal identity and that the Net generates, above all, “mass human prototypes.”

This “social requirement” affects several aspects of Internet-mediated communication, for example turn-taking in synchronous online conversations (Kollock & Smith 1996: 115), thematic maintenance in asynchronous fora (Fernback 1997: 43–44) or the assumptions that are supposed to be mutually manifest to all the members of the community (Bruckman 1996). In short, then, the social context and the personal contribution to the community by using certain discursive forms (of a textual, visual or multimodal kind) define one’s virtual identity. Moreover, the inherently human tendency to form social networks as an *anchorage* of identity (Milroy & Milroy 1992; Milroy 1978, 1992) is also present in virtual communities (Paolillo 1999; Garton et al. 1997).

Several studies have analysed the process of multiplicity (and the parallel effect of fragmentation) of identities in the online/offline divide, and emphasis should be placed on the pioneering research by Turkle (1994, 1996a, 1996b, 1996c, 1997, 1998, 1999, 2011; see also Wortzel 1998, Brody 1996, Davis 1999). Analysts such as Newitz (1995) suggest that, in fact, people do not turn into *different people* in either of the environments (offline/online), but provide a different image, divide their identity into physical and virtual sides of the self. The virtual self may exhibit attributes that the user does not want to show in physical settings, without losing the core identity. This is what happens, for instance, to people for whom the suppression of their “body anchorage” on the Net produces a liberating effect (see Ardèvol & Vayreda 2002, Ellison et al. 2006: 418). This lack of corporeality in virtual scenarios underlies Subrahmanyam & Šmahel’s (2011: 62) claim that users do not have a physical presence when they are online:

> individuals have a “virtual representation” rather than an actual physical presence within digital contexts. A virtual representation is a “cluster” of digital data about a user in a virtual context and includes a name or more accurately, a nickname/username, email address, online history, and status within that virtual setting. In other words, it is simply a user’s face and body within that particular digital context. Individuals can have different digital representations in different online contexts.

Turkle (in Brody 1996) draws a dividing line between people who suffer from split personality, with non-overlapping and fragmented physical/virtual identities, and those who are fully aware of which virtual identities they have created.

These users combine different aspects of their selves and easily shift from physical to virtual identities, thus experiencing a fruitful combination of both that challenges the traditional idea of the self as unitary and unique (see Wynn & Katz 1997, Sweeney 1999). As I have pointed out above, the tendency nowadays is towards an amalgamation or hybridization of physical-virtual interactions with the user as a node in a dense intersection of mixed interactions.[18] The user's identity should also undergo a similar process of hybridization depending on the environment in which it is exhibited.

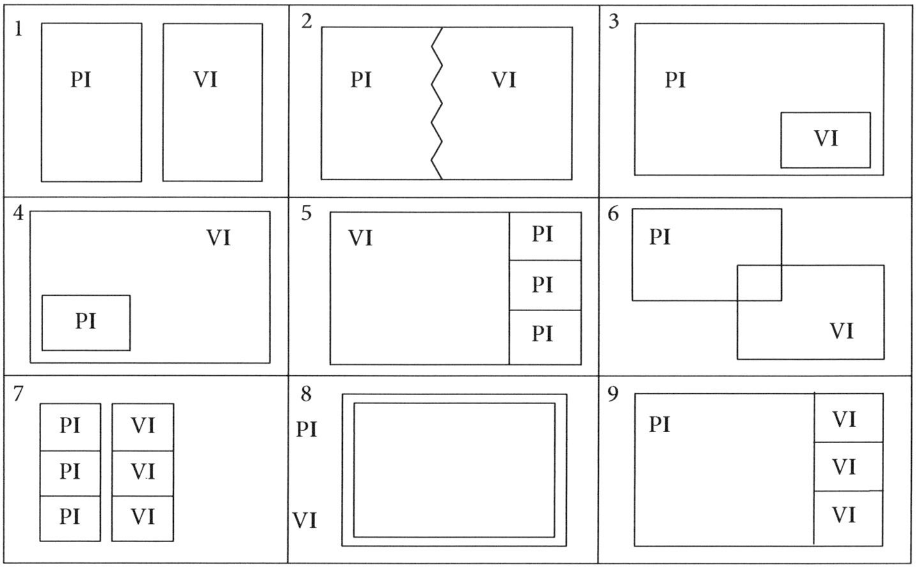

Figure 2.3 Relationships between physical identity (PI) and virtual identity (VI)

18. In this sense, N. Jones (1997: Chapter 3) proposes a classification of the virtual self into (a) *Self*, the human being in front of the computer in the physical world; (b) *Metaself*, the presentation of self in the virtual world, the self that other users perceive, a version of the physical self that the user varies and modifies at will or unconsciously; and (c) *Metafictional self*, a manifestation of a portion of one's self inside a fictional environment such as the MUDs (multi-user dungeons, or more recently multi-player online games), created consciously as a form of alternative (meta)self within the boundaries of the virtual world. Another division is Brewer & Gardner's (1996, quoted in García Gómez 2010: 140) into *the individual self* (those personal characteristics that make the self different from all others); *the relational self* (defined by the dyadic relationship that assimilates the self to significant others; and *the collective self* (the group characteristics that differentiate 'us' from 'them').

The term *faceted identity* also fits this scenario. As Farnham & Churchill (2011) stress, cultural representations of the self tend to favour more individualistic notions of protagonists who struggle to achieve their one "true" identity across situations. But for many people their identity is *faceted*, in the sense that different aspects of identity are performed depending on context, and this is transferrable to this process of physical-virtual hybridization of identities. In this sense, in Figure 2.3 several possible combinations of physical identity (PI) and virtual identity (VI) are provided:

In (1) the user has perfectly delimited physical and virtual identities. There is no overlapping between them and the user values both forms of identity with the same strength. A similar case is (2), where certain aspects of physical and virtual identities *invade* each other, with qualities of each type of identity fitting specific aspects of the user's overall identity.

By contrast, (3) portrays the case of users whose identity is shaped mainly in physical scenarios, with minimal sources of identity on the Internet. This is the case of occasional users of the Net, who still have not benefited from it and hence the role that the Net plays in their identity shaping is minimal. The opposite would be case (4), in which users hardly find any sources of identity in physical contexts but feel at ease on the Internet, where they can "be themselves" and their identities are shaped accordingly.

In (5) the virtual identity occupies most of the identity formation of the users, with several more fragmented and *ad hoc* physical identities. It is on the Net where these users find their main sources of identity. An example could be users who interact in different physical contexts and adopt *ad hoc* instrumental identities for these scenarios, none of which really shapes the individual's main identity. And it is on the Internet where these users find the true medium of expression, consolidation and support of their identities, which are much stronger than those fragmented *ad hoc* physical identities.

In (6) users have physical and virtual identities with similar weight in their daily lives and with a certain amount of overlapping between them. These users are aware that several interactions (and sources of identity formation) are only available online, and that certain interactions among users will probably never happen offline. At the same time, several interactions and several areas of their identity are valid for both scenarios, which justifies the overlapping area.

In (7) users have a number of fragmentary identities that form their global identity. These partial identities apply to both physical and virtual environments. As Androutsopoulos (2006) argues, users do not necessarily have to reproduce their offline identities when they are online, but stress or favour certain aspects of their identities depending on the environment where they are interacting. This opinion is shared by Turkle (in Davis 1999: 72), for whom playing with identities in several computer windows on the screen is a parallel phenomenon to the multiplicity of identities adopted in physical contexts.

In (8) users do not differentiate between physical and virtual sources of identity, maybe because they live both as equally natural and valid sources, and neither of them is prominent. Many adolescents fit this case, since they "jump" from physical to virtual contexts without even noticing that they are changing environments or feeling that there is a loss when moving from the physical context to the virtual one.

Finally, case (9) is the opposite of case (5), since it is now the physical identity that occupies most of the source of identity for the users, with a few fragmented and partial online identities meant for specific purposes. This is a frequent case, since it is difficult to avoid the physical anchorage of the person even when logged onto the Net.

As we can conclude from Figure 2.3, there are many possible combinations between physical and virtual sources of identity, and for many Internet users the virtual sources may be a valid (rather than added) alternative to the physical ones, and they may even overcome the latter in terms of strength or fill the gap of poorly-developed physical identities, as exemplified in the opinion of a user quoted below:

(1) I didn't really have a social life before. But now I've got one, I don't leave my room. (quoted in Welford 1999)

In the past, this multiplication or diversification of identities was clearly supported by text-based communication. Even nowadays typed text is still important in the shaping of users' virtual identities (for example chat room messages, instant messaging, comments and posts in blogs and social networking sites, e-mail, etc.). Of course, the evolution of the discursive properties of *cyber-media* has made visual sources of identity more prominent and important (as in *photologs*), together with multimodal combinations of text, sound and pictures (as in *YouTube*). For example, Schwarz (2010) comments on how self-pictures in *Flickr* have an identity-shaping role for adolescents in terms of public awareness of one's presence there: "*Flickr* may be described as a social space in which users compete for other users' attention (represented by each photo's view-counter); for public recognition of their technical and artistic competence [...] and even for a specific form of social capital (a web of contacts, objectified in each user's 'contacts list')." Similarly, social networking sites contain multimodal profiles that "function as 'digital bodies' which identify a person and constitute the end product of self-reflexive identity production" (Georgalou 2010: 42; see also Kim & Dindia 2011).

A pragmatic consequence of the variability in *cyber-media* is that, depending on the informative richness of the medium and its evolution in the oral-written, verbal-visual and synchronous-asynchronous dichotomies, the addressee users will have to make a greater or lesser effort to compensate, inferentially, for the loss of contextual information in the messages being processed. The presumption of relevance that every text holds must be complemented with a *presumption of honesty* in the way users present themselves to other users on the Net and influence their identity-shaping through sustained interactions.[19]

7. The personal web page

In the late 1990s, one of the most common forms of self-presentation on the Internet was to own a personal web page, with the aim of providing users with information about one's life, interests, hobbies, etc. (see Wynn & Katz 1997). Nowadays, by contrast, personal web pages are being replaced with other forms of self-expression on the Net, such as blogs or profiles in social networking sites, which are easier to edit and with more options for interactions. Personal web pages only remain in academic or scientific contexts. They are used, for instance, by university teachers to list their publications, etc. (see Lamb & Davidson 2002, Thoms & Thelwall 2005).

In general, but to different degrees depending on the options for *real* interaction between authors and readers, the manifestness of information on the personal web page rarely reaches a true level of mutuality, that is, there is no certainty that the information on the personal page will end up mutually manifest to both the author and the reader. An exception would be the e-mail address on a page that allows for certain feedback on its content (Miller & Mather 1998, Jackson 1997, Margolis & Resnik 1999). However, for Miller (1995) this lack of mutuality between authors and readers may have a liberating effect on the users when presenting themselves on the Net: "on the Web you can put yourself up for interaction without being aware of a rebuff, and others can try you out without risking being involved further than they would wish."

On the other hand, the web page is "published," it acquires a certain autonomy from the author, just like novels. This quality allows for the creation of what

19. I agree with Androutsopoulos (2008) when he makes a distinction between the analysis of "static" sources of identity (screen-based), such as self-presentations in blogs and social networking sites, and interaction-centred participatory sources of identity (face-to-face or mediated), and both sources are inherent objects of an ethnography of Internet communication.

has been called the *autonomous media identity*, common to all forms of discourse transferred to other people through media discourses. A web page designer (quoted in Chandler 1997) comments: "my web page [...] mediatively interacts with other people in my absence [...] The images we have of ourselves and which others have of us gain a life of their own independent of our presence" (see Yus 1996a: 24–29, 1996b).

8. The nickname (*nick*)

The nickname (or *nick*) is another form of self-presentation on the Internet. In synchronous *cyber-media* such as chat rooms, nicknames are frequent and often compulsory, and it is logical to ask ourselves what relationship holds between the *nick* and the real user, or whether there are connotations that the choice of a *nick* makes manifest, perhaps beyond the user's will, that is, whether the *nick* plays a role of *opaque mask* behind which it is impossible to guess what the person using it is like or, rather, whether it works as a *translucent filter* that allows for the inference of certain information about the user who has chosen it (see Diago Marco 2002).

The *nick* is, to a certain extent, similar to the proper name.[20] In general, proper names may function referentially ("I've seen *Peter*") or connotatively ("Peter is an *Einstein*"). Within the framework of this book, proper names, in their referential function, entail the formation of a number of encyclopaedic assumptions related to the referent of the name. Besides, if there are several competing referents for the same proper name, the hearer will have to disambiguate them as one of the inferential operations leading to the explicit interpretation of the utterance (*explicature*), and contained in a process of interpretation guided by the search for relevance in the utterance being processed (see Marmaridou 1989). In their connotative function, proper names activate in the hearer a number of implicated assumptions prompted by the information that the name makes manifest. These implications are beyond mere reference, but the hearer will be willing to extend

20. There are intense philosophical debates on proper names. For instance, there is a discussion between the Fregean and the Kripkean approaches. As Rivas Monroy (1996) summarizes, for Frege the referent of proper names is mediated by the sense, and hence any individual or object that satisfies the definite description associated with the proper name is its referent. For Kripke, by contrast, the proper name is a rigid designator, that is, it always designates the same individual in any possible world in which the individual may exist. There are also discussions on the scope of the reference of proper names, with Recanati's (1993) research on *direct reference* as one of the main analyses. However, these discussions go beyond the scope of this heading on *nicks*.

context to yield them as part of his/her interest in obtaining the highest relevance from the speaker's utterance.

In this way, the hearer of (2a), where the proper name is used referentially, will develop its logical form to reach a fully contextualized proposition (2b), often after a process of disambiguation:

(2) a. Peter: "I've seen Tom this morning."
b. [Peter] has seen [Tom Smith?] [during the morning of the day in which he has uttered (2a)].

In (2a), the hearer will take the proper name as part of an ostensive communicative act that carries the presumption of its eventual relevance, an act in which Peter is trying to make mutually manifest to himself and the hearer some information (a set of assumptions) concerning the referent of the proper name, Tom Smith. Similarly, the hearer of (3a), which contains a proper name used connotatively, will extract the necessary contextual assumptions that allow him/her to derive implications such as the ones listed in (3b–c) (adapted from Marmaridou ibid.). The eventual extensions of context and the responsibility for the derivation of these implications will be subject to the relevance-related balance of cognitive effects and mental effort while processing (3a):

(3) a. Peter: "Thomas is an *Einstein*."
b. Thomas is very clever.
c. Thomas is very good at maths.

It should be noted that in this case Peter does not intend his interlocutor to find a referent for Einstein, but hopes that he/she will manage to find the necessary contextual information that makes it possible to derive the intended interpretation of the proper name. Besides *strong* contextual implications (*implicatures*) such as (3b–c) that (3a) makes highly manifest, the hearer may also derive other *weaker* implications, perhaps not supported by Peter, and for whose derivation the hearer would be partly (or wholly) responsible, but which are also initiated by the processing of (3a), such as the implications listed in (4a–c):

(4) a. Thomas used to fail when he was at school but he turned out to be very clever.
b. Thomas' haircut is a mess.
c. Thomas thinks that everything is relative.

As will be commented upon below, the nickname does not seem to fulfil the same referential function as proper names since it does not link the name to the identity/referent of the person who uses it. Instead, it is used with the intention of masking one's identity. However, this is not always the case. Concerning the connotative function, a nickname can convey information on a number of assumptions that the person using it intends to make manifest in a specific context.

Nicks are of course omnipresent on the Internet and are often a requirement for entering conversations in chat rooms. Moreover, nicknames are used in physical scenarios. As de Klerk & Bosh (1999: 2) stress, the nickname allows one to manipulate social conventions when naming people, and therefore it is not surprising that they are particularly frequent among adolescents. For them, the nickname is a symbol of group membership, and provides a feeling of familiarity, of belonging.

On the Net, the *nick* is often used with the intention of concealing the user's identity (Jaffe et al. 1995, Macdougall 1999), but sometimes it is possible to draw conclusions from the choice of a *nick*. For example, in Ruedenberg et al. (1994), Danet (1996a) and Danet et al. (1998), among others, several *nicks* are analysed and several conclusions are obtained from them. My opinion is that this *exuded* information is more these researchers' responsibility than information intentionally made manifest by the user holding a specific *nick*.

Finally, according to Liu (1999) the instability in the use of *nicks* (and parallel instability of identities behind them) comes from the loose rules that govern their choice and use. The lack of restrictions for using them opens up possibilities of which users can take advantage. They can use a different *nick* every time they enter a chat room or keep a single one throughout the sessions. They can change it at will and for any reason (Reid 1994: 35–36). They can do it on purpose (to avoid an unwanted interaction). Finally, although each participant can use only one single *nick* and every *nick* is linked to a single user in one session, it is possible for several users to choose the same *nick* in different sessions.

Nevertheless, there are also chat rooms whose participants have to register their *nicks* and, together with their e-mail addresses, they become linguistic markers of identity that resemble the referential function of proper names (see some users' opinions and comments in Gómez 1998). The software even warns new participants that a *nick* just chosen belongs to another user.

Nicks may also make (mutually) manifest between users the intention to communicate a number of assumptions related to the choice of a certain word as a *nick*, that is, they can also be used connotatively. This use is subject to the existence of contextual information of an encyclopaedic (and often stereotypical) kind that is accessible to all the users in the synchronous conversation (it belongs to their mutual cognitive environment) or else the *nick* might be misinterpreted. But even in this hypothetical case, the users will never be sure of the other users' honesty in using a *nick* or of the underlying intentionality in making these connotations manifest.

CHAPTER 3

Relevance on the web page

1. The web page genre. Intention and manifestness in the interpretation of a web page

Since the vast amount of documents on the Internet were unified under the same "html" protocol (the now ubiquitous *World Wide Web)*, the number of web pages, websites and web portals has increased enormously (to the extent of putting the stability of the whole system in danger). Therefore, the role of search engines such as *Google* has proved to be essential when we surf the Net in order to obtain relevant information and avoid, if possible, the *infoxication* (intoxication due to an excess of information to be processed) that many users suffer nowadays (see 3 below).[1] In short, users surf the Net guided by the psychologically rooted human tendency to search for relevance in incoming stimuli. The different discourses that the Net holds nowadays are really varied (verbal, visual, multimodal) and worth being analysed by the methods of pragmatics.

Schneider & Foot (2004: 16) list four aspects that have aroused the interest of web analysts:[2] (a) which communicative actions are established through the Net and how they change over time, (b) how the web page designers' decisions influence the users' activity on the Net, (3) what kind of user experience is fostered on and between web pages, and (4) how web page authors relate to other authors and to their readers by means of discourses and links. Relevance theory can provide an answer to these questions by focussing on aspects such as the author's (or designer's) intentionality, the content of the page and its arrangement on the space framed by the screen (content would be treated as coded evidence of underlying

1. For an evaluation of the performance and efficiency of the five most used search engines (i.e. *Google*, *Yahoo!*, *Live*, *Ask*, and *AOL*), see Deka & Lahkar 2010).

2. They also list three groups of approaches to the study of the Net: (1) The rhetorical or *discursive analyses* of web pages (more interested in the content of the page than in its design or structure); (2) *structural analyses* (usually focussing on a single page as the basic unit of analysis and on its structure, hierarchy of content, etc.; and (3) analyses that stress the role of hyperlinks. Schneider & Foot (ibid.: 118) propose their own model, called *web sphere analysis*, centred upon communicative actions and inter-relations between the designers of web pages and their users within a certain time span.

intentions), and the users' inferential steps while attempting to interpret the content of the page correctly, guided by the principle of relevance. In other words, there are three perspectives from which a cognitive pragmatics analysis of web pages can be undertaken (similar to the ones for the processing of other types of text, see Yus 2002b):

1. From the author's point of view (*intentio auctoris*). Web page authors offer their readers (make manifest to them) coded information on the page of a verbal, visual or multimodal kind. These pages are normally inter-related by hyperlinks that direct the reader either to another area of the same web page or to other pages on the Internet. Therefore, there is very often no prearranged reading path expected or intended by the author. But in general a cognitive expectation exists: that the content of the page should be interesting enough to compensate for the mental effort that the reader will have to make in order to process it. Hence, the role of the author is to select the content of the page according to expectations of relevance, predictions of context accessibility, etc. The author also has to devise or design its interface (if allowed to do so) aiming at relevance, that is, an interface that provides a sufficient number of cognitive effects (i.e. interest) in exchange for as little effort as possible. This aim is particularly significant on the Internet, where a simple click on a link can take the reader to an immense array of pages.

As in other forms of written or published texts, the author of a web page is not normally present when the reader accesses the page. This is why many discourse analysts have dismissed the possibility of tracking the author's intentionality when he/she created it (e.g. the death of the author claimed by Roland Barthes, or the so-called "intentional fallacy"). Like in literary works, it is the web page authors' responsibility to predict which assumptions are manifest to their readers, but there is no guarantee that these assumptions will in fact belong to their cognitive environments. Similarly, there are fewer chances of achieving mutual manifestness of these assumptions than there are in face-to-face conversations. However, today's web pages offer more options for interactivity that have become conventionalized as part of the web page genre (and hence expected by Internet users) and provide more options for mutuality of assumptions as well. This is the case of forms to be filled online (typical of blogs, see Chapter 4) and applications for synchronous conversations with experts that can be found on pages for e-commerce. In any case, the relevance-seeking pattern invariably applies: a reader's inferential steps intended to turn the content of the page into fully contextualized information that matches the author's underlying intentions, as in traditional literature (Pilkington 2000: 66, Gibbs 1999: 177). Of course, there are additional, more specific issues that are involved in the processing of web documents and that might play a part in the eventual balance of cognitive effects and mental effort.

One of them is the user's ability to command the software for surfing the Net and making the most of link-mediated discourses. If the user is not familiar with this software there might be increased effort without an offset of additional effects in return, to the extent that different readers with different command of software use might get radically different interpretations of the same page.

2. From the textual or discursive point of view (*intentio operis*). For several analysts, texts are stable sources of meaning, and therefore they should be autonomous holders of "the meaning" conveyed (intended or otherwise) by the web page. From the cognitive pragmatics point of view, however, texts normally (if not always) underdetermine the information that their authors actually intend to communicate through them. There is always an information-gap that has to be filled inferentially. Pilkington (2000: 25) argues that the linguistic or structural features of literary texts are significant insofar as they favour specific forms of pragmatic processing. This idea is applicable to web page processing, since these pages are designed and filled with content with an underlying intention to obtain an interpretation (or at least "some interpretation") from the reader (who will use the content plus context as feeds for inferential activity), although very often the reader is free to take full responsibility for this interpretation, even to go beyond the author's intention. Web pages would be examples of "public representations" that, in a more or less faithful way, resemble the thoughts or assumptions that the author intends to make manifest. But of course there is no guarantee that the reader's interpretation will match this intention. Readers often misinterpret authors, texts are typically open to multiple interpretations (all compatible with the coded text), and they also convey feelings and emotions that might well be misunderstood.[3]

However, processing effort and chances of misinterpretation can be reduced by the conventionalisation of the web page genre and its typical strategies for content and interactivity, which authors follow as part of genre conformance and readers expect as part of what in Chapter 4 will be called *interiorized schema*. As Nielsen (2002) summarizes, two main strategies have been used when generating, stabilizing and disseminating the web page genre: "(a) authors or designers tended to copy the form and content of existing homepages when making their own homepage or (b) they consulted the enormous range of handbooks on web design and web writing, which began to swamp the market." So, for these authors

3. Lüders et al. (2010: 952) make an interesting proposal: that web genres are at an intermediary level between the levels of media and text, but influenced by both. In a nutshell, "genres are both enabled and constrained by the actual medium. In some cases, the Internet is likely to influence the level of the genre, as in hypertextual news articles. At other times this influence is marginal, as in the case of research articles in pdf-format available on the net."

we have reached a status in which we can talk about a stabilised web page genre, to the extent that this genre influences how content is provided, structured and arranged on the page, and also how content organization is foreseen by the web page readers. Moreover, web pages are inherently inter-related to other pages and sites. Therefore, even though we can aim at a single web page as a typical instance of genre, the scope of analysis may also be broadened. For example, Santini et al. (2010: 11) propose a three-fold distinction: (a) on the *micro* level, to analyse page-level units and their constituents as self-contained (though not necessarily the smallest) manifestations of web genres. These then enter into websites as more complex web genre units; (b) on the *meso* level, to study single or conglomerate websites and their web-specific structure formation; and (c) on the *macro* level, to deal with the web as a whole from the perspective of complex network analysis and related approaches.

Undoubtedly, the stabilized web page genre has evolved enormously in the last few years from the static quality of the initial stages of web page design. Nowadays, users expect to find certain modern elements of pages such as frames, tags, etc as well as dynamic forms of interaction with the interface of the page. These features have become part of the modern web page genre, so even if there is a constant development in this genre, users rapidly adjust the aforementioned *interiorized schema* to what they can expect when entering a page.[4] Besides, as in more traditional genre theory, the author's expected interactivity of the page is reflected upon a number of "moves" that indicate the stabilization of the web page genre, and these moves will also be expected by the readers of the web page.[5] For example, for corporate web pages Askehave & Nielsen (2005) list the following conventionalized moves: attracting attention, greeting, identifying sender,

4. As such, web pages would fit the features that, according to Berkenkotter & Huckin (1997, quoted in Caballero 2008: 18), constitute modern genres: (1) *situatedness* (the stabilized mental web page schema derives from active participation in web page processing), (2) *community ownership* (web pages exhibit agreed conventions of design and use as shared by a broad community of users), (3) *duality of structure* (web pages constitute social structures and simultaneously reproduce them), (4) *form and content* (engaging in web page production and processing involves the knowledge of the most suitable content or topics to this genre, as well as the lexical and structural resources which suit purposes and needs involved in web page use), and (5) *dynamism* (social changes may prompt changes in the web page genre as well). This last attribute is important for Internet genres, which are under constant (and somehow competing) processes of stabilization and evolution.

5. In order to refer to the different rhetorical strategies in the presentation of information, genre analysis has used the notions of *move* and *step*. As explained by Swales (1990: 228–229) the term *move* refers to a discoursal or rhetorical unit that performs a coherent communicative function in a written or spoken discourse.

indicating content structure, detailing (selected) content, establishing credentials, establishing contact, establishing a (discourse) community, and promoting an external organisation.

In any case, in my opinion web genres are especially interesting as evidences of authors' communicative intentions and interpretive predictions and as key elements influencing the quality of readers' interpretation. Karlgren (2010: 34) points in the same direction: "genres have utility and a purpose in that they aid the reader in understanding the communicative aims of the author; they provide a framework within which the author is allowed to make assumptions on the competence, interest, and likely effort invested by the reader." Authors and readers of web pages, then, constitute the foundations of the web genre, just like any genre: "when the character of text in a typical communicative situation is formed by or based on the bidirectional flow of authors' expectations on their audiences and that or those audiences' expectations on likely behaviour on the part of the authors they are reading, those items, or that family of items, constitute a genre" (ibid.).

3. From the reader's point of view (*intentio lectoris*). If the main objective of relevance theory is "to identify underlying mechanisms, rooted in human psychology, that explain how humans communicate with one another" (S&W 1986: 32), then the quality of readers' interpretations of web pages should, together with the analysis of intentionality, be an interesting issue for research, specifically addressing which inferential steps readers have to take in order to obtain relevant interpretations, or how contextual information is combined with the propositions made manifest by the text on the page (or pictures or multimodal combinations), and which implications can be derived from this combination. In this scenario, the role of the reader is also interesting, because he/she is no longer a passive entity that processes a single text mode in linear sequence. On the contrary, on web pages

> the reading process is not only interrupted because of the graphical frame structure of homepages (similar to newspaper front pages), but also by the users' modal shifts -where they either read, listen, or watch depending on the nature of the media. What is more, the multi-medianess of web texts supplies the texts with a rich polysemous potential where the web user is "invited" to participate actively in assigning meaning in the process of text consumption.
>
> (Askehave & Nielsen 2005: 125)

Furthermore, the specificity of web pages, with their link-mediated inter-relationship, provides additional interest for researchers. For example, it would be interesting to study how relevance is obtained from fragmented chunks of inter-related texts without a pre-arranged reading path, or the role of cognitive effects and processing effort in the time spent on a web page or in the actual reading sequence selected by the user.

Besides, as in any asynchronous discourse, the readers have to compensate for the loss of contextual information that is available in situations of physical co-presence between interlocutors. In personal web pages, for instance, the users' self-presentation is often limited and incomplete (Döring 2002). For other researchers, however, web pages might well resemble oral conversations if users fill in the informational blanks correctly. For example, the *hyperpersonal model* (Walther 1996) predicts that web pages might be even better forms of self-presentation on the Net if the users are able to present and infer the information in an efficient way.

2. Relevance (in information retrieval systems)

This chapter aims to apply relevance theory to Internet-mediated communication, in particular the relation/interaction established between authors (addresser users) and readers (addressee users) with clear-cut roles for both of them:

On the one hand, authors have to devise the content of the web page, link this content to other areas of the same page or to other web pages, predict the readers' accessibility to the contextual information that allows for relevant interpretations, and expect that the eventual balance of cognitive effects[6] and mental effort when processing this content will yield relevant outcomes.

On the other hand, the reader has to assess competing interpretations that are compatible with the text (or pictures, video, etc) coded on the web page and calculate the balance between cognitive effects and mental effort resulting from this processing, paying special attention to the information that fruitfully combines with the reader's cognitive environment.

So far, relevance theory has not been extensively applied to users' interpretations of web pages created by other users. However, several analysts have applied it to the outcomes of search engines such as *Google* and other computational agents and information-retrieval systems (IR systems), or have used a broader concept

6. As I have already pointed out, when I refer to "cognitive effects" I mean *positive* cognitive effects, beneficial to the user. Indeed, initially S&W only mentioned "cognitive effects," but in the second edition of *Relevance* (1995) they specified that the information that leads to negative effects should not be labelled as relevant. As Higashimori & Wilson (1996: 3) summarize, with this differentiation S&W wanted "to capture this intuition by distinguishing two types of cognitive effect: positive cognitive effects (e.g. true beliefs), which contribute to cognitive efficiency, and hence to relevance, and others (e.g. false beliefs), which are not worth having, and hence irrelevant."

of "relevance" as central in their research.[7] In this case, the object of analysis is the estimation of relevance measured when a system used for extracting and indexing information lists results according to the users' typed text (query). It is, hence, a "system-to-user" interaction and not a "user-to-user" type of interaction and therefore not an inherent object of study for pragmatics. Nevertheless, the application of relevance theory to IR systems is interesting, because it provides insights as to how users combine cognitive effects and mental effort in their interactions with IR systems.

In this sense, Kalyanaraman & Sundar (2008: 219) assert that the cognitive activity carried out when users interact with other users is not radically different from the one performed when using a computer system: "individuals respond automatically or 'mindlessly' when cues that elicit social responses are present in media messages and form impressions when responding to computers or other new technologies, just as they would in human-human interaction." In other words, they claim that *mediated person impression formation* is similar to *mediated technology impression formation*, as long as the system interface is capable of producing an illusion of interactivity (see Patel et al. 2006: 30), or an illusion of non-mediation (Kumar & Benbasat 2001). Of course, obtaining relevance is different in system-to-user and user-to-user communication, especially because of the influence of the design of the interface in the estimation of relevance (see White 2007a: 536). In fact, computer systems are "idiots savants" when compared to human users, "tremendous at remembering millions of bibliographic details, but pea-brained at critically evaluating them. Even the best systems may respond poorly or not at all to people's questions" (White 2007b: 584). Perhaps the latest trend in search engines such as *Google*, with interesting algorithms for ranking results by relevance, are advancing really towards more intuitive outputs fitting the user's search for relevance.

Many studies, which can be traced back to the 40s, have addressed the role of relevance in the output results of a system.[8] All of them share, to a greater or lesser extent, the emphasis on the idea that relevance entails the analysis of users and systems, but there are blurred boundaries in the application of this emphasis. For example, Greisdorf (2000: 67–68) stresses that it is the relationship between users and output results (more than the acceptance or rejection of these results) that has been ignored in the bibliography, and this relationship allows for

7. See Park (1994), Harter (1992), Spink, Greisdorf & Bateman (1998), Greisdorf & Spink (2000), Greisdorf (2000, 2003), Brouard & Nie (2004), Saracevic (1996, 2006, 2007), Spink & Gresidorf (2000, 2001), and White (2007a, 2007b), among others.

8. An interesting *time line* about the use of the term "relevance" can be found in Mizzaro (1997).

degrees of relevance instead of being assessed as an absolute judgement. It is the users' intuitions of relevance that have to be studied; not the presence or absence of relevance, but the degrees of relevance as ranked by the users (see Saracevic 2006: 11). Besides, Bodoff (2006) holds the opinion that it is more important to distinguish between the relevance obtained when one is surfing without a precise objective (*browsing task*) and the one obtained with a specific purpose (*focussed search*). This distinction has important consequences, for example in the field of e-commerce (Detlor et al. 2003: 82).

From a relevance-theoretic perspective, the most important element in the triangle system-output results-user would be the user, whose judgement is guided by a search for relevance when clicking on one of the links from the output results of the IR system. In other words, instead of *objective* (or *topical*) relevance, centred upon the extent to which the output results match the search query typed by the user, relevance theory would be more interested in a user-centred approach, analysing whether the user estimates that the results are relevant to his/her search for relevance (Harter 1992: 604).[9] Schamber et al. (1990: 774) also stress the users' role when they treat it within the three possible approaches to relevance.[10] Finally, Saracevic (2006) lists a number of attributes of the term "relevance" that, again, emphasize the user's role in interactions with IR systems:

1. *Relationship*. Relevance arises from the relationship of certain attributes that include people besides information. In this relationship, there are degrees of relevance in the combinations of the information from search outputs and the user's cognitive environment in order to produce certain effects.

2. *Intention*. Relevance has to do with intentions, which are also essential in the relevance-theoretic model (for S&W the only real communication is the *ostensive-inferential* one, based on the fulfilment of *communicative* and *informative*

9. By this I do not mean to dismiss the importance of *objective relevance*. In White (2007b: 585), for example, this is defined as the situation in which a document provided by a system and a query typed by a user are, to a certain extent, similar or related to the same topic. White stresses its importance in the sense that the results should not depart radically from what the user has in mind when typing the query. However, if there are different levels of acceptability depending on how closely the output result matches the words typed, it is difficult to foresee which results will end up being relevant if we do not incorporate into the analysis the user's specific needs and cognitive environment to the analysis at the moment of making the search.

10. (a) Relevance as a multidimensional cognitive concept whose meaning depends on the users' opinion about the information and their needs in a specific situation; (b) relevance as a dynamic concept that depends on the users' judgement about the relationship between the information and their needs in a specific situation; and (c) relevance as a complex concept but systematic and measurable if it is assessed from the user's perspective.

intentions). Similarly, web page authors (or designers) intend to make manifest a number of assumptions. This information will be relevant to varying degrees depending on the readers' cognitive environments and on the way this information is (or is not) fruitfully combined with these environments and the users' specific needs. However, search engines such as *Google* do not distinguish degrees of relevance, but automatically offer results based on certain algorithms.

3. *Context.* Relevance is obtained in a context and makes no sense outside it. Saracevic divides it into *internal context* (of a cognitive or affective quality) and *external context* (centred upon situations, tasks, problems, etc.), to which social or cultural issues can be added. In my opinion though, there is no need to distinguish types of contexts. Of course, context is essential because the same information can be very relevant in one context and utterly irrelevant in another. But according to relevance theory, context is information ("a set of assumptions" following its terminology) that is used in the interpretation of stimuli. There may be differences as to how easy or hard it is to retrieve this information, but not in its mental quality. In the context of searching on the Net, it is obvious that the context in which the search is performed (the user's specific needs when typing the query) affects the subsequent choice of one of the output results, irrespective of the position that it occupies on the list retrieved by the system (even though users do tend to select the ones at the top of the list).

4. *Inference.* Relevance requires an inferential assessment in the form of "information gap filling" between what has been coded (the content of a web page, the text typed by a user) and the interpretation that is intended. And this inferential activity starts at the moment of choosing one result from the output list given by a search engine. The selection of results is, of course, not a simple trade-off between effects and effort, but a rather complicated task that also involves feelings and emotions. As Flavián-Blanco et al. (2011: 540) correctly put it, "the affective states or emotions experienced during the search activity, conscious or unconsciously, can determine to a great extent how an online search is performed and what alternatives are chosen."

5. *Choice.* The inference in the processing of output results provided by a search engine also requires the choice among competing sources of information in order to maximize interest and minimize mental effort. A great deal of effort is devoted to running through the results before selecting one, which may alter the eventual relevance. In Jamali & Asadi (2010: 290), some complaints made by users in this direction are quoted, for example: "I find Google a bit annoying because no matter what you put in, you get 20,000 answers back. Half of them are referring to the same thing, linked through different ways and you've got to be very, very careful what sort of search words you use."

6. *Interaction.* Inference is carried out as a dynamic and interactive process in which the interpretation may change as happens with contextual parameters as well. The same applies to the users of search engines, who possess different cognitive environments and engage in different inferential estimations of relevance from the list of results.

7. *Measurement.* Relevance involves weighing the cognitive effects that a choice offered by the search engine produces against the mental effort involved in this selection. There has been a great deal of discussion among relevance theorists (for instance inside relevance theory e-mail List)[11] about whether it is possible to quantify numerically this balance of effects and effort, but S&W opt for a more qualitative approach to it (see 4 below). "Effects" are taken to mean information that productively combines with the user's cognitive environment and satisfaction at finding (via the search engine) what he/she is looking for. The mental effort refers to the negative outcome of the search, the frustration at being unable to find relevant results or taking too long to get them. In this sense, Al-Maskari & Sanderson 2010: 866) included "user effort" as one of the variables affecting user satisfaction, and observed that

> user effort – as measured by the number of queries submitted to obtain relevant documents and the rank position in the results list accessed to obtain the relevant documents – was inversely correlated with user satisfaction. As users exert an increasingly greater effort to complete a given search task, it was very likely that their satisfaction decreased.

Curiously enough, effort may be on some occasions positive to the user. As Flavián-Blanco et al. (2011: 542) point out, "the perceptions of the effort exerted on a search process could have positive effects on users' feelings. By engaging in a search process, consumers acquire knowledge and learn from the experience, which can be used for future searches. Thus, the effort exerted on a search task could have a positive impact on consumers, as they value this effort as leading to enhanced levels of expertise."

It can be concluded that, in general, the bibliography on the term "relevance" applied to search engines and IR systems tends to focus on proposing types of relevance, rather than on a coherent (and unitary) proposal of what really goes on when a user searches for information. Most studies propose a dyadic division into:

---

11. On the web page [http://www.csmn.uio.no/homepages/nick/relevancemailinglist/] there is an archive of messages sent to this relevance theory mailing list and instructions on how to join it.

a. *Objective* or *system-based relevance*. It refers to "a correspondence between the user's query terms and the terms that are indexed and stored in the retrieval system" (Maglaughlin & Sonnenwald 2002: 328). Labels such as *topical relevance* and *logical relevance* would fit this type, because they emphasize the fact that the relevance of the output results from the system is based on algorithmic computations and not on the user's needs when typing the text.[12] This kind of relevance is of little interest to cognitive pragmatics specialists, except for the influence of the system and its ranking procedures on the user's eventual estimation of relevance. For example, it would be interesting to analyse why search engines such as *Google* are preferred by users on the basis that they provide more relevant outcomes, and to determine whether the search criteria match the users' search for relevance.
b. *Subjective* or *user-centred relevance* (see Borlund 2003: 914). This would indeed arouse the interest of cognitive pragmatics and relevance theory specialists by focussing on the user's inferential activity when offered a list of output results from a search engine. Labels that one way or another fit this kind of relevance include *subjective relevance, situational relevance* and *psychological relevance*.[13] Sometimes the label *cognitive relevance* is also proposed, referring to the link between the users (in terms of knowledge, cognitive needs, etc.) and the information that they interpret. As described by Cosijn & Ingwersen (2000: 539), cognitive relevance is inferred from criteria such as informativeness, novelty, or preferences for information, among others. This kind of relevance links the system (which has to be efficient in indexing information and offering interesting results) and the user (who has to be efficient in choosing the search text so that it leads to the expected results). Cosijn & Ingwersen admit that research on relevance is shifting more towards a more user-centred emphasis and less towards a system-centred one.

12. *Topical relevance* was proposed by Park (1994) as context-free and centred upon fixed assumptions regarding the relationship between the topic of a document and the text typed, ignoring the particular context of the users or their needs. *Logical relevance* was proposed by Cooper (1971) to describe a relevant result for the computer system, even if it has little to do with the user's needs.

13. Swanson (1986) explains *subjective relevance* as "what the user says that is relevant is what turns out relevant": the user is the one who decides, because an information retrieval system is designed to help him/her. Wilson (1973) suggests that *situational relevance* deals with the user's perception of his/her informational needs and underlines the fact that there are aspects of relevance that only the user can identify. Besides, in Wilson's view (ibid.: 458) *psychological relevance* refers to the real uses and effects of information, how people use information and how their opinions change (or not) when they are given output results.

Harter (1992:606) also makes a distinction between the *objective* and *subjective* varieties of relevance:

> relevance is thought to be a relation between a document and a question, where the question is usually a representation of an information need. Alternatively, pertinence or another aspect of subjective relevance has been used to refer to the relation between a document and the need itself. In the former sense, relevance is objective, and is measured by asking experts in the discipline to make relevance judgments. In the latter sense, relevance is subjective and personal, and must be assessed by the user himself.

Park (1994:136) stresses that a shift towards the user is the right track, since the users are the major interest of any system that provides information. And nowadays the design of IR systems seems to follow this direction, because the system interface is increasingly user-friendly and devised by and for the users. Any research on objective or topical relevance is subject to the emphasis on the users' behaviour when searching for information.

A parallel issue concerns the final relevance of the output results proposed by the IR system. In general, it can be stated that the user will tend to trust these results (Pajares Tosca 2000) by focussing on the one at the top of the list. Undoubtedly, one of the reasons for the popularity of *Google* is that its indexing system (*PageRank*) yields results that very often match the best balances of effects and effort in the user's search for relevance.

On the other hand, there are aspects of information-searching that can generically be called *exogenous features* and that may influence the user's estimation of the relevance of output results provided by the system. One of them is the interface of the IR system. Search engines such as *Google* are simple, but other interfaces have proved to alter the user's estimation of relevance. An example is cited in Kalbach (2006): several design problems were found in one of the search engines. Firstly, all the navigational elements of the page were preserved after the search was performed (utterly unnecessary), as well as flash-animated advertisements (*banners*) that made it difficult to sort out the list of results. Secondly, the output results occupied a very small area of the screen. Features such as these may discourage and "scare off" users.

An additional *exogenous* feature that should be borne in mind is the distinction between expert and neophyte users of search engines. As Greisdorf (2000:68) stresses, expert users can compensate for incomplete information by filling the informational blanks and thus making the most of all the options that the search engine offers, while neophytes will face effort-producing challenges when working out what the really interesting information is (see also Greisdorf 2003). For example, expert users of *Google* not only decide which link is the most potentially relevant, but also pay attention to the number of bytes that the document has (to determine whether it is a long or short document), to the web address at the bottom of the result (to determine whether the document is part of a collection of items, for example if it belongs to Proceedings of a Conference as in the "www.../proceedings/article.pdf" nomenclature), and to the possibility of accessing the "cache area" of the document, etc.

3. Relevance in the user who is surfing the Net

A more interesting research area for cognitive pragmatics (and within that, relevance theory) is the interaction that takes place between an "addresser user" who uploads content on a web page and "addressee users" who access the page and process this content. It is user-to-user communication and therefore a suitable object of analysis for relevance-theoretic pragmatics.

3.1 The role of the "addresser user" and the role of the "addressee user"

Web-mediated communication has traditionally been asynchronous, with the author and the reader being online at different times and unable to engage in real-time interactions. Relevance theory emphasizes that effective communication is the one in which it is *mutually manifest* to both interlocutors that the author intends to make manifest a number of assumptions (i.e. information). Upon accessing a web page, by contrast, the user is faced with information that is manifest, but without an adequate opportunity for mutuality with the author. This entails a lack of certainty regarding the interpretation or intended reading patterns intended.

Nevertheless, there is no doubt that web interfaces have evolved enormously, allowing for a much higher level of interactivity with the authors (e.g. with online comment forms, a typical feature of blogs).[14] Therefore, two scales of web page communication may be proposed:

14. Heeter (1989, quoted in Chung & Zhao 2004) proposes that interactivity on the Internet is subject to six attributes: (a) *complexity* of choice (to what extent the user is offered the possibility to select the information available); (b) *effort* that users have to make in order to access the information; (c) *responsiveness to the user* (to what extent the web interaction resembles everyday human discourse); (d) *potential* to monitor system use; (e) *degree of ease of adding information* (when the user can add information to the system and other users can access it, regardless of whether they are registered or not): and (f) the *degree of interpersonal communication* that the medium allows for.

The first scale would be centred on the author, who can provide readers with a clear-cut and linear reading sequence or can decide to simply upload information with no expected reading paths. Furthermore, some non-propositional aspects may be communicated by the author. For example, when designing a page many users resort to typically human tendencies (but also contradictory ones to some extent) such as differentiating themselves from other web authors or mimicking them, both of them potential anchorages of the user's identity (see Sundar 2008). Kalyanaraman & Sundar (2006) studied the first of these possibilities and, predictably, concluded that web page personalization (customization) is an identity marker and an effect of the need for differentiation, for shaping a unitary self against the others' selves: "Because users can recognize their own preferences in the customized output, the self as source criterion will likely result in users perceiving a greater sense of ownership of portal content, leading to increased liking for the portal (e.g., *MyYahoo!*), especially when compared to other Web interfaces that offer generic, noncustomized content (e.g., *Yahoo!*)" (ibid.: 113).

An additional feature of customized pages has to do with the user's ability to interact and obtain supplementary effects beyond the processing of the content on the page. This can be described as a process: (a) the illusion of interactivity leads to (b) a feeling of control in the user, which produces (c) the user's positive attitude towards the content of the page, and generates (d) more involvement in the user and (e) more motivation to process the content. This produces the overall effect of (f) a positive attitude towards the page or portal, which eventually may lead to (g) the development of affinities among users of the page and building up (h) a greater feeling of group or community membership.

The second scale is user-centred, depending on the level of interactivity that the page allows for and on whether the reader is looking for specific information (*scanning*)[15] or is surfing the Net with no clear objective (i.e. for the fun of it). There are, hence, many possibilities ranging from effortless processing of information to the willingness to expend considerable cognitive resources in order to obtain the expected reward.

The picture becomes more complicated if we take feelings and emotions into account. Indeed, non-propositional effects such as emotions are harder to de-

15. Choo et al. (2000) propose four types of information scanning (a–d) related to four parallel types of information seeking (a'–d'): (a) *undirected viewing*, realized as (a') *sweeping* (to scan a variety of sources, choosing which one is more accessible); (b) *conditioned viewing*, realized as (b') *discriminating* (to search in pre-selected sources or topics); (c) *informal search*, realized as (c') *satisfying* (the search is focussed on a specific topic, but a minimally satisfactory search is enough); and (d) *formal search*, realized as (d') *optimizing* (systematic retrieval of information by following a method or procedure).

scribe within pragmatics than wholly propositional thoughts (Yus 2002b).[16] This happens to the qualitative properties of emotions, which are subjective and elusive (people have a rather personal – and often poor – picture of what feelings such as sadness or sorrow are really like). However, emotions also refer to specific beliefs and desires, to states and events of the world and, therefore can be characterized in terms of intentionality.

Singh et al. (2005) proposed a model of users' experience of web pages that can be re-interpreted in relevance-theoretic terms. Two theoretical assumptions underlie this model: (a) accessing web pages generates feelings, and (b) these feelings influence the user's opinion about the content of the page. The first one deserves commenting upon. For these analysts, the user's initial reaction when entering the page is to try and make sense of it, and in this process several feelings and emotions arise. These feelings and emotions should exhibit a certain regularity across users and it should be possible to measure them (ibid.: 33). But this claim is not valid if we adopt a relevance-theoretic approach, since users possess very different cognitive environments and, at the same time, very personal feelings and emotions that are difficult to generalize, measure or comment upon.

Relevance theory has addressed the analysis of feelings and emotions mainly through the term *weak implicature*. Because feelings and emotions are often less supported by the speaker, in the sense of intending a specific interpretation of them, the implicatures that are generated tend to be weaker and require the hearer's responsibility in their derivation. Pilkington (2000: 66) applies this term to what are called *poetic effects*, which arise in the interpretation of literature and, above all, of poetry or innovative metaphors. The wide exploration of context that characterizes poetic effects produces a particular kind of brain activity with a certain aesthetic feeling. The cognitive correlate of this feeling is a subtle increase in the prominence of several assumptions, impossible to control or describe in an exact way. It would be a matter of affective mutuality, rather than cognitive mutuality, between the author and the reader.

16. Downes (2000: 100) also comments that traditionally linguistics has not addressed the analysis of experiences that are not proper thoughts, that have no propositional form, but which can also be manifested through language. Emotions and intuitions constitute a complex and varied store of human experience. Downes wonders if language possesses systematic resources to deal with emotions and intuitions, to mould and express them linguistically. Similarly, we can picture a *funnel effect* (Yus 2009f), according to which the words we have in a language (the narrow end of the funnel) are insufficient to communicate all the range of concepts, feelings, attitudes and emotions that we store in our minds (the wide part of the funnel).

3.2 Levels or patterns of interactivity

Users' ability to obtain a satisfactory level of interactivity (or *illusion* of interactivity) may influence their willingness to engage in future interactions and foster or sustain contacts on the page or portal (see Niekamp 2003, Trammell et al. 2006). It is not surprising, then, that many studies have either stressed the importance of interactivity on the web or included it on the list of aspects to take into account in the analysis of web-centred users' behaviour.[17]

From a relevance-theoretic approach, interactivity is important insofar as it favours an adequate mutuality of assumptions between interlocutors. It would also be interesting to check the different attributes that indicate the existence of true interactivity. For example, Kiousis (2002) proposed, among others, the following: (a) bidirectional communication, where (b) the roles of sender and receiver are interchangeable and (c) the speed of communication is closer to "real time." These aspects may influence the estimation of relevance during a web-based interaction. Besides, as will be studied in Chapter 4, interactivity in social networking sites is an essential element of its current popularity. This is why Sachdev et al. (2010: 591) propose a specific definition of interactivity in these sites (*social computing interactivity*) as follows: "the degree to which the interaction (user-medium and user-user) is perceived to: (a) enable control; (b) exhibit responsiveness; (c) enable reciprocal communication and social presence; and (d) provide capabilities for self-presentation and deep profiling."

Furthermore, Solanilla (2002) suggests three criteria for analysing interactivity on the Net. Firstly, an important factor is whether the interaction is established with another user or with a computer system. For this criterion, she proposes the distinction between *interactivity* (user-to-system) and *interaction* (user-to-user). Secondly, a quantitative analysis should be carried out to determine the intensity of the interaction. Finally, it needs to be established whether the interaction is public or private.[18]

---

17. This is the case of Burbules (2002), who includes interactivity in the five types of activities that can be found on the Net: (1) *movement* vs. *statism* (the structure of the page allows, facilitates or inhibits movement); (2) *interaction* vs. *isolation* (the design of web spaces also communicates assumptions and expectations of interaction); (3) *public* vs. *private* (to what extent the design allows or prevents the users' expression of their identities and activities); (4) *visibility* vs. *concealment* (transparency of the web structure or capacity to restrict access to information on users); and (5) *inclusion* vs. *exclusion* (what one decides to include on the page).

18. In a similar fashion, McMillan (2002) proposes three types of interactivity: *user-to-system* (the user interacts with the interface, for example by clicking on a link), *user-to-user* (the user engages in a conversation with another user, for example in a chat room), and *user-to-document* (the user can modify the content of a document on the web page). Again, only the second type

3.3 Availability of information on the Internet and infoxication

Another interesting aspect of today's web pages is the vast amount of information that they hold and the consequences for a correct estimation of relevance. No doubt, the attempt at processing more information than one's cognitive systems allows for (the inability to overcome the increasing *infoxication* we suffer nowadays) can have negative consequences for the eventual relevance, for example increasing processing effort without an offset in cognitive effects. Using the two conditions of relevance (W&S 2002c: 602),[19] two cognitive limits for information processing on the Net can be proposed, which satisfy these conditions for relevance:

a. everything else being equal, the greater the user's capacity to process information on the Internet at a given moment and to filter information from different sources available on the Net, the greater the relevance of that information to that individual at that time; and
b. everything else being equal, the smaller the processing effort expended by the individual in obtaining and filtering information, the greater the relevance of this information to that individual at that time.

There are, of course, many elements to bear in mind when we consider the link between the information available online and the user's willingness to process it. Among them, the following can be listed:

1. When texts that were initially published outside the Net are transferred to the electronic format, very often they have to be adapted so as to maintain similar balances (of effects and effort) to those achieved when they were first published. It comes as no surprise that, according to Giltrow & Stein (2009), this transference plays a major role in the issues that a genre theory for the Net faces nowadays:

a. Does a new medium automatically make for a new genre? Is it possible for a traditional genre in the spoken or written media to migrate into the Internet without any loss of identity? How much can be lost or changed for a genre to retain its identity? Is it possible to have stability of genre across the medium change?

---

would interest cognitive pragmatics, even if interesting conclusions on how inference works can be drawn for the other two types.

19. These are: (a) everything else being equal, the greater the positive cognitive effects achieved in an individual by processing an input at a given time, the greater the relevance of the input to that individual at that time; and (b) everything else being equal, the smaller the processing effort expended by the individual in achieving those effects, the greater the relevance of the input to that individual at that time.

b. Do genres on the Internet systematically possess properties that the traditional spoken and written genres do not have? Are there systematic changes to genres in cases where genres are held to have felicitously migrated into the Internet?

c. If "new" genres are to be found, what is their relationship with previous genres? Do all genres have some sort of ancestry? How is such ancestry to be conceptualized?

The importance of this "genre transference" has made it necessary to make distinctions such as the one proposed by Shepherd & Watters (1998), who divided *cyber-genres* into two classes of sub-genres:

Extant sub-genres, based on genres existing already in other media and which have been turned into a digital format. As will be shown in the case of *cybernewspapers* (see 5 below), when a genre is transferred to the Net it is initially faithfully replicated, without taking advantage of the new possibilities of the web, and processing it yields similar balances of effects and effort to the ones obtained offline. At subsequent stages, however, the medium evolves, exploits the new possibilities of the Net (hyperlinks, multimedia, etc.), and processing it yields different outcomes of relevance in terms of effects and effort. Similarly, the very notion of an *extant web genre* has to accommodate the evolving quality of web discourses and their increasingly multimedia and multimodal quality (Paolillo et al. 2010).

Novel sub-genres, wholly dependent on the new medium. They may be initial extant sub-genres that have become completely differentiated from the initial offline genre, or brand-new genres created on and for the Net. Again, *cybernewspapers* would be an example, since today's news portals (e.g. *ElPais.com*) differ completely from the printed counterparts (e.g. the printed *El País* newspaper) and from the initial transference of the newspaper to the Net with no genre alteration (e.g. the now extinct *ElPais.es*).

Besides, the estimation of relevance also depends on how much information is made available in the new medium and how it combines with the users' cognitive environments and with the search for relevance they are engaged in at a given moment. If a lot of information is simply made manifest but is not effectively combined with this cognitive environment to generate relevant outcomes, the user will tend to become *infoxicated* with information.

2. Internet users are now familiar with the non-linear and link-mediated quality of online information (even though they still cognitively expect that processing it will compensate for the mental effort involved). This is particularly noticeable in the multimodal quality of web content, parallel to the users' increasing reluctance to process traditional text-only and linearly encoded information. And users also

reject passive processing of discourses uni-directionally generated on the Net. It has been demonstrated, for instance, that the rigid format of television does not suit today's *digital natives*, accustomed to a different kind of online communication *à la carte* (see Grau 2010). In the case of television news, Pitts (2003) points out that today's users do not want a unitary perspective on a news event. Rather, they want to build up their own news programmes by using the initial information and complementing it with video, archives, etc.

3. Furthermore, it is likely that a systematic reiteration, on the Internet, of certain balances of cognitive effects in exchange for very little mental effort (as happens with very short messages in instant messaging, chat rooms or Twitter) might lead to a reluctance in users to process discourses which require more elaborate processing in terms of cognitive resources for the derivation of optimally relevant effects (see Yus 2007c, 2008b, 2009b). Indeed, many users – and especially adolescents – are used to exchanging very short messages and the mental effort that these demand is minimal, due to their brevity and simple construction. Therefore, it can be predicted that a reiterative processing of this kind of text will lead to a cognitive unwillingness to expend supplementary resources for the processing of longer texts or discourses with deferred relevance. This hypothesis may have important consequences in a future in which a high percentage of human communication will be Internet-mediated (see Grau 2008).

4. Another aspect of processing that has consequences on the estimation of the relevance of Internet documents is multitasking, that is, the tendency – again, especially among adolescents – to be engaged in several simultaneous tasks while they are online, for example using instant messaging while writing a school paper on the word processor, downloading a film and sending messages to a chat room (see Salvucci et al. 2011). Baron (2008a, 2008b) claims that significant conclusions can be drawn from this habit (which affects relevance assessment), but these depend on whether we are dealing with *cognitive multitasking* or *social multitasking* (mental or social consequences of multitasking, respectively). Predictably, information-processing worsens when two or more activities are being performed simultaneously, but it is possible to specify that this deterioration of processing takes place especially when both activities involve similar discourse formats (e.g. two simultaneous visual activities entail worse processing than one visual and one verbal activity).

5. On the Internet it is also possible to find surprising balances of cognitive effects and mental effort when the content of a web page is processed (Yus 2010c, 2010d, 2011a). As can be seen in Table 3.1, there are many possible combinations (and with different degrees in each case). In general, relevant outcomes occur when the processing of the web content enlarges the user's cognitive environment

(i.e. combines fruitfully with it), and this can happen not only by adding new information to that already present in the cognitive environment, but also by forcing a revision and erasure of previous assumptions and, above all, by combining with already-stored information to yield relevant conclusions.

Table 3.1 Possible combinations of cognitive effects, mental effort and resulting relevance

Case	Positive cognitive effects	Mental effort demanded	Eventual relevance
1	high number	high	positive to the user
2	high number	high	negative to the user
3	high number	low	positive to the user
4	high number	low	negative to the user
5	low number	high	positive to the user
6	low number	high	negative to the user
7	low number	low	positive to the user
8	low number	low	negative to the user

Case 1. *High number of cognitive effects, high mental effort, with positive relevance.* Sometimes the user will be willing to devote substantial cognitive resources in order to obtain the expected relevance. For example, a user who fears that he has the symptoms of an illness and reads a long, dense journal article on the Net will be willing to spend as much effort as necessary to check whether he has contracted the disease or not.

Case 2. *High number of cognitive effects, high mental effort, with negative relevance.* If the user spends a lot of effort in processing the content of the page with no offset in cognitive effects (for example because it does not combine effectively with the user's cognitive environment), the resulting relevance will be negative.

Case 3. *High number of cognitive effects, low mental effort, with positive relevance.* This is a prototypical case as predicted by relevance theory: high number of effects in exchange for the least effort leading to positive relevance to the user.

Case 4. *High number of cognitive effects, low mental effort, with negative relevance.* This case occurs when the user easily processes a lot of information but this information does not interact fruitfully with the user's cognitive environment.

Case 5. *Low number of cognitive effects, high mental effort, with positive relevance.* By contrast, this case is difficult to explain in relevance-theoretic terms, since it is not easy to find situations in which difficult-to-process information yields few cognitive effects and paradoxically ends up being relevant to the user.

Case 6. *Low number of cognitive effects, low mental effort, with positive relevance.* This case is covered by relevance theory, this time as a typical example of irrelevance: many cognitive resources are devoted to processing web content that does not produce enough cognitive effects for the user.

Case 7. *Low number of cognitive effects, low mental effort, with positive relevance.* This is one of the most intriguing cases in web content processing. Apparently, it can be stated that no user will be willing to process information on the web page that demands little mental effort but, at the same time, yields few cognitive effects. However, this combination is surprisingly very frequent on the Net (see Yus 2007c, 2008b, 2009b, 2010b: 84, 2010c, 2010d, 2011a). It is the case of pages whose content is utterly uninteresting but also demands no processing effort, which we can label as "boring," but which arouse the interest of thousands of users. An example was a web page where we could see a Cheddar cheese rotting (at http://cheddarvision.tv/, now closed), and another page where we can see hens moving about in a farm (at www.hencam.co.uk). In all of these cases, the effort demanded to obtain the information is minimal (or zero), but the information provided is also minimal. It may be, as mentioned in Burkeman (2007), that we suffer a lag in which the slow horse of human comprehension is unable to keep up with the fast horse of the information that is available on the Net, and maybe dull websites are popular because they are a rebellion against information overload, a space for our slow horses to graze.

Case 8. *Low number of cognitive effects, low mental effort, with negative relevance.* Finally, this case can also be found in the processing of web content. Although the mental effort demanded is low, the offset in cognitive effects s is also low, and therefore the eventual outcome is irrelevant.

As in other forms of input processing, the eventual relevance of cases 1 to 8 is affected by a number of aspects that should also be incorporated into the analysis. Indeed, there are myriad forms of cognitive satisfaction that may offset the effort involved in processing the information and that have often been dismissed, including a whole range of feelings, emotions, empathy, phatic connotations, community membership, socialisation, etc. I group all of these 'alternative' sources of user satisfaction under the generic label of *cognitive rewards* (Yus 2010d, 2011a).

The combinations of cognitive effects and mental effort, as predicted by relevance theory, are also influenced by the kind of Internet navigation. Indeed, the search for relevance, the expectation of cognitive effects, and the willingness to devote mental effort to the surfing activity will vary enormously depending on whether the user is simply browsing the Net without a specific purpose (e.g. to kill time) or is engaged in a focussed search for specific information.

Besides, the design of the interface plays an important part in the amount of cognitive effects obtained by the user and the mental effort involved in obtaining relevant conclusions from the web content (e.g. Hasan & Ahmed 2007 concluded that the interface style affects the user's willingness to accept and use this interface). Needless to say, the user's familiarity with the interface, even if it is not user-friendly, and also the user's intuitive ability to interact with the interface also affect the balance of cognitive effects and mental effort involved in processing information from this interface, generating multiple outcomes of (in)efficient interpretations. And frequency of use is a parallel variable affecting relevance.

Furthermore, the eventual relevance is also influenced by the type of cognitive task that the user is currently engaged in while surfing the Net, which also affects the user's willingness to devote mental effort to the task at hand. In this sense, Amichai-Hamburger et al. (2007) applied Cacioppo & Petty's notion of *Need for Cognition* to Internet communication. It refers to the individual's readiness to engage in and enjoy effortful cognitive endeavours, and it ranges from those users who have a 'low need for cognition' (i.e. those who do not enjoy cognitive efforts and acquire information by using simple cues offered by the environment) to users who have a 'high need for cognition' (i.e. have motivation to seek knowledge and will devote whatever effort it takes to get the information they need).

All of these non-propositional aspects of Internet communication that may influence the eventual relevance positively or negatively can be generically labelled *environmental constraints*. Among these, the type of surfing, the position of the interface in the different scales (oral/written, visual/verbal, synchronous/asynchronous), the user's familiarity with the interface and need for cognition, the frequency of use, the suitability of the interface to the user's navigational pattern, and the quality of the interface design.

In conclusion, a general equation for effective web page processing can be proposed, in which the basic conditions for relevance are present but with the addition of *cognitive rewards* (positive to the user) and *environmental constraints* (positive or negative to the user), as follows (Yus 2010d, 2011a):

> the cognitive effects obtained from information processing
> plus
> cognitive rewards
> plus
> positive environmental constraints
> [should exceed...]
> the effort demanded for the derivation of these cognitive effects
> plus
> negative environmental constraints

In Table 3.2 some examples of effective equations of Internet communication are listed and commented upon (see Yus 2011a for a more detailed discussion). Not all of them are web-page-related but are nevertheless illustrative of the elements involved in effective relevance outcomes.

Table 3.2 Examples of equations of effective Internet communication

Equation	Comments
The cognitive effects from a text that is difficult to process [+] cognitive rewards [+] positive environmental constraints [exceed…] The mental effort needed to process it [+] negative environmental constraints	This equation fits case 1 in Table 3.1. The user faces the processing of a text that is dense, difficult to read, but whose information combines fruitfully with the user's cognitive environment producing relevant outcomes. Even though mental effort is high (and some environmental constraints may also be involved), the user's reward in terms of cognitive effects offsets the effort.
The cognitive effects from an 'oralized written text' full of unusual orthography and creative use of spelling [+] cognitive rewards [+] positive environmental constraints [exceed…] The mental effort needed to process it [+] negative environmental constraints	This oralization of text will be analysed in Chapter 5 as part of the users' tendency to connote text with oral features of discourse. For neophytes, innovative text-mediated conversations will require additional cognitive rewards that compensate for the increased effort demanded, so that the interaction turns out effective, for example in terms of feelings of empathy, sociability, community membership, etc.
The cognitive effects from an interaction within a virtual world using a 3D avatar [+] cognitive rewards [+] positive environmental constraints [exceed…] The mental effort needed to process utterances in virtual worlds [+] negative environmental constraints	The environmental constraints in this equation will be analysed in chapter five as part of the problems involved in managing verbal interactions and nonverbal behaviour of avatars in 3D virtual worlds. But users can also obtain cognitive rewards from these 3D interactions, for instance from the comfort at being able to sustain fruitful interactions, or from the attention-drawing potential of the avatar designed by the user.

3.4 Cognitive effects, mental effort and estimation of relevance

Under this heading, a prototypical situation will be analysed, consisting of: (a) link-mediated web discourses and (b) users who click on these links to obtain the specific information that they are seeking. In the situation comprising (a)–(b), relevance can be studied both from a *quantitative* and a *qualitative* point of view, as will be shown below. The premise is that users have a *cognitive need* of information, and will try to get it by clicking on the right links. Reduced effort will be obtained by as few clicks as possible (quantitative approach) or by as much inter-link coherence as possible (qualitative approach).

Other authors have addressed similar situations in other terms. For example, Wirth (2002: 163) uses a different terminology but with similar underlying premises: "The user wants to save time and money, and therefore needs an economical guessing instinct that will offer a hint as to which link is worth following. The interaction between 'guessing instinct' and 'economical navigation' constitutes a standard of relevance." Similarly, the research on *uses and gratifications* has concluded that users enter the Net because they have a psychological or cognitive[20] need and expect a reward – a gratification – in terms of psychological effects (see Barnes 2003: 82–83). As Leung (2003: 112) summarizes, *Uses and Gratifications* theory claims that the media and their messages are sought in order to satisfy a variety of social and psychological needs. The quality of these needs (and values or beliefs) provides reasons for behaviour, which is oriented towards obtaining varied gratifications in the consumption of the media. Internet would no doubt be a rich medium in the sense that it offers users multiple and immediate options for gratification.

From a relevance-theoretic perspective, this kind of navigation can be accounted for in quantitative and qualitative terms. As I have already mentioned in passing, there are discussions on whether this theory can provide a quantitative explanation of how individuals assess relevance. W&S (2002a, 2004: 610; S&W 1986: 129–132) opt for a qualitative (or comparative) approach for the analysis of the estimation of relevance, since it is really difficult for analysts to assess relevance in purely quantitative terms and also for people in general to select the most interesting inputs. As W&S (2002: 253) stress, it is highly unlikely that individuals have to compute numerical values for effort and effects when assessing relevance 'from the inside'. Such computation would itself be effort-consuming and therefore detract from relevance. Moreover, even when individuals are clearly capable of computing numerical values (for weight or distance, for example), they generally have access to more intuitive methods of assessment which are comparative rather than quantitative, and which are in some sense more basic.[21]

20. Katz et al. (1973, quoted in Witmer & Taweesuk 1998: 292) include *cognitive needs* (important for information searching and knowledge) in the five categories that represent people's needs from the media. The other four are: (2) *affective needs*, (3) *personal integrative needs*, (4) *social integrative needs*, and (5) *escapist needs*.

21. The problems in assessing balances of relevance in quantitative terms increase if we intend to analyse not only the objective role of information content on the web page, but also other features that play a role in attracting the user's attention. For instance, Willis (1999) lists aspects such as *memory retention* (of web content), *interest* (capacity of web content to draw and keep the user's attention), *emotional bonding* (capacity to give the user some emotional support), *aesthetic satisfaction* (the design offers a pleasant audio-visual experience), *clarification of the message* (arrangement of content so that it is easier to process), and *immediacy of understanding message* (desire to present information in a clear and direct manner).

In any case, for the analysis of how specific information is interpreted by the user on a web page, two types of relevance conditions can be proposed, one in quantitative terms (condition (b_1)) and one in qualitative terms (condition (b_2)) (see also Yus 2008b: 638):

Condition a. The information in link-mediated content of web pages is relevant to an individual to the extent that the cognitive effects achieved when it is optimally processed are large.

Condition b_1. This information is relevant to the individual to the extent that the number of clicks that the user has to make in order to obtain these effects is small.

Condition b_2. This information is relevant to the individual to the extent that the level of coherence obtained from linking different texts is optimal despite the non-linear arrangement of the link-mediated texts.

Certainly, one of the most typical complaints by Internet users is their frustration at being unable to find the information that they are looking for. This difficulty, which is directly related to processing effort, can be measured as the number of clicks required and also as the level of coherence maintained after clicking on links and processing non-linearly arranged chunks of text in sequence (see Smith et al. 1997: 69). The user will expect that the new piece of discourse just accessed after clicking on a link will combine, in a relevant way, with the information already processed (from previous chunks of discourse) and which is still active in the user's short-term memory store.[22] Besides, the user will react negatively if the clicks do not lead to the expected information, to the extent that the user might even get lost in the array of link-mediated texts.

It is obvious that certain qualities of the interface of the web page may influence the number of clicks and what can be labelled *inter-link coherence* while reading the content from several link-mediated sources.[23] One of these qualities has to do with the design of links and their inter-relationships. Some web authors expect a more or less linear reading sequence and offer their readers a predictable

22. See Salmerón et al. (2010) for an analysis of three main strategies for the user's link selection based on (a) link screen position, (b) link interest, and (c) the semantic relation of a link with the section just read.

23. This *inter-link coherence* is similar to the label *internodal coherence* proposed by Engebretsen (2000). He distinguishes three types of coherence between links: (a) *intranodal coherence*, applied only to one link and which corresponds to the traditional notion of coherence used in text linguistics, since the text is read in a linear way; (b) *internodal coherence*, referring to the relation between two nodes of text that are link-mediated and read in sequence; and (c) *hyperstructural coherence*, reflecting upon the structure that governs the whole system of nodes and links.

structure of texts, while other authors offer total navigational freedom, leaving the full responsibility for the processed content and the order in which it is processed to their readers. In this sense, Reitbauer (2006) suggests three possible structures of links: the *linear structure* (a conversion of traditional texts into a link-mediated format but with linear arrangement); the *axial structure*, characterized by a number of central nodes that work as organizing axes and recommend a determinate reading sequence (often with the aid of frames at one side of the screen); and the *network structure*, which offers the user the highest level of navigational freedom. Besides, Danielson (2002: 9) has analysed the most typical links and groups them into dichotomies: (a) *associative* (relates content according to content similarity and relevance) vs. *structural* (relates content according to the structure of the page or portal); (b) *embedded* (visually and semantically included in the interface) vs. *isolated* (separated from it); and (c) *static* (all the attributes are maintained in different sessions and with several users) vs. *dynamic*. Needless to say, as the reading sequence becomes less predictable and less supported by the author, the content will be more open to personal interpretations and to increased effort if personal choices of reading paths and links do not lead to the expected outcomes.

Links are, in short, a promise of increased relevance, of supplementary interest, in exchange for the effort that clicking on them demands. This is why Pajares Tosca (2000) redefines S&W's principle of relevance as the condition in which "every link communicates the presumption of its optimal relevance." In her view, links entail a kind of "suspended meaning" that will not be confirmed until the user really sees where the link leads.[24] As such, links do not possess a fixed and definite meaning, but are mere indices of a possibly deferred relevance. Mitra & Cohen (1999: 186) point in the same direction when they write that, unlike other texts, Internet content is constantly inviting the reader to move to another chunk of text by using a link. The traditional presumption that the reader will process the whole text is substituted on the Net by the expectation that the user will explore and surf the page by clicking on the links. Of course, against the author's expectation, the link (and subsequent text related to it) may end up being irrelevant to the reader due to increased mental effort and frustration at the lack of inter-link (or inter-node) coherence. Furthermore, Mobrand & Spyridakis (2007: 44) state that

24. Pirolli & Card (1999, cited in Murphy et al. 2006) suggest the analogy of a user who is exploring a web page to an animal foraging for food. For animals, the calories spent foraging for food must be less than the calories that the food provides. Similarly, the wording of a hyperlink promises a potential benefit to a site visitor, while the time that the visitor spends reading and choosing a link, and also finding an adequate inter-link interpretation, represent a cost.

Being unable to predict the reader's path, hypertext authors can find it difficult to provide effective signals that facilitate understanding. Hypertext authors may, however, address coherence between nodes in several ways: by explicitly indicating the semantic relationship between linked nodes, juxtaposing the content of the preceding and current nodes in order to connect existing (or 'given') information to new information, or providing orientation cues that help readers identify their current position and obtain an overview of the hypertext structure.

4. Usability: A relevance-theoretic approach

Usability refers to the effort that the use of a computer system demands. However, as Casaló et al. (2008: 326) correctly qualify, nowadays the term is mainly associated with the ease of use of web pages or interfaces. Because usability affects users' processing effort when using an interface, accessing the information on a web page or surfing the Net, it is clearly linked to estimations of relevance.

Casaló et al. (ibid.) propose several factors that are related to usability: (a) how easy it is to assimilate the structure of a web page, its functions, the interface and the content; (b) how simple it is to use the page at initial stages; (c) how fast users can find what they are searching for; (d) how easy the user feels it is to navigate the page; and (e) how much users can control what they are doing and where they are at a particular stage of navigation. All of these factors play a role in the eventual relevance of web content.

Besides, several methods to measure usability have been proposed. In the same way as a quantitative model of relevance assessment was deemed necessary by several analysts, even if difficult to put into practice, a quantitative measurement of usability has also been attempted. But Van Schaik & Ling (2006: 872) acknowledge that currently the tendency is to use subjective techniques to explore user satisfaction, sometimes in conjunction with more objective tests. However, the latter are typically carried out in laboratories and this prevents the study of "natural" behaviour of users when assessing web page usability.[25]

25. Alby & Zucchermaglio (2008: 496) list some deficiencies of these laboratory tests: (1) the informants often have different motivations from the ones for which the technology had been designed; (2) the group or community practices of end users are not taken into account; (3) the interaction with the technology is supposed to be individual and mental, but the complex phenomena that have to do with local, social and organizative mechanisms are not taken into consideration; and (4) usability tests are usually performed at the final stages of the design of the interface, when it is almost finished and there is little room for changes.

4.1 Users and interfaces

Web pages are normally designed so as to provide easy access to the most interesting information and arrange it in relevant ways. In this framework, the visual, verbal or multimodal design of the interface is essential in web usability and user satisfaction. Furthermore, satisfaction with usability can have implications for future visits to the web page. This is essential, for example, for customer loyalty in e-commerce.[26]

Nowadays web pages include text, pictures, flash animations, graphics and videos, among other elements. The combination of these elements has an impact on the user's satisfaction (and the user's desire to return to the page in the future), and also on the eventual relevance of web content. However, this assertion does not entail that pictures invariably play the major role in the interpretation of web pages, as we could intuitively conclude. Although several studies claim that information is processed (and retained) more easily when it is presented visually than when presented only textually (Casaló et al. 2008), for a pragmatic analysis the importance of these elements lies in whether these are combined effectively to generate relevant interpretive outcomes.

What are the user's expectations and actions when they surf the Net? There may be no foolproof way to predict patterns of users' behaviour on the Net because the users' cognitive environments and expectations of relevance differ enormously. Nevertheless, some generalizations can be made about their needs, search for relevance and prominence of certain information from informative sources that "compete" for the users' attention. In the next heading some comments on usability are summarized (see also Yus 2010c).

4.2 Designing for relevance

Designing a web page entails a prediction of readers' needs and actions, e.g. that they will follow certain interpretive steps with the aid of inference, that the way information is arranged will be positively valued by the readers, etc. As summarized in the cognitive principle of relevance (S&W 1995: 260–266), the human cognitive system tends to pay attention and devote cognitive resources to potentially relevant stimuli. In the same way, Internet users tend to pick up, often unconsciously, the information from the web page that is likely to be relevant. Those who design pages or fill them with content foresee that certain information and

26. For example, Casaló et al. (2008) and Chang & Chen (2008) concluded in their studies that there is a relationship between user loyalty and usability.

the way it is arranged on the page are more likely to be selected and processed, together with a prediction of which conclusions the users are likely to draw from this information.

S&W (2002: 14–15) stress this human ability to predict the mental states and inferential steps of others, as part of the general human tendency to maximize relevance. Specifically, individuals (and also web page authors and designers) can predict:

1. What stimulus in an individual B's environment is likely to attract B's attention (i.e. the most relevant stimulus in that environment). In the context of web page processing, the author (or designer) is aware that certain ways of designing a page and putting information in it are likely to interest the user.

2. What background information from B's memory is likely to be retrieved and used in processing this stimulus (i.e. the background information most relevant to processing it). In the context of web page processing, the authors (or designers) assume that certain information will already belong to the reader's cognitive environment and that the web content will combine effectively with this background information to yield relevant conclusions.

3. What inferences B is likely to draw (i.e. those inferences which yield enough cognitive benefits for B's attentional resources to remain on the stimulus rather than being diverted to alternative potential inputs competing for those resources). The same would apply to the processing of web page content.

In other types of communication, relevance theory has proved to be useful for showing that speakers are indeed able to predict (1–3) above. An example is the structure of jokes and other humorous texts.[27] Humorists are aware that some of their jokes (or parts of them) can lead to different interpretations (some of them not noticed by the audience) and can predict which of these interpretations is more likely to be picked up as the intended one (since it offers the best balance of cognitive effects and mental effort). The audience will then discover that this interpretation just picked up is not the intended and eventually valid one, which produces an incongruity. It is not until the end of the joke that the audience are given the resolution of the incongruity and the humorous effects are produced. In the bibliography there are examples such as (1a), in which the humorist predicts that the most likely interpretation is (1b), more accessible and relevant, but which is later invalidated and replaced with a more unlikely but eventually correct (1c):

27. See Yus (1997c, 2002c, 2003e, 2004, 2005d, 2008c, 2009c, 2010e, 2011b, forthcoming b) for some applications of relevance theory to humorous discourses (jokes, stand-up monologues, etc.).

(1) a. A doctor thoroughly examined his patient, and said, "Look, I really can't find any reason for this mysterious affliction. It's probably due to drinking." The patient sighed, and snapped, 'In that case, I'll come back when you're damn well sober!"
b. The reason for the patient's affliction is alcohol (which the patient drinks).
c. The reason for the patient's affliction is alcohol (which the doctor drinks).

In a similar way, authors or designers of web pages make predictions of relevance, of the reader's manifest information, inferential steps and conclusions that they can draw from web page content and background information. But there is no guarantee that these predictions will be successful, since cognitive environments and manifest information vary enormously among users (see Nielsen 2008). Livingstone (2007: 167) also stresses that web pages are open to interpretations that are not necessarily predicted by their authors or designers. She illustrates this with a page for adolescents, *The Epal Homepage*. Its creators tried to make it attractive to its potential young readers by designing an avatar-mediated interface that looked like the famous video-game character Lara Croft. But contrary to the predictions, adolescents did not find it interesting at all.

Perhaps it would be more sensible to assume that certain designs and web content will surely satisfy certain users and discourage others. An example can be found in Wu et al. (2008). They concluded that there is a dilemma between the advantage of being guided in web navigation and the freedom of navigation that users take for granted. These analysts conclude that the best web page design is the one based on a multi-faceted categorization system, that is, a structure where several categories are offered on the page and all of them are equally prominent and independent. In relevance-theoretic terms, this arrangement minimizes the presumption of an intended interpretation or reading path; instead, the author simply makes manifest the information, organized in categories, and leaves all responsibility of choice and final satisfaction to the reader.

In any case, Wu et al (ibid.) propose that some control over navigational options should be exerted, with an emphasis on predicting users' needs, so that some options have more prominence over the others in the interface. Again, this control and these predictions do not guarantee eventual relevance for the users, who are very different and with differing cognitive environments. This unpredictability is clear in the two types of interface that these authors analyse. On the one hand, in *hierarchical structure* there is a relationship of subordination among elements and nodes, but the user might not understand the structure that underlies and provides coherence to the arrangement of these elements, even if the users are guided in their navigation. On the other hand, *multiple categorization* adopts a uniform structure that treats each category as independent, but this extreme freedom might generate unwanted effects, for example it might cause the user to get lost among the array of options (which increases mental effort and the eventual relevance of the content of the page).

In conclusion, the design of the page, the choice of what information it contains and the selection of items to be turned into links are subject to hypotheses about their relevance. To these hypotheses other factors can be added, for instance referring to the people that are involved in web-page-mediated communication[28] or to those technological aspects that may influence the user's satisfaction or lack of it. For example, Rajani & Rosenberg (2000) list what they call *computational aspects* of web page usability and which may influence successful navigation: (a) *maintenance* (the page has to incorporate recent technological advances and the user has to be up to date on how to use a technology in constant evolution); (b) *speed of access* (pictures and animations may slow the access to the content of the page and frustrate users); (c) *WYSIWYG* ("what you see is what you get," the printed output and what one sees on the screen might differ, even if nowadays there is a presumption of equivalence); (d) *navigational aids* (among them, frames, bars, content indexes, site map, etc., which help to reduce mental effort when processing the page); (e) *anonymity* (web page designers do not usually know the future users of the page, nor do they have access to their cognitive environments or how the web content will interact with these environments); (f) *design traits* (clarity, accessibility, consistency, simplicity, navigability, etc.); and (g) *limitations of html* (html is the language of web pages, but it is still limited when complicated layouts are intended on the page).

Given the variety of elements influencing web processing and navigation, it is difficult to make recommendations that will invariably lead to relevant interpretive outcomes. An attempt is found in Éthier et al. (2008: 2773). They classify their pieces of advice into three categories. The first one comprises opinions and recommendations made intuitively by professionals and web design experts. The second category comprises typical recommendations for web page design intended to reduce the user's disorientation when surfing the Net. Finally, the third category feeds from the actual users' experience in order to guarantee a usability-oriented web page design.

Other suggestions include those made by Abels et al. (1998). They group them in criteria such as (a) *use* (the page should be easy to use, with navigational

---

28. See Chevalier & Bonnardel (2007), who in the field of e-commerce differentiate clearly between the client of the web page designer (the owner of the page) and the future users of the page. These users can have radically different motivations and expectations of relevance from the owner's.

aids, etc.), (b) *content* (information should be updated, useful, etc.), (c) *structure* (it should be intelligible, with clear organizing principles, hierarchy of texts, etc.), (d) *links* (they should relate pages in a relevant way, etc.), and (e) *search* (there should be some area on the page for searches).

Fuccella & Pizzolato (1998) propose a web design in three steps that are meant to be integrated in the creative process. The first step entails a *definition of the audience*. Then, a *collection of data and requirements* of the page takes place in order to have a clearer picture of the intended content. Finally, there is an *organization of the information*.

Lim (2002) prefers to list specific pieces of advice for web page designers, such as: (a) *Follow a sequential progression*: "information on websites unfolds in a sequential manner where one hyperlink is an elaboration of a previous interconnected hyperlink and so on" (ibid.: 165). Users are used to this form of processing information and get annoyed if the structure of the page does not fit it, with a parallel increase in the mental effort involved. (b) *Mimic real-life scripts*. As humans, we make sense of the environment and extract stereotypical mental scripts that we store in our encyclopaedic knowledge. Later, we use this store of information to make predictions on what is going to happen on a particular occasion. In a similar fashion, Internet users expect certain scripts in their processing of web content, thus saving mental effort. (c) *Provide visual indicators*, especially useful for e-commerce. (d) *Place functionality above aesthetics*, also useful for e-commerce websites (but see Mitra et al. 2005 for a different position on this).

5. Transferring discourses to the Internet: The printed newspaper

Newspapers are a good example of how users search for and evaluate the relevance of information and how the design has to be altered so as to meet the expectations of relevance in the transference to a different medium (from the printed page to the Net). At the beginning, printed newspapers were simply scanned and uploaded to a website without changes in their format. Nowadays, however, *cybernewspapers* have turned into news portals, even if the scanned printed newspaper can still be accessed on a different area of the portal (often under subscription). Crucially, the evolution of the news format from "paper" to "electronic" on the Net has huge consequences on how relevance is assessed and how much attention the user pays to the news in both media (printed/electronic). There are also variations in the level of participation expected from the user to complete the information initially provided by the news article (see Yus 2003d). As Mobrand & Spyridakis (2007: 43), stress,

> traditional print-based signals do not translate easily to such hypertexts, and the reader finds no comforting boundaries or sense of closure in the malleable web environment. The reader must share his or her limited pool of cognitive resources available for reading between comprehending the text, manipulating the reading tool, and making decisions about the path to take through the text.

Online newspapers have gone through a number of phases in their evolution. Each phase has had an impact on the readers' interpretations and their relevance (see Cabrera González 2001 for a proposal of stages). Differences have become notorious as the online format has evolved into an autonomous genre with hardly any resemblance to the printed counterpart and with more options for interactivity with the reader.[29] The following differences are particularly worth commenting upon:

1. *Immediacy*. The printed newspaper cannot compete against the fast "real-time updating" of Internet news portals. Maybe the only advantage of printed news is to offer the reader a more elaborate and detailed account of the newsworthy event. In any case, the authors of printed news have to predict, at every stage of the composition of the text, whether the event already belongs to the reader's cognitive environment or not, and to what extent the information it conveys combines in relevant ways with this cognitive environment, because the relevance obtained in each case will vary enormously. Certainly, if the reader already knows about the event, the information in the news article will be irrelevant. This explains why authors try their best to make relevant predictions of manifestness. Take, for instance, the following headlines, both referring to the same newsworthy event:

(2) a. El Columbia se desintegra poco antes de tomar tierra.
(*Información*, 2-2-2003, p. 1)
[*Columbia disintegrates before landing*].

b. Bush promete continuar los vuelos al espacio a pesar del desastre del 'Columbia'. (*El País*, 2-2-2003, p. 1)
[*Bush promises that space missions will continue despite the 'Columbia' disaster*].

29. Beyers (2004: 12) proposes two senses of the term "interactivity" in *cybernewspapers*. Firstly, interaction is part of communication with the reader, for example with a readers' forum, chat rooms included in the news portal, or e-mail. This is called *communication interactivity*, *interpersonal interactivity* or *audience involvement*. Secondly, there can also be interaction when readers select content. Internet allows for a choice of "my own newspaper," which Beyers labels as *selection interactivity*. On the other hand, Bucher & Schumacher (2006) distinguish two other senses of the term: (a) a technical sense of interactivity as *responsiveness*; and (b) a sociological term that refers to *reciprocity*.

The headlines in (2) mention an accident involving the Columbia space shuttle that took place on February 1st, 2003, at around 3 o'clock in the afternoon (Spain's local time). Although it was very likely that the next day the reader would know about the accident from other media, the author of (2a) offers the accident as "new information" for the reader's cognitive environment, and its relevance will lie in adding information to this environment (and it will be irrelevant if this information is already manifest to the reader). By contrast, the author of (2b) presupposes that the news on the accident is already part of the reader's cognitive environment and seeks relevance by offering additional information about the event.

With the Internet, these predictions of manifestness of information are not so much an issue due to the "real time" quality of information on news portals. As Baehr & Schaller (2010: 36) correctly state,

> In the real-time unfolding of news events online, context is often lost and never recouped as other events break and the early news is stowed at the expense of current events that draw the readers' and viewers' interest. In the past, people would find out about a tsunami on the evening news or the next day. In real time, the news unfolds literally as it happens. The analysis and contextual pieces in the newspaper arrived on the doorstep the following morning. Now, users are left to find their own context.

2. *Ubiquity*. The *cybernewspaper* can be accessed from any part of the world with Internet connection, while the printed paper is normally bought and read at specific physical locations (but there are interfaces for reading the printed paper online; see below).

3. *Scroll factor*. The text of the news on the screen disappears from the top as we scroll down while reading it. Many users are reluctant to move the text so that the unframed part of the text appears on the screen and prefer to click on links that lead to smaller chunks of screen-sized texts (Armentia et al. 2000a, 2000b). As a consequence, when printed news is transferred to the Net it is often necessary to "fragment" it and link the chunks of text together so that they fit the limits of the screen.

4. *Interactivity*. The *cybernewspaper* offers the reader a level of interactivity that cannot be found in printed newspapers. It is a system-to-user kind of interaction, what Chung (2008: 660) calls *medium interactivity*. But the constant advances in the design of news portals and the rise in the trend towards user-generated context under the Web 2.0 umbrella (e.g. *Menéame* and *Fresqui* in Spain)[30] make interactivity increasingly "natural," bi-directional and even user-to-user, besides e-mail and online forms to contact the newspaper (Kenney et al. 2000). The kind of interaction that the user engages in depends on the user's needs when accessing

30. The first one in [http://www.meneame.net] and the second one in [http://fresqui.com].

the *cybernewspaper*: those who have a high information seeking motive are more likely to expose themselves to mediated contents or messages by clicking hyperlinks (i.e. *medium interactivity*). On the other hand, people with a high social-interaction motive will tend to participate in discussion with other readers, for example by leaving comments in a journalist's blog or sending e-mails (i.e. *human interactivity*) (Yoo 2011:72–73).

5. *Multiple formats on the same page*. News portals offer information in multiple types of discourse (visual, verbal, graphics, audio, video…) whereas the printed page has a more limited range of formats.

6. *Updating*. Internet portals can be constantly updated, but the printed newspaper has a time limit beyond which it has to be published and delivered to newsagents.

7. *Trans-temporality*. In *cybernewspapers*, articles are frequently archived but are, at the same time, always accessible to users. This results in a new form of news processing that combines old and new information to generate more fine-grained outputs of relevance.

8. *Personalization*. Sometimes users are allowed to alter the content of the web portal according to their informational needs or search for tailored sources of news that directly appeal to them and combine with their cognitive environments effectively.

It is clear that both sources of news, printed and online, are so different nowadays that the eventual relevance will necessarily be affected by how (and how much) information is accessed and processed. Nevertheless, for both formats there is an activity that always applies: the reader's inferential gap-filling of information to turn the schematic words into fully contextualized and relevant interpretations. In Dor (2003:704) Example (3) (from the Israeli national newspaper *Ma'ariv*) is proposed. I have added in italics several questions that readers would ask themselves in their search for a relevant interpretation. A short and simple paragraph such as (3) demands a lot of inferential enrichment, and in this process, the reader has to determine which information is new and which information is already manifest, plus combinations of both:

(3) The bodies of John Kennedy Jr. [*who is he? what do I know about him?*], his wife Caroline and his sister-in-law Lorraine [*who are they? what do I know about them?*] were discovered yesterday in the ocean [*were they dead? what were they doing there?*], at a depth of 30 meters [*is this information relevant?*], 10 kilometers away from Martha's Vineyard Island [*what do I know about this island?*], where they were headed on Saturday [*which Saturday? why were they going there?*]. Senator Edward Kennedy, John's uncle [*what do I know about him?*], arrived at the site where the bodies were found, in order to identify them. Kennedy Jr. will be buried in NY in the coming days [*is this information relevant?*].

The text of the news article always *underdetermines* the information that its author intends to communicate with it (i.e. it is always less informative than the intended interpretation) and the reader's collaboration is thus essential. But again, the variability of cognitive environments, attitudes, needs for information, interests, etc among readers is so wide that authors can only make more o less accurate predictions about the potential interest of the news article.

It is also interesting for a relevance-theoretic study of news to analyse the reading pattern of the printed newspaper and compare it to that of *cybernewspapers* so as to determine if the expectations of relevance (and ways to obtain it) differ between the two formats. Figure 3.1 shows a prototypical reading sequence (of course, not the only possible one) for a printed newspaper. The reader will probably start by browsing the pages or paying attention to the front page, or going to a specific section that he/she likes.

When the reader stops browsing and reads an article whose content is potentially relevant, the reader will tend to process each part that the article is composed of (headline, lead, first paragraph…) and will stop when his/her expectations of relevance are satisfied. Many readers will stop processing at the very headline, because most headlines are informative enough to optimize relevance (see Dor 2003).

The online format, on the other hand, alters enormously the aforementioned reading sequence. Firstly, in *cybernewspapers* we can frequently see how the authors have also filtered the information from the Net, which might be interesting as a complement to an article. Secondly, today's users are fully aware of all the possibilities that the Internet offers and are more demanding in their need for immediacy and inter-connection of information. They are used to multi-tasking and willing to click on any link that offers a presumption of potential relevance. As Orihuela (2000) states, nowadays we have to learn to read and write all over again, to obtain and spread information. Today, information tends to be built up in the form of navigational spaces, as networks in which the different formats (text, audio, video, graphics, animations…) are inter-connected, open to the users' decisions and their own contributions. Knowledge in the information age is fragmented, dispersed, and hyper-specialized.

Besides, Tewksbury (2003:694) claims that the Net provides the audience with a more substantial control over the selection process of information than with traditional media. Internet users are particularly prone to following their own interests and reluctant to follow the wishes of editors or producers of news. However, the availability of information comes at a cost in terms of mental processing effort. D'Haenens et al. (2004:365) conclude that users devote too much effort to working out the structure of the page and to creating their own pathways through the content, and there is no guarantee that this effort will be offset by supplementary interest.

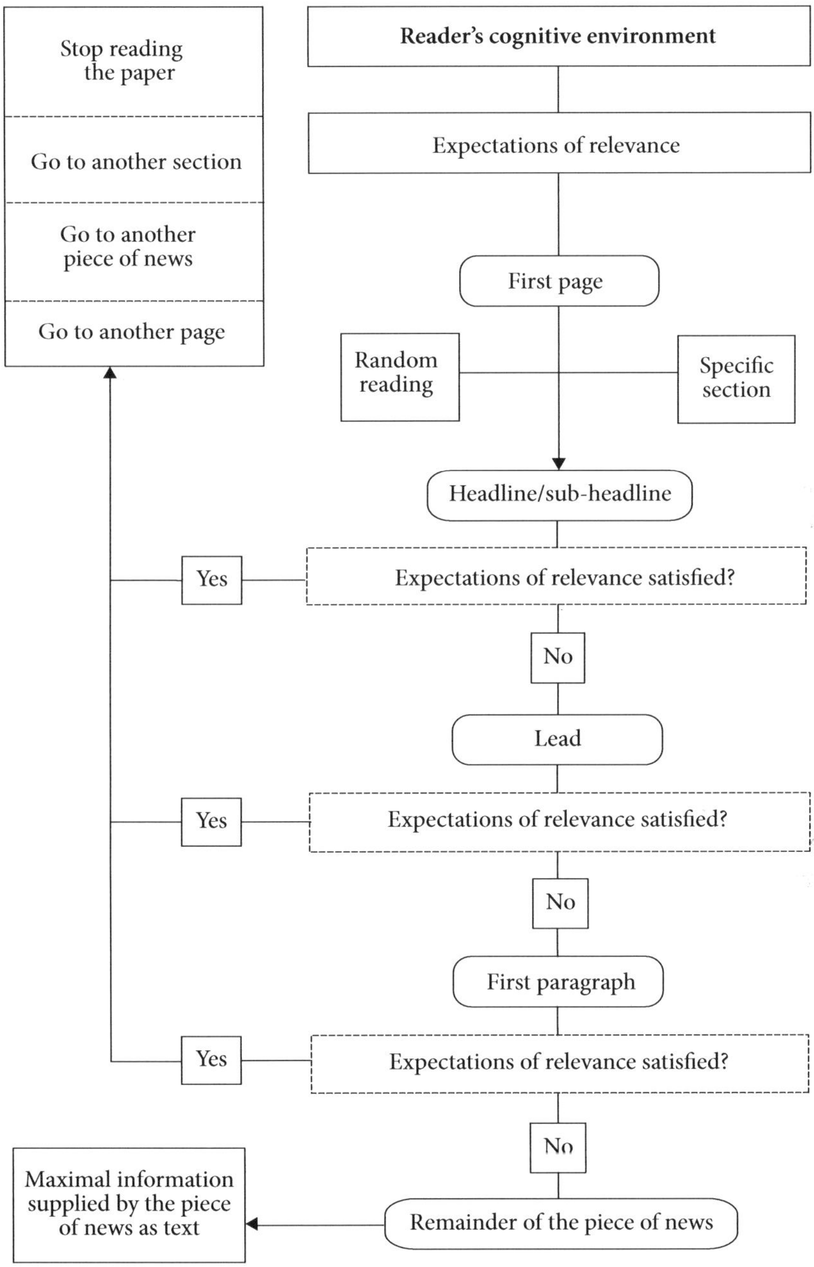

Figure 3.1 Relevance in the printed newspaper

Nowadays, *cybernewspapers* bear little resemblance to their printed counterparts, and both designers and users benefit from alternative sources of relevance within the portal or in other areas of the Net related to the text they are reading. Reading *cybernewspapers* is therefore different from reading printed papers. Figure 3.2 shows a prototypical sequence (again, not the only possible one) for the processing of news in *cybernewspapers* which differs from Figure 3.1. The inferential steps start, again, with the user's cognitive environment and expectations of relevance. Normally, the search for relevance starts at the homepage of the *cybernewspaper*, from which a wide array of options are offered to the user. It should be noted that starting the reading activity at the front page of a printed newspaper is very different from starting it at the online newspaper's counterpart: "interactivity, multimodality and dynamic content make the online frontpage different from a paper frontpage. While the paper frontpage has the physical boundary of the first page in a newspaper, and one can dwell on it, the online frontpage is a gateway, i.e. a navigational page providing access to other pages" (Santini et al. 2010: 10).

Among other options offered to users, in Figure 3.2 some are listed: to type words in the "search" option, to access the site map, to click on the link of a specific news article that has been highlighted at the home page, or to enter a section (these are normally listed in a side frame or in an arrangement of parallel tags). After choosing one option, the user will probably read the most appealing headline. In *cybernewspapers* headlines are mere stretches of text that link to the non-visible content, and it is very unlikely that the user will be satisfied only with the information that it provides and hence he/she will tend to click on the link to check the presumption of relevance that the link creates. From this point the pattern is similar to that for printed news: accessing different sections of the news article and stopping when one is satisfied; but there are differences in format. The online news article may be fragmented in link-mediated chunks.

An important difference between newsprint and *cybernewspapers* is that, while reading the latter, the user can leave the page of the news article and look for alternative sources of information in other areas of the Net (represented as dotted lines in Figure 3.2). For example, the reader can use the "search option," or check multimedia content related to the article, as well as graphics, or access the general archive of news in the portal to see what has been published before on the topic of the news article. All these inter-connections of information play an important role in the eventual relevance obtained from reading the online article.

On the other hand, and at the present stage of the evolution of *cybernewspapers*, there are in general three versions of the same news article. The first version bears no resemblance to the one that was printed and, as such, it could be labelled *novel cybergenre* in Shepherd & Watters' (1998) terminology. The second contains the same content as the printed paper, but with the addition of several options

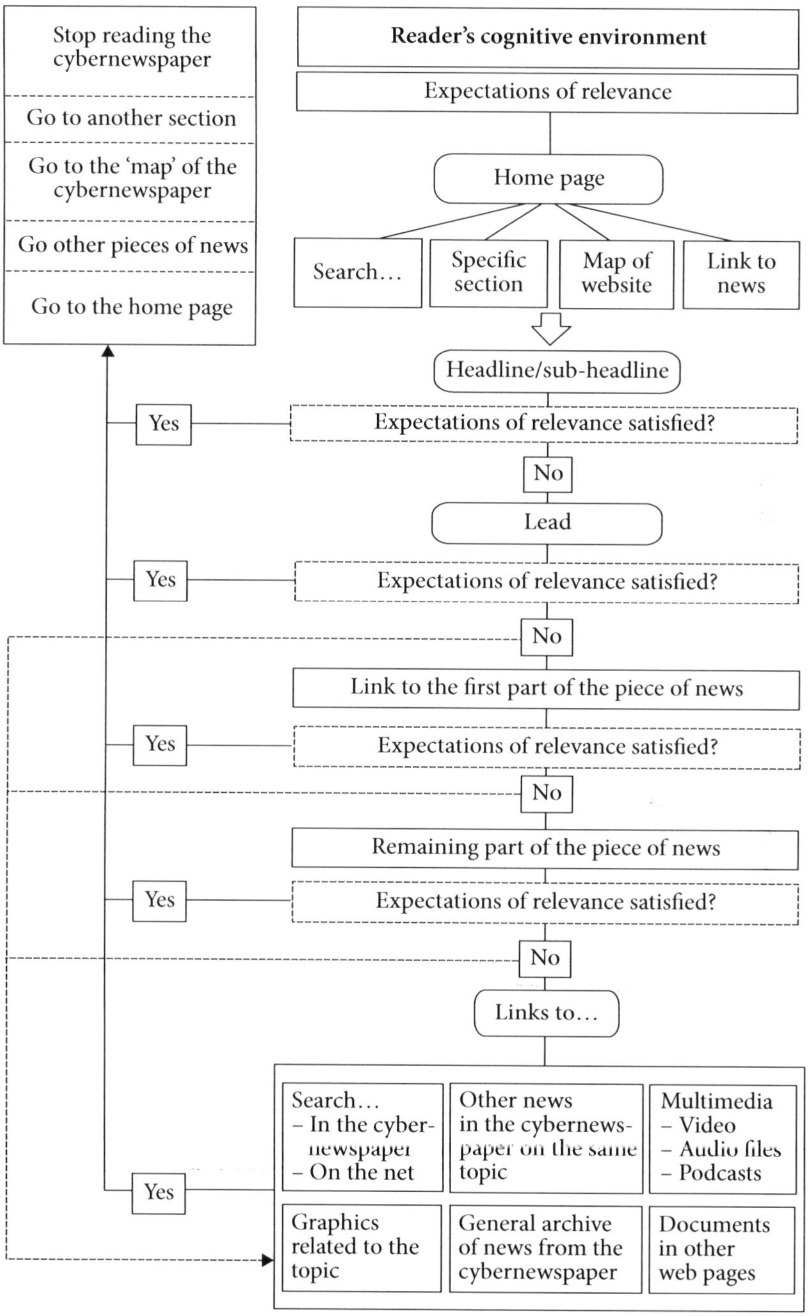

Figure 3.2 Relevance in the *cybernewspaper*

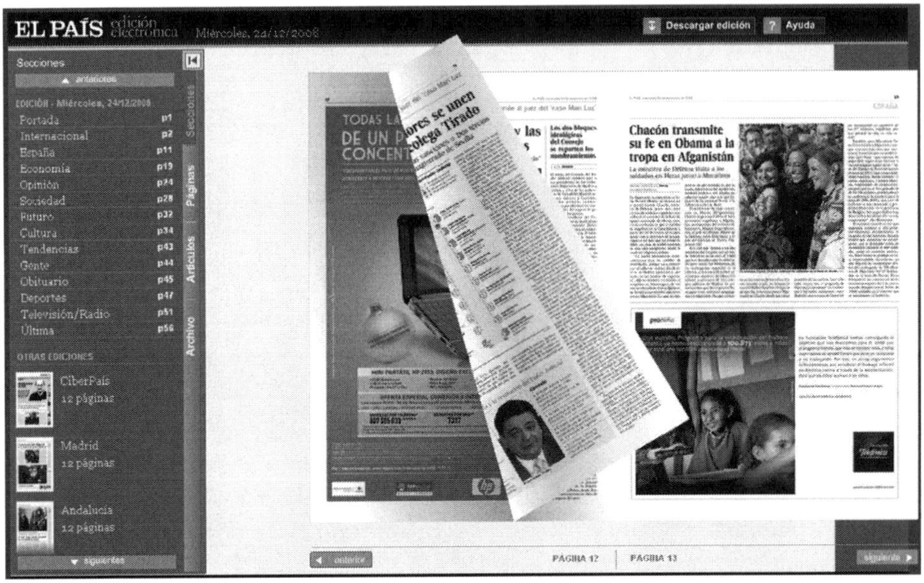

Figure 3.3 The interface of *El País* for reading printed news online

(video, links to other pages, etc.). And the third is a faithful reproduction of the news article just as it was printed, such as the ones that we can find in news portals like *El País* in Spain and *The Guardian* in England.[31] These versions fit the label *extant cybergenres* (Shepherd & Watters ibid.), especially two variants: the second would be a *variant genre* because it is "based on existing genres but [has] evolved by exploiting the capabilities afforded by the new medium" (Shepherd & Watters ibid.: 99), whereas the third one would be an example of a *replicated genre*, since it is similar to the already existing printed genre.

In the case of *El País*, the reader can choose between an "html version" of the article or a "pdf version" with the same content and for the latter there is a special viewer (under subscription) that reproduces the printed paper faithfully and the user can even turn its pages (Figure 3.3).[32] Obviously, it is in the "html version" where we can find the most striking differences between the two formats. In this one there is additional information on the characters mentioned,

31. The first at [http://www.elpais.com/diario] and the second at [http://guardian.newspaperdirect.com].

32. My thanks to *El País* for granting me permission to reproduce Figure 3.3.

plus *RSS feeds*, links to pages in other newspapers where the same event is addressed, access to multimedia content, etc. Specifically, *RSS feeds* are an interesting way of accessing information that inevitably alters the final relevance of the processing of the news article. Instead of clicking on potentially relevant links, it is the site that *feeds* the user with pre-established topics of interest. The *RSS feed* is re-written automatically whenever some updating takes place in the content of a website. In this way, the *RSS* file makes it possible to know whether the website has added new content or texts, but without having the need to go to the actual website unless one wants to read the "extended version" of the potentially interesting content.

In any case, I think it is interesting to analyse *cybernewspapers* in comparison with their printed counterparts in order to delimit the qualities of this *cybergenre*. Orlikowski & Yates (1994) point in the right direction when they propose the notion of *genre interdependence* so as to capture the dependencies and connections that may exist between genres when they are enacted in communicative action. Ihlström & Henfridsson (2005: 175) use this term for investigating the potential sequential dependencies that exist between the online and the printed newspaper genres: "Given the long-established genre elements and rules of the printed newspaper genre, it is plausible to assume that the evolution of the online newspaper genre to a large extent overlaps with its printed counterpart." In the meantime, the medium keeps evolving and generating adjustments in the assessment of relevance. At present, the *cybernewspaper* has jumped from the computer screen to smart phones, demanding from news producers and readers information that is

> comfortable in a new format, on a smaller screen, where the issue of where you are may also become part of how you are reading or reporting or watching. This is [...] a permanent shift in communications consumption. It is ripping the format of journalism and potentially other media so far away from the page-centric world we all grew up reading and writing so that it raises the question of how long it will be before even the concept of a website becomes old hat. (Bell 2010: 4)

And within the context of the computer screen, traditional formats fight against new developments over ways to present information and generate relevant outcomes. For example, Gómez (2010: 34) suggests that in the near future journalists will have to develop at least five parallel formats for the "electronic newsagent's": alerts, articles, podcasts, video-reports and news on social networking sites. Each of them will provoke variations in content accessibility and processing, and hence on users' balances of effects and effort.

6. Transferring discourses to the Internet: The printed advertisement

Internet advertisements (most of them banners or pop-up ads) are another example of a genre that has been transferred to the Net from the initial printed format. But they also exhibit an informative underdetermination when compared to the interpretation that they intend to communicate. Therefore, the user's participation is essential to compensate for the informational gaps.

The effectiveness of Internet advertisements is measured by the number of clicks obtained. As they have to share the space of the screen with other discourses, they are typically schematic, as in (4a), and demand the user's cooperation in turning them into fully contextualized and relevant interpretations, such as (4b), with the aid of context (see Yus 2005c):

(4) a. Special offer, 3X2 in milk shakes. Free delivery over $70.
b. We are currently running a special offer in milk shakes: if you buy three bottles at our chain of supermarkets you will only pay for two of them. Besides, we are offering free delivery of the product if the purchase of milk shakes is over $70.

Internet ads are combinations of text, pictures, video or Flash animations that are placed on a web page in order to draw the user's attention (*communicative intention*) towards the advantages of a product and of purchasing it (*informative intention*). Normally, the fulfilment of these intentions is evident when the user clicks on the ad to be taken elsewhere and buy the product. But nowadays there are innovative techniques for the dissemination of the qualities of the product, such as *viral advertising*,[33] in which the users play an essential role in telling other people about the qualities of the product and spreading them through their networks, as will be briefly commented upon below. In any case, advertisements on the Net have to fight for a share of the users' attention that typically focuses on other areas of the screen while they are engaged in multi-tasking activities with other users. Bulkley (2008) correctly states that the online space is increasingly about bloggers, twitterers (mini instant messages) and online forums. Websites are places for communities to congregate, share and chat. In this digital space brands need to engage with people by being relevant to what they are already doing or offering them something they want or need. Unlike TV, the advertiser does not have the luxury of taking over the entire screen for handfuls of time.

33. Also called *viral campaign*. It refers to the "mouth-to-mouth" dissemination of information on the product among friends. For Porter & Golan (2006), though, the two terms are different. A *viral campaign* is a marketing strategy with several components and a "mass media" attribute, whereas *viral advertising* is a technique for selling products on the Internet based on the participation of users, who are persuaded to spread the information among their peers.

An Internet advertisement is relevant when the number of cognitive effects obtained from processing it is high and the mental effort involved in deriving them is low, as in the interpretation of any stimulus. But many users are willing to expend considerable cognitive resources on the Net. For example, Carter (2004) claims that, once the users decide to participate in the interaction with the brand, they exhibit a higher involvement with it than in traditional media.

Banners are placed on web pages and "invite" readers to click on them. To achieve that, the designers predict that a certain type of ad or verbal-visual composition inside it will be regarded as potentially relevant and will lead to its interpretation and eventual click. What is more difficult to determine is what makes a user interested in a banner, i.e. how to determine why the banner communicates a presumption of its eventual relevance. This question is particularly pertinent since banners are normally intrusive, and pop-up ads (those that suddenly appear on the screen) are utterly irritating (Bahr & Ford 2011). Therefore they are not, on paper, very effective and new formats of *netvertising* have been devised to compensate for this burden (e.g. attractive Flash animations) and avoid, if possible, the user's tendency to get used to their presence or even avoid them altogether, what Benway (1998) called *banner blindness* (see also Janoschka 2004).

This negative reaction in the user is clearly explained by relevance theory: since human cognition tends to focus cognitive resources on what is potentially relevant, these advertisements cannot fight against other areas of the web page for the user's attention. Designers of Internet ads should strike a balance between the capacity of the banner to draw the user's attention and its level of intrusiveness in the user's current navigational goal that led him/her to the page where the banner is located. This balance is influenced by the kind of navigation. In an experiment, Calisir & Karaali (2008) concluded that the content of the banner is not interpreted in the same way (or the banner is not detected in the first place) if the user is simply surfing the Net with no specific purpose or is looking for specific information.

Clearly, all advertisements are intrusive to a greater or lesser extent and interfere with the task that the audience has at hand (printed ads in magazines, TV ads interrupting a film, etc.) but on the Internet the level of interference is even higher because of the size of the screen and this burden has consequences on the estimation of relevance (see Bruner & Kumar 2000). The key to a successful advertisement on the screen is to obtain the user's contextualization of the ad by combining its information with the user's cognitive environment so that relevant outcomes are produced. Of course, the design of the advert may play a part in drawing attention and leading to this contextualization.

An effective strategy for the contextualization of the advertisement is to attach it to search engines (as in *Google* ads) or to place it on web pages whose content is somehow related to what the advertisement is offering, or to the user's current task at hand, his/her preferences, etc. When such a connection is made between the content of the ad and the user's cognitive environment, then the format of the ad is a secondary issue. This is the case of *Google* ads: they are text-based but are very effective in their combination with the user's search for relevance, as G. Stuart, former president of *Internet Advertising Bureau* corroborates:

> Most of the time when text ads are used, it's based on a search mechanism. Naturally, [users] will read all the results given because they are on a search for something. From that standpoint, text ads are really valuable. In terms of the lower production costs, the speed [in which they can be put online], the ability to change things; you can develop a text ad for every single keyword at a low cost in a way that you can't do with graphics. (quoted in Honan 2001)

There are several ways in which a banner may end up being relevant. One of them is thematic similarity between the content of the page and the content of the ad. An example would be a reader of the online website of *The New York Review of Books* who has found the review of a book particularly interesting. This reader will find it relevant to see on that page a banner that leads directly to an online bookshop like *Amazon* or *Barnes & Noble* where the book can be purchased.

But certainly the most interesting way of attracting interest is when the stimulus is combined with contextual(or already manifest) information to yield interesting conclusions that can only be obtained from this combination. This is why the most efficient way to obtain relevance for an Internet ad is to place it on a web page whose content favours this combination and subsequent conclusions, as in (5):

(5) An Internet user with a low budget for buying a car remembers that some friend of his told him once that there were brand new cars on sale for less than six thousand euros. He is reading a piece of news on cars in his favourite online newspaper and comes across a banner leading to a section of an online magazine in which there is an article comparing a number of low-price cars. He clicks on it and discovers that two out of five cars are under six thousand euros, and one of them also has air conditioning and power steering. This banner is relevant in the current context (willingness to buy a car) and this expectation of relevance is confirmed: on the one hand, the user strengthens his previous assumption that there were cars under six thousand euros. Moreover, and crucially, the new information about these cars is combined with the user's previous intention to buy a low-price car, leading to the implication that the one with air conditioning and power steering is the one which he should buy. As a side effect, his satisfaction with this article on cars may lead, in the future, to a subscription to the online magazine.

The usefulness of placing the ad on thematically related pages is important, but of course the design of the advertisement also plays a part in relevance outcomes.

It is undeniable that certain designs are attractive and draw the user's attention, setting in motion the search for relevance (see Alamán 2003).

Furthermore, placing the banner at the top of the web page is often suggested as a tip for ad success. However, in my opinion, this attribute is counterproductive, because by placing the ad systematically in the same area of the page, it is inevitable that the user will end up constructing a "banner placement stereotype" (part of what in the next chapter will be labelled *interiorized schema*, in this case "banner schema") and users will even end up not noticing that the banner is placed on the page at all (the aforementioned *banner blindness*).

Several studies have empirically tested the effect of placing banners systematically on the same area of the page, the extent to which the user's attention varies when the banner is placed elsewhere, and banner blindness in general. One of them is Pagendarm & Schaumburg's (2001) study. Several informants were asked to find certain information on a web page. This information was easier to find by clicking on a banner than by looking for it in a different part of the page. In the first experiment, the informants found it more difficult to find the banner-mediated information than to find it through the ordinary links. In the second experiment, several attributes of the banners were changed (colour, animation or lack of it…). More than 70% of informants were unable to find the information that would have been easy to access simply by clicking on the banner. They did not expect the banners, typically disruptive and non-related to the content of the page, and also placed in identifiable places of the page, to be the source of the information they were searching for.

Another interesting study is that of Benway & Lane (1998). The informants were asked to find information on English courses. The web page contained a prominent banner with the text "*New! English courses*" and, in smaller letter size, the text "*Click for more info*." To the authors' surprise, the informants did not see that information and clicked, instead, on a small link with the text "Courses" (but the relevant information was not there). When the informants were shown the banner where they should have clicked, they were surprised at having missed it. Again, the explanation is that the informants unconsciously avoided the expenditure of cognitive resources for the processing of the banner, since the mental schema for banners includes the assumption that they are useless and disruptive in the context of web content processing.[34]

34. Other studies mentioned in Pagendarm & Schaumburg (2001) and also in Bayles (2000) yield apparently contradictory conclusions, because in some experiments the informants paid attention to the banner for up to a second. But actually there is no such contradiction, because even if the informants did notice the banner on the page, that is, even if they did not suffer from *banner blindness*, a deeper level of processing is necessary to determine the really useful information beyond the link in the banner and conclude that it is potentially relevant to click on it, a level that the informants did not reach.

This tendency to the user's dismissal of pop-up advertisements was empirically demonstrated by Bahr & Ford (2011:782), where users rapidly developed as sort of mental schema of uselessness for these ads: "users operated using a mental model for future pop-up stimuli to be perceived as meaningless and disruptive and promptly eliminated with a heuristic considered relatively safe." But this research should be complemented with analysis of how the structure and overall content of the web page where the banner is allocated affects banner identification and recall, as in Hsieh & Chen (2011). One of the main conclusions was that

> Web pages with different information type of browsing contents do affect users' attention on advertising. Video-based webpages and picture-based webpages have a better advantage in advertising attention than text-based webpages and text-picture-based webpages do. Among these information types, the video-based webpages are the best for drawing users' advertising attention. (ibid.: 943)

In conclusion, the design of the banner with attractive colours, animations, etc. is useful to draw the user's attention, but it does not guarantee that the user will actually click on it to purchase a product. The design is in fact subject to the general cost-benefit cognitive assessment that all web page processing entails. *Google* has demonstrated that it is possible to be effective with plain text-based ads (also called *microads*). According to Omid Kordestani (Vicepresident of *Google* at the beginning of this century) the goal was to make something that did not subtract from the *Google* user experience. The idea was that the commercials should be complementary to the standard search, be clearly set apart as ads by colour, service the client to learn how to better target the ads, and be designed around relevancy. These ads allow fast load time, clear distinction, and tools for relevancy (quoted in Honan 2001).

The most positive way to make banners more efficient is to obtain the users' involvement, to get them to participate through expectations of relevance. This can be achieved by two means: (a) by asking the users to solve some inferential or metadiscursive puzzle suggested in the banner; or (b) by getting them to spread the qualities of the product to other users in a virus-like way.

An example of (a) is to challenge users by creating some kind of incongruity in the banner and expecting them to be willing to solve it as part of the general human tendency to make sense of the world we live in and sort out inconsistencies. This strategy is typical of printed advertisements, where their texts often include word-plays on polysemy, homophony, etc., as well as text-image incongruities (see Tanaka 1994). Internet banners can also exploit this strategy, but the way this is sustained on the Internet necessarily differs from printed ads. As Gustavo Núñez, from *Nielsen Online España*, correctly asserts (in Sevillano 2010: 12), the challenge for the advertiser "is not so much the format as the way communication ends up being successful." It is useless to simply transfer the TV spot to the Net; the language has to be different, not unidirectional. "The user expresses an opinion, replies, talks to other users [...] the brand image is constructed in real time." User participation can also be fostered by asking them to compete for a reward. This is what the advertiser for a new film on Sherlock Holmes did by asking *Facebook* users to find Watson with a number of clues and whoever solved the puzzle got free tickets to see the film. Similarly, *advergames* (defined as "computer games specifically created to function as advertisements to promote brands, where the entertainment content mimics traditional game forms," Kretchmer 2005, quoted in Cauberghe & De Pelsmacker 2010: 5) are effective means to obtain user interest, participation and, eventually, the spread of the brand or product advertised.

In the case of (b), users may also be willing to play an active part in the viral spread of the content of the product advertised if they think that it is going to be useful for their circle of friends and acquaintances. An example is the billboard that appeared in the streets of Spanish towns in 2008, described in (6):

(6) a. Advertisement: "Cambio mi sexto sentido por un instinto básico... Quién fuera hombre."
[*I exchange my sixth sense for a basic instinct... I wish I was a man*].

b. Brand: A company that sells men's underwear.

c. Visual context: The text in (6a) is superimposed on the photograph of a topless woman who is wearing one of the male underwear garments.

During this campaign, the users could enter the brand's website and send this ad (the woman with the text) to their friends by e-mail, but changing the text into whatever message they wanted to type (for example, "I'd like to introduce you to my new girlfriend"). By humorously sending this ad to their friends, the users were unconsciously spreading the qualities of that brand's underwear. This virus-like spread may even be more effective than the traditional Internet banner.

Another example is described in Fleck et al. (2007: 232) and reproduced in (7) below:

(7) An advertising campaign by Volkswagen used short video clips which could easily be shared and distributed across the World Wide Web. The story was trivial, showing an old man (Horst Schlämmer) preparing for the driving licence examination. Thanks to the performance of the well-known German comedian Hape Kerkeling, the clips were amusing and entertaining.

At the heart of *viral advertising* lies the concept of *meme*, a cultural unit disseminated among the population and which, in the context of Internet-mediated communication, ends up being manifest to a number of users and online communities.

From an anthropological perspective, there is an interesting debate between those who defend a cultural dissemination based on the *meme* (with a *genetic* conceptualization of cultural spread) and those who defend an *epidemiological approach* (with a more virus-like conceptualization), the latter proposed by Sperber (1996), among others.[35] Similarly, mouth-to-mouth (or, rather, "keyboard-to-keyboard") communication of advertisements on the Net may also obtain a cultural, communal status thanks to the users' obsession with spreading and sharing information in their social networking sites, chat rooms, etc.

35. An application of this epidemiological approach to humour was suggested in Yus (2002c, 2004, 2005d), specifically to stand-up comedy monologues. When the comedian addresses the audience, they communicate (make manifest) a number of assumptions related to cultural representations, often of a stereotypical quality, which are already stored in the minds of the audience. The comedy venue is an ideal scenario for the propagation and dissemination of cultural information, a version of which ends up being shared by all the members of the audience (with subtle variations deriving from the combination of this cultural information and the personal cognitive environments of all the members of the audience). In short, cultural information – typically made of stereotypes – spreads in the comedy venue with the aid of the comedian's monologues.

CHAPTER 4

Social networks on the Internet: The Web 2.0

The Internet is under constant evolution and development. One of the most strikingly successful environments for virtual interactions and information transmission is the popularization of a new form of production and reception of information that avoids the traditional "pyramidal media communication pattern" based on an authority that uni-directionally filters and delivers Internet content to the mass of users. Instead, this new trend of informational dissemination feeds from the users through special interfaces for interactions and content sharing. This phenomenon, now consolidated, has been given different labels, such as *social networks*, Web 2.0 (see O'Reilly 2007), *wiki phenomenon*, *participatory culture* (Jenkins et al. 2006), *user-generated content*, *Me Media* (Garfield 2006), and *social software*, among others. It implies a new form of conceptualizing the Net that is interesting for a pragmatic analysis of the information exchanged in these networks and the way it is interpreted, contextualized and transmitted. Besides, analysts within pragmatics would also be interested in the quality of interactions therein and how these are sustained in these virtual scenarios. Furthermore, it is also worth studying the role that the interface (e.g. blogs, social networking sites, *Twitter*) plays in information transmission, identity shaping and collective actions. For example, Androutsopoulos (2010: 208) claims that any analysis in this direction should take into account processes such as *integration* (the co-existence of various communication modes on a single platform), *embedding* (the ability to place digital content, especially videos, on a web page, or combinations of text and multimedia), and *modularity* (the way in which web pages are composed of a number of different elements in terms of origin, authorship, affordances, conditions of production and so on).

A pragmatics-oriented analysis would initially focus on the "addresser users," i.e. the producers of information, whose task of uploading information has to be compensated for by an offset of cognitive effects that are more related to "contributing to the collectivity" than to obtaining self-oriented cognitive reward, as can be deduced from these two addresser-centred conditions of relevance:

Condition a. Socially generated information is relevant to an individual to the extent that the social benefit achieved when it is produced is large.

Condition b. Socially generated information is relevant to an individual to the extent that the effort required to produce it does not threaten the user's satisfaction at being engaged in collectively generated content.

These two conditions are influenced by inherent features of the Net. For example, condition (b) is affected by several factors, as concluded by Li (2011), to the extent that in that study effort was not a significant predictor of the willingness to contribute information to online communities. Specifically, effort is typically minimized by the quality of digital discourse and users' surfing habits:

> the regular information contributors are likely to be ritual visitors to the community so, when they take time to visit, the extra time needed to contribute information may not be substantial. [...] The effort needed to contribute could also be significantly reduced by using computer technology. Since information is often stored in digital format, contributing information could require little more than the ability to copy and paste. Considering these factors, it is conceivable that the cost of contributing information to online communities is negligible. (ibid.: 291)

Of course, there is also a recipient-centred estimation of relevance. Addressee users have to make sense of the vast amount of user-generated information on the Net and select the potentially relevant one, often without the aid of an authority that filters out irrelevant content.

In this chapter, I will analyse several forms of Internet communication and networking that clearly emphasize sociality, interactivity and mutuality of information within the generic label of Web 2.0.[1] Firstly, blogs (or weblogs) will be studied from several points of view (author, content, reader and interactivity). Although blogs are a development of the traditional personal web page studied in Chapter 2, they possess an explicit social orientation and a purpose of interaction with other users. Secondly, the trend of social networking sites such as *Facebook* or *MySpace* will be addressed. They are also developments of the personal web page but the new interface allows for a great number of interactions and the management of shared information with friends or acquaintances. Finally, a brief analysis of the *microblog Twitter*, a short-messaging service with an explicit social networking orientation, will be made.

---

1. Chiang et al. (2009) argue that the qualities of social networks on the Internet are, in fact, scalable, that is, a website exhibits a greater or lesser quality of "Web 2.0ness" depending on how many prototypical parameters of social networks they exhibit.

1. Blogs

In the last few years, web pages have evolved into more interactive forms of Internet-mediated communication. Unlike the static quality of traditional web pages, which only made manifest information to passive readers, the level of interaction that blogs achieve today makes it possible to obtain a mutual manifestness of this information. Besides, one of the reasons why users abandoned the web page and created their blogs is that the latter are easy to use and update (and social networking sites are even easier to manage, thus reducing blog popularity, see Arthur 2009).

Blogs are web pages that have evolved into an identifiable genre (see Yus 2008d, 2008e). Among the many definitions of blogs that can be found in the bibliography, I have selected the following:

> A blog is a website that consists of short entries made by a writer, or a blogger. The entries are arranged in a reverse-chronological order (latest entry first) by time and date, much like on a message board or a website guestbook. Usually the entries consist of the entry text itself, a title and a time/date stamp. Only the newest entries are displayed on the main blog page while older entries are usually arranged in archives where they can be accessed on a later date. Many blogs nowadays also allow readers to post comments to individual entries, much as they would do in threads on a discussion forum. (Vuorinen 2005: 5)

> A frequently updated web site consisting of personal observations, excerpts from other sources, etc., typically run by a single person, and usually with hyperlinks to other sites. (Oxford English Dictionary, 2002, quoted in boyd 2006a)

> Blogs are online publications that are characterised by short entries which are usually written in an expressive and authentic style and are arranged in reverse chronological order. (Fleck et al. 2007: 228)

As was suggested for web pages in a previous chapter, blogs can be studied from three main points of view. Firstly, the author's intention when uploading information on the blog is essential in a pragmatic analysis. Secondly, the qualities of blogs as a stabilized genre can also be studied, insofar as they are evidence of the blogger's intentions. And thirdly, the analyst can use the content of the blog as a tool to predict the quality of readers' interpretations. To these three perspectives a fourth can be added that focusses on the blog as a medium to sustain interactions and as evidence of group or community ties.

1.1 The blogger's intention

Blogs are verbal-visual discourses that work as evidence of the blogger's communicative and informative intentions. These intentions are typically focussed on a desire to filter out and select the information which, on paper, is potentially relevant to the readers (*filter blogs*), or a desire to provide personal information about likes, dislikes, daily events, etc. (*diary blogs*). But other types of blogs have been suggested in the bibliography according to their format or other criteria.[2] In any of these types, tracing the author's intentionality is important for efficient blog communication. Gibbs (1999: 16) stresses that our interest in sharing other people's intentions is such an important aspect of how people construct meaningful interpretations that sometimes we get the feeling that intention ascription is somehow optional and can be discarded if one wants to do so. However, the explicit search for the intentional foundation of human actions reveals the extreme importance of communicative intentions in many aspects of our experience of meaning.

Initially, blogs appeared as some users' attempts to filter information and select the most interesting content for other users, who had to trust the filtering criterion and the blogger's "authority." The outcome of this filtering tended to be relevant because the negative condition of relevance (mental effort) was minimized through a reduction in the time and effort required to access interesting information. But if the selective criterion was unsatisfactory, the outcome was likely to be irrelevance.[3] Over the last few years, however, another type of blog has become popular: *the diary blog*, where users make manifest information about their lives, opinions, beliefs, etc. It is sometimes difficult to understand why certain

2. For example, Andreevskaia et al. (2007) add the *notebook* to *filter* and *diary* blogs. Biz Stone (2004, quoted in Chesher 2005) differentiates between *technology blogs*, *political blogs* and *diaries*. Lankshear & Knobel (2003) propose *links with commentary* (that work as filters), *journalling*, *hybrids* (between the first two types) and *meta-blogs* (blogs about blogs). Fleck et al. (2007: 231) make a format-based classification: *blog*, *photoblog*, *moblog* (entries created from a mobile phone), *audioblog* (mostly audio files) and *videoblog*. Herring et al. (2004) propose a classification based on the purpose of the blog: *filter*, *diary*, *k-log* (knowledge log, highly specialized), *mixed purpose* and *others*. Finally, Holbrook (2006: 7) suggests two types of blog: those which comment on information available on the Net (*epiphytic blogs*) and those which create their own content (*generative blogs*).

3. Umberto Eco (quoted in Origgi 2002) has mentioned the danger of a lack of filtered information on the Internet: "With the Web, everyone is in the situation of having to filter information that is so vast, and so unsustainable, that if it isn't filtered it cannot be absorbed. It is filtered unsystematically, so what is the primary metaphysical risk of this business? That we'll end up with a civilization in which every person has his own system of filters, in other words where every person creates his own encyclopaedia. Now a society with five billion concurrent encyclopaedias is a society in which there is no more communication."

information about bloggers' lives might ever prove relevant, but as we have seen in the previous chapter, on the Internet some combinations of cognitive effects and mental effort are surprisingly beneficial to the readers. Similarly, it does not seem to be worth the effort to trace the exact intentionality underlying these diary blogs (Miller & Shepherd 2004). One source of relevance for users may be what Thompson (2008) calls *ambient awareness*, the feeling of being physically near the blogger with the aid of the information posted about his/her feelings, moods, ordinary activities, etc.

But in general it can be stated that simply uploading vast amounts of information about oneself is counterproductive in terms of cognitive effects and mental effort. This is why several applications have been designed to help users control the flow of information that their blogs generate. For example, Kendall (2007) mentions that one of the portals for blogging, *LiveJournal*, offers the possibility of using "cut tags" that allow users to link part of their blog entries. The hidden part of the entry will not be visible on the screen until the reader clicks on it.

The attempt to understand the phenomenon of diary blogs is even more difficult if we take into account the fact that bloggers are often unable to explain why they uploaded the information about themselves on their blogs. A possible explanation is that, by updating their blogs, the bloggers shape, strengthen and develop their identities, and the blogs acquire a certain corporeal quality for them (boyd 2006a, Efimova et al. 2005). Efimova & Hendrick (2005) note that

> what makes weblogs different is not the publication of content *per se*, but the personalities behind them. Weblogs are increasingly becoming the online identities of their authors. Most weblogs are not formal, faceless, corporate sites or news sources: they are authored by individuals (known as webloggers or bloggers), and perceived as 'unedited personal voices' [...] Often a weblog is written as a narration of its author's thoughts and feelings, [...] allowing personality and values to emerge from the words. Even weblogs that are little more than collections of links and short commentaries say something about their authors. The selected content a weblog author finds interesting enough to link to and to comment on functions as a public record of personal interest and engagement.

Holbrook (2006:7) uses Genette's famous terminology on literary narratives in order to look into the bloggers' presence (and identity) on their blogs in more detail. When the *implied blogger* tells a story in which he/she is also the main character, the blog is *homodiegetic*, and when he/she does not participate in the story, then the blog is *heterodiegetic*. A parallel interest lies in determining whether the bloggers are "transparent" in the construction of their identities on the blog or play with some sort of fictitious identity, which has enormous consequences for the extent, intensity and eventual relevance of blog entries.

Other authors have suggested underlying intentions for bloggers when they upload information. A good example is the work of Nardi et al. (2004, 2005), who list the following motivations for blogging: (a) to update others on activities and whereabouts; (b) to express opinions and thus influence others; (c) to seek others' opinions and feedback; (d) to "think by writing"; and (e) to release emotional tension. Efimova (2003), for her part, concludes that, among bloggers' motivations, some stand out, namely curiosity, the will to improve the management of information, and learning, plus an overall interest in content sharing. Bortree (2005: 26) also proposes reasons for blogging, specifically related to self-presentation: *ingratiation* (the goal is to be appreciated by others), *competence* (also called *self-promotion*, where the goal is to be perceived as skilful and qualified), *intimidation* (being seen to exert power), *exemplification* (attempting to be perceived as morally superior or possessing high moral standards) and *supplication* (being seen as requiring nurturance or appearing helpless so that others will come to one's aid).

Diary blogs should be intended for a more restrictive audience in mind, and this is very often the case: such blogs are usually read by an intimate and previously selected audience.[4] Holbrook (2006: 9) defines a diary blog as "any blog that generates its own content rather than commenting on other content, presents a narrative that is presumed to be reflective of the implied blogger's real experiences, and is tied together by a focus on one or more characters rather than themes."

Another interesting issue is that many bloggers value more the effect that the blog has on themselves than the one it has on its readers, which can be intuitively explained as a complement to the achievement of mutuality of assumptions and the alteration of the readers' cognitive environments, the main reasons for keeping up a blog. In a certain sense, the blogger's subjectivity is in a state of constant updating in parallel to the updating of the blog content and the current state of interactions and comments. Lu & Hsiao (2007) also conclude that personal rewards such as consolidating one's image and obtaining other bloggers' praise play an important part in the blogger's desire to share information on the blog (see Ko et al. 2008, Lenarcic & Sarkar 2008).

An example of the difficulty that tracing bloggers' intentions entails is the *photolog*. Cohen (2005) describes how *photo-bloggers* experience two types of feelings when they upload photos. On the one hand, the picture recalls some aspect of the moment when it was taken. On the other hand, there is a certain feeling of surprise that the *photo-blogger* experiences after some time has elapsed. Indeed, these bloggers are very often unable to explain the intentions that underlay the creation or uploading of the picture. Besides, "the fact that digital photographs are free of cost

4. See, for instance, boyd (2004a), Mortensen & Walker (2002: 209–210), and van Dijck (2004).

allows them to take pictures whenever they want, of whatever they want [...] And this proliferation of photographs and the situations in which photographs are made creates the conditions under which surprise is possible" (ibid.: 889). Although many *photo-bloggers* write texts at the bottom of the pictures to explain what they had in mind at the moment of taking the picture, sometimes they admit that they do not have a clear picture of the motivations. That is, when coming to explain a *photo-log*, instinct seems to play a greater role than explicitly communicative intentions.

1.2 The blog genre

A second pragmatic approach to blogs focuses on the verbal, visual and multi-modal attributes of the blog; it studies the existence (or not) of a stabilized *blog genre*, and analyses the role that blogs and their genre can play in the effective transference of bloggers' communicative intentions and the (in)correct interpretation of their intentions.

Among the possible approaches to the blog genre, it is important to determine to what extent it has evolved into a clearly identifiable and conventionalized genre that can be differentiated clearly from other competing discourses such as the traditional personal web page[5] or social networking site profiles. This status of "genre autonomy" is interesting for cognitive pragmatics, since the immediate identification of the blog genre may affect the quantity of effects obtained, the mental effort involved in interpreting the content of the blog, and the relevant conclusions that might be derived from its processing.

In this sense, overlappings with other genres have been mentioned, besides web pages. Herring (2003) points out that blogs have inherited features from personal diaries, from opinion essays in the 17th century, and they also share similarities with newsgroups (see Chapter 6) and chat rooms (see Chapter 5). For Lawley (2004), blogs are unique in the way they blend the temporal quality of e-mail distribution lists and newsgroups with the stability of the web page stored on *Google*. Furthermore, Herring et al. (2005: 143) conclude that

5. Among the differences, boyd (2006a) points out that blogs are not as complex as web pages. Chesher (2005) finds differences in format and options for navigation. And Karlsson (2006: 10) stresses that the web page does not contain so many interactive elements or the immediacy of instantaneous publication. Besides, Lu & Lee (2010: 22) list a number of clear differences between web pages and blogs: the latter exhibit update frequency, clear ownership, optional links to other blogs, support by tools and freeware that can automate certain functions, entries that are displayed in chronological order, and readers that "are given the option to leave comments, which help them interact with particular bloggers, thus it is more suitable than traditional web sites for online relationship building."

> the blog is neither fundamentally new nor unique, but that it – along with other emergent genres expressed through interactive web technologies – occupies a new position in the internet genre ecology. Specifically, it forms a de facto bridge between multimedia HTML documents and text-based computer-mediated communication, blurring the traditional distinction between these two dominant internet paradigms, and potentially contributing to its future breakdown.

Genres have been defined as "instances of conventionalised or institutionalised textual artefacts in the context of specific institutional and disciplinary practices, procedures and cultures" (Bhatia 2001: 5). This is a rather static definition that does not explicitly account for the fact that genres are processed in specific contexts, identified as evidences of communicative intentions and stabilized inside a community of people. A better definition is Swales' (1990: 58): "a class of communicative events, the members of which share some set of communicative purposes. These purposes are recognised by the expert members of the parent discourse community, and therefore constitute the rationale for the genre."

Besides, genres are typically identified as fulfilling specific goals and demand mutual accessibility (between interlocutors) to their qualities. The goal of a genre is not the personal motivation to communicate, but the need to be constructed and recognized socially by the pertinent community of users, and it is bound to be used in typical situations where the genre is inherent and necessary (Orlikowski & Yates 2002). The interesting part of this function of "recognition" that genres entail is that the mental effort required to process the genre tends to decrease when its features become clearly distinguishable from other discourses and hence become exponents of the specificity of the genre. Starting our processing with the identification of the genre (and the corroboration of our expectations about its conventional features)[6] allows readers to generate specific interpretive patterns that save mental effort and which would not be created if the genre was not conventionalized. As Santini et al. (2010) stress,

> genres can be seen as sets of conventions that transcend individual texts, and create frames of recognition governing document production, recognition and use. Conventions are *regularities* that affect information processing in a repeatable

6. These expectations will tend to be interiorized and the reader will use them by default in the interpretation of a blog, due to their accessibility. Dillon & Gushrowski (2000) have corroborated that adequacy to genre conventions helps to recall discourse and increases the reader's satisfaction. Moreover, analysts in the area of hypermedia and web page design have concluded that Internet orientation and navigation are influenced by the user's identification of rules about how information should be presented and, therefore, the absence of genre conventions in the digital world are a source of potential difficulty for user navigation.

> manner. Regularities engage *predictions* about the "type of information" contained in the document. Predictions allow humans to identify the *communicative purposes* and the *context* underlying a document.

Therefore, even if it is true that blogs are hybrid genres (or *inter-genre-al*, following Devitt 2009: 44), in my opinion they have acquired a clearly identifiable status within the range of *cyber-media* that allows for their initial effort-saving processing of layout, etc. Lüders et al. (2010: 956) add, specifically for "diary blogs," that

> whereas the personal blog differs in significant ways from the paper diary, its resemblance to the paper diary explains how users nevertheless approach this new genre (or any new genre) based on generic knowledge, crucial for making sense of specific texts. New genres never emerge without a context. Hence, whereas users need to internalize conventions, they already possess generic knowledge derived from antecedent or similar genres. This knowledge enables communication, and is thus crucial in the emergence and stabilization of a new genre.

An analysis of 100 blogs was carried out in Yus (2008e) to determine the discursive features that are so conventionalized in the blog genre that users invariably expect to find them every time they read a blog. These features shape the user's *internalized weblog schema*. This mental schema, based on the establishment of the "blog genre," has been facilitated by the availability of companies such as *Blogger*,[7] whose simple, easy-to-use templates make the blog genre even more identifiable, to the extent that other users, faced with the challenge of creating a blog, will tend to use these default templates, spreading the weblog schema to other users in a virus-like way. This picture of a mental schema of blogs fits the view that "genre is not only something manifested in texts, but also a knowledge which users must have to be able to interpret and act in accordance within a given communicative context" (Lomborg 2009).

Initially, the identification of conventions that belong to the store of properties of blogs that the readers possess (and expect to find every time they enter a blog) entails a double analysis of the visual (iconic) and verbal (symbolic) content of the blog (or multimodal combinations). In practice, though, blogs tend to a mixture of iconic and symbolic signs in which, very often, the iconic content acquires a symbolic quality and the verbal content becomes *iconized*. This is a phenomenon

7. Evan Williams, one of the creators of this famous software to create blogs, proposes a definition of blog that is worth quoting because it underlines the qualities of the genre above the importance of the information that the blog contains: "To me, the blog concept is about three things: Frequency, Brevity, and Personality. […] This clarification has evolved over time, but I realised early on that what was significant about blogs was the format -not the content" (quoted in Mortensen & Walker 2002: 249).

that blogs share with other discourses such as comics, where this blend of iconic and symbolic signs is also typical (see Yus 2008f).[8] The discursive features of blogs are not, therefore, simply verbal or visual. During the processing of these features, texts and pictures are not processed as purely symbolic or iconic, respectively, but as a mixture of properties that activate the identification of the blog genre. For example, readers use their store of prototypical features of blogs and visually spot the textual elements of the page (categories, entries, comments…) even before these texts are actually processed.

As I pointed out above, the stabilization of the blog genre has been accelerated by the availability on the Net of free software for designing blogs such as *Blogger*, with pre-determined steps to create a blog and a fixed interface layout. And inside the blog it is possible to determine different categories. For example, Lomborg (2009) proposes three axes on which we can place all blogs. The first one, the *content axis*, comprises blogs ranging from personal experiences (*internal*) and events which are external to the user (*topical*). The second one, the *directionality axis*, classifies blogs between *monologism* (user's own stories and little interactivity with readers) and *dialogism* (reader-oriented and filled with interactions). Lastly, the *style axis* can be either *intimate* or *objective*, depending on the style of the text in the entries of the blog.

An analysis of blogs also leads to the conclusion that bloggers are, in general, more interested in the information that they want to communicate (i.e. make manifest or mutually manifest if the interface allows for this level of mutuality) than in a more or less creative design of the blog (see Scheidt & Wright 2004), maybe because most bloggers are not trained in programming or because the range of default options is sufficient for their communicative purposes. Moreover, these default options are easily identifiable, save processing effort and alert readers to the content that they expect to find inside the blog. This was corroborated by Lu & Lee (2010), who concluded that content quality is the most important factor in making users stay longer and revisit the blog. The context quality – how the blog is presented – turned out not to be so important in retaining readers. These authors also mention the list of popular blogs provided by *Blog Look* (http://look.urs.tw), and the fact that most of these famous blogs only slightly modify the default options provided by blog service providers.

The *interiorized weblog schema* (Yus 2008e: 125) comprises those features of blogs that have stabilized in the readers' minds and save processing effort because

8. In comics the reader can find highly symbolic visual signs (e.g. lines to show movement, a light bulb that symbolizes an idea, etc.) and highly iconic verbal signs (e.g. text that is deformed to show the character's emotions). This parallelism between visual and verbal attributes also exists in the processing of verbal and visual metaphors, as claimed in Yus (2009d).

of their high accessibility and capacity to generate expectations. The choice of the term "schema" indicates that the information stored about blogs does not include all their possible features, but only those that readers invariably expect and that facilitate blog identification. After all, schemas contain archetypical information and some blog features do exhibit this quality. The idea that underlies this *interiorized schema* is the certainty that the blog genre and its recurrent qualities play a part in the general search for relevance, against opinions such as boyd's (2006a):

> The prototypical blog has many of the features supported by the most popular tools: commenting, links, trackbacks, time stamps, reverse chronological posts, and syndication feeds. While prototypes have communicative efficacy, they should not be the basis upon which analysis is built. The properties of the prototype do not define the boundaries of the medium nor do they convey value or normative practice. As technology changes, the properties of the prototype will also change.

Needless to say, this *interiorized schema* is not fixed, but changes as the blog evolves due to the users' demands or because companies introduce new options that end up being used massively by bloggers (and expected by readers). The stabilization of the schema is a gradual process of permanent updating of blog features, some of which may disappear from the schema because of a recurrent lack of usage or presence in the blog, while others are incorporated into the blog schema after stabilization.[9] Devitt (2009:41–42) correctly writes that "bloggers and their guests do not encounter genre forms in isolation but rather as collections and absences of features in specific blogs [...] neither bloggers nor readers require a single, closed set of unchanging forms to participate in blogging. If we abandon trying to define genres through closed, static sets of forms, we can permit forms to be what they appear to be, multiple, fluid, and yet constructive of generic actions." Schmidt (2007) adds that "by incorporating shared expectations and routines into their individual ways of handling the format, bloggers not only fulfil their communicative goals, but also reinforce and reproduce the sets of adequacy and procedural rules." Hence, readers expect, as part of their *interiorized schema*, a number of blog features, but this does not mean that the schema is static, unchangeable, nor does it imply that not comprising *all* of these features prevents successful blog interpretation.

9. See Mariottini (2011a, 2011b, forthcoming) for interesting analyses of the quality of this *interiorized schema* in the context of specialized discourses (blogs for lawyers and blogs of tourism). In her research, several questionnaires were given out that revealed the exact quality and extent of this schema, which users invariably expect to find when entering these specialized blogs.

Several insights into the quality of this blog schema were derived from an analysis[10] of 100 blogs (see Yus 2008e). Among others, the following aspects deserve a few comments:

1. *General blog layout.* Readers expect a layout of a single column of text (normally with the entries) and one or two side columns or frames with other elements. Besides, 66% of blogs contain a rectangle at the top with the title of the blog or a distinctive picture. This rectangle probably belongs to the reader's *interiorized schema.*

2. *Background colour and wallpaper.* It does not seem likely that there is an expectation of a specific background colour, but 50% of blogs have a white background. This suggests an interest in legibility, which is corroborated by the lack of wallpapers (97% do not have one) that also reduce text legibility.

3. *Reference to the blogger.* Most blogs contain references to the blogger, and their readers expect to find this reference when entering the blog. Contrary to my expectations, many bloggers avoid the use of *nicks*, but provide their real names (55%) and pictures or drawings of themselves (27%). Frequently, most of the information that the blogger wants to convey to the readers is found after clicking on "see my complete profile" or "about the author" (63%).

4. *The word "blog."* In theory, a useful means to alert readers to the fact that they have just accessed a blog is to place the word *blog* or *weblog* on the front page. 49% of blogs do contain this word, but in my opinion it does not play a crucial role in the readers' efficiency at identifying the blog, nor is it likely that the word will belong to their *interiorized schema.* The explanation lies in the evidence that, once the blog has stabilized, many bloggers feel that they no longer need to label their pages as "blogs" so that their readers can identify them, since other elements in the blog will perform that function (Yus 2008e: 130).

5. *Blog sections.* Most of the blogs analysed (97%) contain links to sections and these links are placed on a side frame or column (*sidebar*). The reader will invariably expect these "sections as links" on the main page of the blog. Specifically, they will expect a neutral letter font (81% of blogs), in bold (65%) and black colour (56%).

6. *Letter font.* The analysis revealed a great variety of letter fonts in different sections of blogs. The most frequent font type is *Verdana* (in 28% of headings, 39%

10. A similar analysis was carried out by Klamma et al. (2007), who compared sites such as *MSN Space*, *Blogger*, *Squarespace* and *MySpace*. They concluded that only a few features such as comments, the archive of entries, the blogger's profile or permanent links (*permalink*) were found in all blogs. But, in my opinion, being present in *all* the blogs is not a necessary condition for a feature to belong to the *internalized schema.* A high percentage of presence suffices for the reader to expect its presence on the blog.

of entries and in 30% of date and hour of posting), but its usage is not so frequent as to predict that it will be expected by the readers when they enter the blog.

7. *Elements in the entry.* For several analysts, the entries are essential in a blog, together with their reverse chronological order.[11] Hourihan (2002) comments that, in fact, entries or posts favour a specific type of reply or comment from readers: "Blog posts are short, informal, sometimes controversial, and sometimes deeply personal, no matter what topic they approach. They can be characterized by their conversational tone and unlike a more formal essay or speech, a blog post is often an opening to a discussion, rather than a full-fledged argument already arrived at." For boyd (2005), entries are written with a lot of suppositions and expectations incorporated to them. It is assumed that the reader knows the blogger's motivations and beliefs. There are entries previous to the one being processed and which may have been archived but are essential in order to interpret the latest entry correctly, and the blogger expects their information to belong to the reader's cognitive environment at the moment of interpreting the latest entry.

If we analyse the posts or entries in detail, we can deduce which of their elements are candidates to belonging to the reader's *internalized weblog schema*. For example, most of the entries contain the date (97%) and hour (64%) of publication. Showing the number of comments that the entry has received is also common (85%). And more than half of the entries include their location, either by category or by permanent link (*permalink*).

Readers will also expect a specifically devised area in the blog to send comments on an entry. Most of the blogs analysed (88%) offered a blank form to send these comments, often below the text of the entry. This form can also be accessed after clicking on the link "comments." Additionally, the reader can be offered a list of the most recent comments or related entries (in 55% of blogs).

8. *Links.* They are essential in the blog genre, and readers expect to find them. Links are used in entries, comments, categories, archives, etc. They also create a network of inter-connected blogs by relating one another with the aid of *trackbacks*.[12] In this case, since readers are "invited" to surf through the inter-linked documents, more responsibility is demanded from them when finding a reward in the eventual congruency that they might achieve after processing all the chunks of information that are scattered but link-related on the Net.

Ali-Hasan and Adamic (2007) distinguish different types of links that indicate relationships between blogs: (a) *blogroll links*, typically found on a sidebar,

11. See Mortensen & Walker (2002: 249), Blood (2003: 61), and Orihuela (2005: 18).

12. According to Li (2005: 42), *trackbacks* help to fill a page with relevant information by linking entries that share the same topic but are located in different blogs. *Trackbacks* offer an innovative solution to collect entries that are dispersed all over the *blogosphere*.

provide connection to other blogs that the blogger recommends; (b) *citation links*, normally located inside the posts and referring to the entire blog or a particular post on that blog; and (c) *comment links*, which occur when a reader adds a comment to another blogger's post (see also Luzón 2009).

9. *Archive*. The reader's *interiorized weblog schema* also includes an expectation to find some form of archive of past entries and their comments. The most recurrent form of archiving is by month-year (72% of blogs) and more than a half of blogs also contain a search option. Besides, readers are often offered categories where the entries are archived permanently (43% of blogs).

Another perspective of analysis of the blog genre involves its ability to spread across the community of bloggers if this genre has become sufficiently conventionalized and interiorized, to the extent that all users have a more or less faithful version of what the blog genre is like. There is a cumulative process in which bloggers tend to use the basic options that they are offered in the templates and hence the *schema* ends up being increasingly similar in the *blogosphere* and in the readers' minds, thus facilitating the identification of blogs and their content. This would be a kind of *epidemiological stabilization* of the blog genre, but, as pointed out above, this *blog schema* is under constant re-shaping and updating among the community of bloggers depending on which options are discarded and which ones are incorporated to this schema (Yus 2008e: 137; see also Yus 2003c, 2005b, 2007b). Similarly, several discourses under the "blog umbrella" exhibit different levels of conventionalization. This is the case of *vlogs* (video-blogs). Frobenius (2011: 816) comments that *vlogs* "constitute a genre so young that the conventions are still in a process of negotiation." He compares *vloggers* to television news presenters ("a vlogger is an independent (usually, but not necessarily) unpaid, private and untrained individual, while a TV news presenter is a journalist representing a broadcast network") and to traditional blogs, but the former are still far from a genre that has become conventionalized within the community.

As Crowston & Williams (2000: 203) assert, since the members of a community extract their knowledge from a range of genres in order to interact with one another, these members strengthen the use of these genres, making them more appropriate for a given situation. That is, the group of genres in use (i.e. the repertoire of genres) is both a product and a shaper of communicative practices within a community. Scheidt & Wright (2004) add that new bloggers tend to incorporate fewer new options to their blogs. As they share an increasingly fine-grained picture of what blogs should be like, they avoid innovations that are inconsistent with the stabilized genre. Of course, some users do design innovative blogs and their attributes might end up conventionalized and part of the *interiorized schema* if a substantial number of bloggers incorporate these innovative features into their blogs.

1.3 The reader's interpretation

One of the main objectives of cognitive pragmatics is to predict the addressees' inferential steps and accessibility to contextual information when they interpret utterances (or texts). Specifically, cognitive pragmatics is interested in determining why readers select (or not), among the range of possible interpretations of a coded utterance or text in a specific context, the one that the speaker or writer intended to communicate. In the case of blogs, there is a huge amount of information available to readers (made manifest by the bloggers) and lots of links to click on.

One problem that blog readers face is the initial lack of *mutual manifestness* with the blogger (although options for interactivity in blogs do facilitate mutuality of assumptions). In addition, bloggers expect in their readers the desired context accessibility and thus leave all the information that they expect to be already manifest to these readers implicit (not coded). In this way, a sort of *scale of readers* is created depending on the level of mutuality with the bloggers and their greater or lesser ability to fill in the information blanks of blog discourse and reach relevant interpretations. For example, some blogs contain jargons or specialized vocabulary that only certain readers can understand.[13] As Hanley (2005) states, "surely the idea of blogging – that is, writing about things you're interested in without the tiresome presence of an editor or censor – is to communicate; but when you're making up words without explaining what they mean, aren't you immediately alienating most of your audience?" Tony Thorne (in Hanley ibid.) thinks that part of the appeal of blogs lies in the fact that they are "geeky, anoraky and self-referential. All slang and jargon is essentially about exclusivity." A good example is found in Myers (2010: 91). He mentions *Sepia Mutiny*, a blog for people of South Asian origin (desis) living in North America. In one of the posts, a user is commenting on a previous post about dating:

> for some vague unexplainable reason, I tend to do much better with the dbd grls… I do better with DBD Mallus of any religion, than ABDs. I've decided that I will only marry someone fobulous. Yeah, I said it. But I'm going to marry one so I can totally do that.;)

13. Jargons are essential in identity formation and group identification. As is argued in Yus (2002a: 3729), jargons provide a feeling of belonging and entail the use of discursive features that are sources of *intra*-group identity, as much as sources of *inter*-group differentiation. Typical examples would be scientific discourse and specialized languages (see Yus 2007d, Alcaraz et al. 2007, Mateo & Yus 2009).

The "outsider reader" of this post will have to devote additional cognitive resources to understanding it and concluding that ABDs are "American Born Desis" and thus DBDs are "Desi Born Desis." The phrase FOB (Fresh Off the Boat), used by many immigrant communities, is here made into "fobulous." And "mallu" is a colloquial term for someone from Kerala, the writer's home state.

Another interesting reader-centred trend of blog research is to study the ways in which readers influence authors and vice versa, especially in discourses such as blogs, where the traditional passive role of readers has become more active and participatory (see Baumer et al. 2008, Karlsson 2006: 2, Kendall 2007). Hollenbaugh (2010: 1659) stresses that "regardless of whether or not a blog is private, bloggers' perceptions of who their audiences are may also impact their choices of what information to disclose. If bloggers believe that it is predominantly their close friends who are reading their blogs, they may disclose more intimate information than bloggers that believe their readers are relative strangers."

Readers may also feel overwhelmed by their role as commentators and readers of blog entries, whose relevant information is often scattered across fragmented link-mediated texts (and it is the reader that has to make sense of them). For instance, Mishne & Glance (2006) point out that readers often get annoyed when they realize that a discussion is fragmented in many entries and their related comments.

1.4 An emphasis on interaction

Other analysts prefer a more interaction-oriented approach to blogs, particularly how they foster and sustain dialogues between bloggers and readers. These interactions are often fragmentary and additional cognitive resources have to be devoted to making sense of the (intra- and inter-blog) interactions and to determining the portion of bloggers' and readers' cognitive environments that is mutual. De Moor & Efimova (2004) add that the multimodal quality of blog conversations is a supplementary source of fragmentation in blog interactions. Certainly, it should be borne in mind that nowadays several alternative channels of Internet communication are used to complement blog-based interactions, such as e-mail or chat rooms or even Internet-supported phone calls (e.g. *Skype*). Interactions on blogs have become multi-channel, rather than text-based. But this multiplicity may also aid blog interlocutors to achieve a more fine-grained sense of mutuality (see Efimova & Ben Lassoued 2008: 137).

Therefore, the interactive attribute of blogs is becoming a rather complex phenomenon with multiple links and threads, and sometimes with online and offline overlapping and the complementation of other *cyber-media*. Besides, interactions in blogs depend on entries whose connection is the blogger's responsibility (Yus

2008d: 29). As a consequence, the eventual relevance of the reading paths chosen by the readers is constrained by the authors' choices of what content is to be uploaded and how different entries are to be linked. In this sense, Efimova (2004, quoted in Luzón 2009: 77) proposes the term *distributed weblog conversations* for these conversations that are scattered across many blogs. Moreover, Lin et al. (2006) list the following aspects of blogs that are oriented towards interactivity: (a) *Temporal dynamics*. Entries can be created, edited and commented upon in a very dynamic way. (b) *Event locality*. The information provided by entries is dependent on the moment when they are created, and these have to be interpreted within a specific time-span beyond which they are no longer relevant. (c) *Link semantics*. Some link-related blog elements (*blogrolls, trackbacks*) are much more than simple links to other pages, but exhibit interactivity. And (d) *Community centric*. The purpose of a blog is to share information, and this purpose normally leads to community bonding. It should be noted that this fourth aspect illustrates the capacity of blogs to create and foster social gatherings and is, therefore, an indicator of the suitability of studying blogs and social networking sites in the same chapter (see Furukawa et al. 2007).

Several elements of blogs facilitate interactions and aid in obtaining a relevant degree of mutuality between bloggers and readers, whose cognitive environments are enlarged as a consequence of satisfactory interactions.[14] This mutuality is also necessary for an adequate feeling of (blog) community membership (see 1.5 below).

Among the elements of blogs that are oriented towards interactivity, the following deserve some attention:

1. *Entries or posts*. Inside each entry there is a clear indicator of interactivity in the attached comments. Other aspects that suggest the presence of interaction are the inclusion of links inside the text of the entries (which indicates inter-relation among them). For Estalella (2006: 26), "multi-situated" interactions, built up by links between blogs, are able to generate a space for shared communication.

2. *Trackbacks*. They make it possible for a blog to notify bloggers if their entries have been discussed, mentioned or commented upon. This is a very useful tool for readers, who can follow several threads of conversations and obtain a more fine-grained mutual manifestness of information.

14. Interactions would be one of the main purposes of blogs. However, several authors minimize the role of interactivity in blogs. As is summarized in Lenhart (2005: 37), for these authors bloggers only expect a sufficient level of interactivity so as to be aware of the readers' presence. In a similar fashion, instead of being an opportunity for mutual manifestness, readers' comments are often taken as a "threat" to the blogger's control over the quantity and quality of communication.

3. *Blogrolls*. These are links placed on one sidebar of the blog and refer to other blogs that the blogger either visits regularly or recommends, creating a feeling of inter-blog interactivity.

4. *Permalinks*. They make sharing of information easier, as they create stable links to a web page or blog. This link may be shared between users via e-mail or instant messaging, facilitating the spread of information.

5. *Tagboards*. They are defined as "little messages attached to your blog, where your readers can leave you notes. They differ from comments, which are attached to individual posts, in that you just have one tagboard for your whole site, and visitors can read the messages right on your homepage" (by *Blogger*, quoted in Lenhart 2005: 76). In this way, tagboards can foster feelings of interactivity around a blog.

6. *E-mail*. Probably the oldest means to foster virtual interactions, and also incorporated into blogs.

1.5 Communal bonding through blogs

Blogs can also create and sustain feelings of community membership arising from the imbrication of social networks and blogs. To understand how blogs may foster communities it is necessary to relax the rigid criteria that are often applied to the definition of communities in offline scenarios. If we do that, we will discover that there are elements in blogs that indicate the presence of communal groupings that have this genre as the main foundation. For example, in a study by Kervin et al. (2010), the comments posted in response to other blog entries provided evidence of the growing rapport and sense of community felt by bloggers and readers, with exchanges of comments such as 'I have had exactly the same bad experience!', 'I can definitely relate to what you say…', or 'Nice point, I hadn't thought about [that] before'.

Efimova et al. (2005) propose the following symptoms of blogging communities: (a) a *meme*-like dissemination (or, as proposed in Yus 2007b, an *epidemic*-like dissemination) of information among blogs; (b) the patterns of blog processing, which can be analysed with the aid of *blogrolls*, subscriptions to *RSS feeds*, etc.; (c) link patterns, which reflect to what extent blogs are positively valued (e.g. recommended); (d) "blog conversations," when a blog provokes the feedback of other blogs; (e) event indicators, in the sense that mentioning online or offline meetings of bloggers indicates the communal relationships among them; and (f) "tribal" marks, group spaces, blog directories.

As was mentioned in Chapter 2, nowadays there is a tendency to a hybridization of offline and online sources of community, with the user as an intersecting "node" of all the personal networks in which he/she participates (Yus 2005b, 2007b, Willson 2010: 494). Blog communities would fit this hybridization where users can obtain feelings of interactivity, reciprocity, belonging, etc. that compete in intensity with traditional sources of communal satisfaction in physical settings. Of course, relationships initiated and maintained in blogs may, in theory, "jump" from the Net to physical scenarios as a complement to communication on the blogs, although several studies indicate that few blog communities have a counterpart in offline gatherings (Furukawa et al. 2007). For example, Nardi et al. (2004) concluded that blogs are more like a refuge from the intense interactions that users carry out through other forms of Internet-mediated communication. Blogs allow for the expression of identities without the requirement of immediate feedback. The picture of blogs is that of diffusion, rather than reiterative communal interactions.

2. Social networking sites on the Internet

In the last few years there has been a revolution in Internet-mediated communication with the popularisation of portals such as *Facebook*, *MySpace* and, in Spain, *Tuenti*. These portals include "the profile" as the basic unit for content sharing and communication with other users. To this profile, applications such as chat rooms or RSS feeds (from "Really Simple Syndication") have recently been incorporated.

Although in the bibliography these sites are usually labelled as *social networks*, I think it is necessary to distinguish between social networks on the Internet, which can be developed and sustained in different ways and in different environments (not necessarily in these portals) and what from now on will be called *social networking sites* (henceforth SNSs), which undoubtedly offer a user-friendly interface for interactions, uploading content, etc., but which are only a sub-group of all the possible scenarios available for Internet-sustained social networks.

2.1 Definition, attributes and types

SNSs have been defined in several ways. Many of the definitions tend to equate them to the more general term of Web 2.0, as Beer (2008: 519) criticizes, among others. Possible definitions include the following:

> Web-based services that allow individuals to (1) construct a public or semi-public profile within a bounded system, (2) articulate a list of other users with whom they share a connection, and (3) view and traverse their list of connections and those made by others within the system. (boyd & Ellison 2007)

> Web-based services that allow users to interact, share information, coordinate actions and, in general, keep in touch. These applications are the new way in which our social network is represented, but also the place where our online identity is constructed and the means by which our activity on the Net is created and shared. (Orihuela 2008: 58)

> On-line environments in which people create a self-descriptive profile and then make links to other people they know on the site, creating a network of personal connections. Participants in social network sites are usually identified by their real names and often include photographs; their network of connections is displayed as an integral piece of their self-presentation. (Donath & boyd 2004: 72)

There are also studies that delimit the features or attributes of these SNSs. All of them seem to share the assumption that one of their inherent qualities is the role that profiles play as the "basic nucleus of social networking," where users make a self-presentation of themselves and make manifest potentially relevant information. As will be analysed in 2.3 below, the profile, the information it makes manifest and the possibility of making this information mutually manifest play a major part in the users' identity shaping, and provide some clues about the socio-cultural context in which these SNSs are inscribed (boyd & Heer 2006). Being relevant on SNSs means capturing the other users' attention in an environment where readers can focus on multiple sources of satisfaction. As can be observed in the prototypical schema of a profile on SNSs (adapted from Joly et al. 2009, see Figure 4.1), these profiles typically contain a photo, a short description and general information about the user, a list of friends, a number of applications and a wide area for entries and comments (both by the owner of the profile and by other users). SNSs users are, hence, *produsers*, a term coined by Bruns (1998b, 2006) as a blend of *producer* and *user* that describes this kind of user who, far from being the classic passive consumer of content, plays now an active role both in the production and consumption of information. An analogous term is *prosumer*, coined back in the 70s by McLuhan & Nevitt, who anticipated the advent of a new kind of media consumer, able to assume the roles of *producer* and *consumer* of content (see Islas 2008).

A number of authors have suggested the following elements or qualities of SNSs:

It has been proposed by boyd (2007a) that SNSs are characterised as having: (a) *persistence* (communication between users can be stored indefinitely); (b) *searchability* (with a "search form" we can find information inside these sites); (c) *replicability* (we can copy and paste content from one area to another); and (d) *invisible audiences* (many strangers can access the content of the profile, although the software can filter out information that is intended only for pre-selected friends).

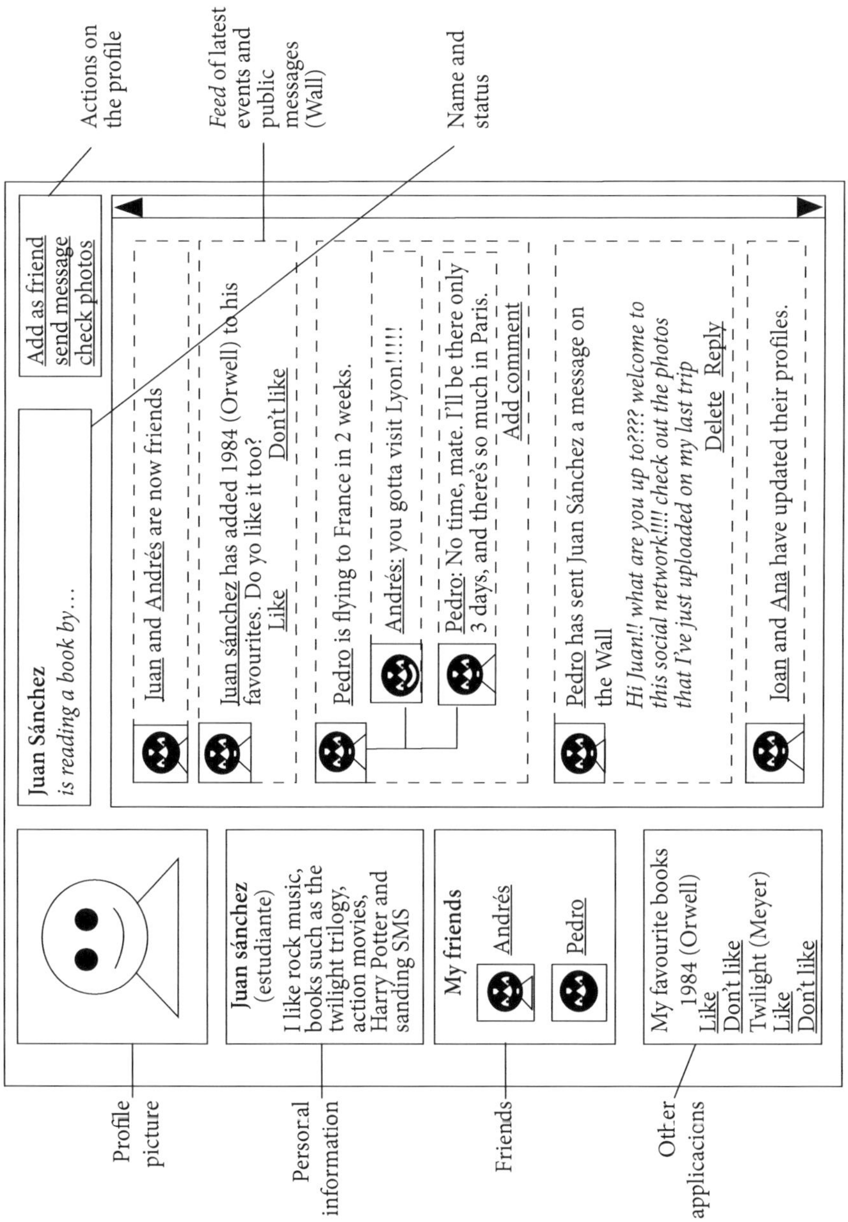

Figure 4.1 Prototypical SNS profile (adapted from Joly et al. 2009)

Cormode & Krishnamurthy (2008) propose, as distinctive features of SNSs, (a) that the users are the key element in the whole system; (b) that they exhibit a great capacity to generate connections among users; (c) that they offer the possibility of uploading content of multiple formats; and (d) that they make it possible to integrate other technologies and applications into the system.

Finally, Golbeck (2007) lists a number of requirements for SNS labelling: (a) they are accessible from the Net without the need of a special software; (b) inside them, users express the links that bind them with other users clearly; (c) the software has to foster the creation, maintenance and development of interactive connections among users; and (d) these connections need to be visible.

One attribute that is rarely listed in the bibliography is the capacity of these environments to alter or blur the neat dividing line that, in the past, separated interactions and networks in physical scenarios from the ones sustained in cyberspace. Within the picture proposed in Chapter 2 of a current process of hybridization between physical and virtual interactions with the user as an intersecting node (Yus 2007b), SNSs play an important part in the management, development and perdurability of these hybrid networks. Willson (2010: 498) points in the same direction when she writes that

> People (as nodes) are seen as able to access their social networks largely according to their own individual temporal, spatial and material needs and desires. These are constrained only by proximate and embodied demands and by the particular spatial and temporal rhythms of the various social networks in which they are involved. According to this understanding, the individual experiences her/himself as largely in control of her/his sociability through the possibilities of the technology.

In this scenario, it is possible to devise a scale or continuum of SNSs depending on whether the site is mainly an extension of the users' physical interactions and, therefore, their contacts are people that the users already know in physical environments and communicate with on an ordinary basis, or these sites are settings that favour virtual interactions with users that will never meet face-to-face, and with mixed options in between.

In any case, the combinatory possibilities for online/offline interactions on SNSs are rather limited. For example, Antheunis et al. (2008) only picture three possibilities: (a) *online friendships*, initiated on the SNS and always kept within the boundaries of virtual communication; (b) *mixed-mode friendships*, which start on the SNS and extend to physical settings; and (c) *physical friendships*, initiated outside the Net but transferred, at a later stage, to the virtual scenario. For Jarrett (2008), this third possibility is the most frequent one and qualifies SNSs with its most inherent function: namely, the maintenance of interpersonal relationships that already exist in physical settings, that is, a role of "extension" of physical relationships into the virtual realm (see Martín 2009: 29, Marwick 2005, Ellison et al. 2011).[15] Furthermore, Lampe et al. (2006) make an interesting distinction between the use of SNSs, specifically *Facebook*, for *social searching* (finding out information about offline contacts) and for *social browsing* (the use of the site to develop new connections, sometimes with the aim of offline interactions and perhaps hybrid ones). The former is the primary use of this site, according to this study.

Of course, this does not mean that SNSs cannot create ties and interactions whose strength may even compete with the ones we can obtain and foster in physical settings. The picture of increasing hybridization that I proposed in Chapter 2 makes this levelling of ties not only possible but predictable. Research by McKenna et al. (2002) and Tidwell & Walther (2002), among others, concluded that SNS interactions possess a surprising strength and inside these sites users tend to display more personal and intimate information and develop friendships that may even become more solid than offline ones. For example, a teenager describes in Holland & Harpin (2008: 123) how his relationship with a friend he has never met offline is more intense than the ones he has with people he sees on a daily basis:

> She used to be friends with a mate… so at least she is not entirely random. Me and [my friend] talk a lot, and I think she's a really good friend – she knows more about me than other people I see everyday and she knows how to cheer me up… she['s] a real friend (Charlie).

Finally, several researchers have proposed typologies of SNSs that can shed light on the attributes of these sites. One of them is by Fraser & Dutta (2008: 4–5), who distinguish between (a) *egocentric networks*, platforms for massive networks of friends based on inter-connected profiles and, as will be analysed in 2.3 below, important sources for users' identity shaping; (b) *community networks*, whose members share very tight identity linkages based on nation, race, religion, class, etc.; (c) *opportunistic networks*, whose members join for "rational" reasons, for example to look for professional connections; (d) *passion-centric networks*, that bring together users who share interests, hobbies, etc., also called "communities of interest"; and (e) *media-sharing sites*, defined not by their membership, but

15. However, Golder et al. (2007, quoted in Joinson 2008: 1028) report that, while the vast majority of messages are sent to friends (90.6%), a large proportion (41.6%) is sent to friends outside one's local network. This suggests that messaging is used to maintain and create social ties across distances. And it further confirms today's tendency to a hybrid physical-virtual quality of social networking, as proposed in Yus (2007b).

rather by their content (as in *YouTube* or *Flickr*). Another proposal is made by Thelwall & Stuart (2010: 265–266), who divide SNSs into (a) *socializing SNSs*, which support informal social interaction between members (e.g. *Facebook* and *MySpace*); (b) *networking SNSs*, which support non-social interpersonal communication (e.g. the business networking site *LinkedIn*); and *(social) navigation SNSs*, which support finding resources via interpersonal connections (e.g. *Flickr* and *YouTube*).

2.2 Some theoretical approaches

Although SNSs have been popular for just a few years, a lot of literature on the subject is already available. Nevertheless, this bibliography basically tends to apply pre-existing theoretical models (or models already applied to other media) to interactions and communication on these SNSs. Before providing my own approach to the subject, some of them will be reviewed below.

Firstly, one of the theories offering a direct applicability to SNSs is the *Social Network Theory*. According to this theory, both social behaviour and interpersonal communication are influenced by the qualities of the ties that bind people. In general, it is stressed that the more people get connected with one another, the more likely it is that these people will intensify their connections by using different forms of communication, including the Net. Therefore, Internet-mediated communication complements and extends traditional interactive networking behaviour carried out in physical contexts (Birnie & Horvath 2002). From this approach, what interests most is the interactions and strength of the ties that are achieved by means of SNSs, rather than what the user individually does inside these sites.[16]

A second theory that has been applied to SNSs is the *Technology Acceptance Model*. It is based on two variables: the user's perception of how easy it is to use some technology and the parallel perception of its usability, both regarded as key elements that affect the regular use of this technology (see de Souza & Dick 2007). This theory has mainly been applied to e-commerce and general uses of the Internet, but it can also be applied to SNSs, especially if we also include in the typology of variables the concept of "social pressure" to use a certain technology. In terms of SNSs research, social pressure would be conceptualized as the pressure and influence of friends and contacts of a user to create a profile on these SNSs.

16. See Haythorntwhaite (2009: 127), Hinduja & Patchin (2008: 127) and Papacharissi (2009: 201 ff), among others.

Thirdly, there is the *Signalling Theory*, used initially in biology and economics and centred upon the hypothesis that part of the information that we collect from others is not directly observable, but comes from signals that they *exude*, that is, "more or less reliably correlated with an underlying quality" (Donath & boyd 2004: 72); "we cannot directly observe others' beliefs, experiences, or what they really think of us; instead we rely on signals such as facial expressions, consumption patterns, or the statements they make on their profiles in order to infer these qualities" (Donath 2007). In the case of SNS profiles, this theory would explain why, for example, *Facebook* users tend to be more realistic when they describe themselves: since this is a SNS made up of users who, to a large extent, also know one another in physical settings, it is easier to extract signals that corroborate that what the user is writing on the profile is true.

Fourthly, the *Social Identity Theory* explains SNS behaviour from the premise that human beings have an inherent necessity to label themselves inside a group with which they feel some form of connection or identification. Indeed, people label themselves according to similarities with the archetypical features of the social group to which they want to belong. According to this theory, everybody needs both a feeling of being unique and a feeling of group membership. In other words, users shift between *independent self-construal* (constructing one's self by reference to one's personality, beliefs, etc., regardless of what others think) and *interdependent self-construal* (emphasis on blending with the group and mimicking its qualities, regardless of what one thinks) (DeAndrea et al. 2010: 427).

This double source of self- and group-connoted identity would explain why certain users prefer to belong to one SNS and not to another (Ferebee & Davis 2009). And it would complement more socially connoted theories such as the *Social Capital Theory*, which focusses on the value – or capital – that is obtained from interpersonal interactions inside a collectivity and which is also produced inside SNSs (see Ellison et al. 2007, Valenzuela et al. 2009).

A fifth theory, *Uses & Gratifications,* has already been mentioned in this book. It justifies the use of a certain technology depending on the benefit or reward that its use provides and depending on its capacity to satisfy the needs that motivate this use. Taken to the field of SNS research, it is easy to conclude that the application of this theory is directed towards the extent to which these sites are used regularly for the gratification or satisfaction of personal needs (see Shao 2009: 8–18). Joinson (2008: 1035) also applies this theory to SNSs and concludes that "the different uses and gratifications relate differentially to patterns of usage, with social connection gratifications tending to lead to increased frequency of use, and content gratifications to increased time spent on the site." Besides, the typical function of SNSs, namely "keeping in touch," comprises two main functions according to him:

> The first is a surveillance function [...] *Facebook* is used to see what old contacts and friends are 'up to', how they look and how they behave. In keeping with this use, there is evidence that *Facebook* profiles serve an important self-presentation tool [...] Associated with this use is the social capital building gratification, where *Facebook* is used to build, invest in and maintain ties with distant friends and contacts (ibid.).

2.3 Profiles, entries and (mutually) manifest information

SNSs share some properties with instant messaging (see Chapter 5) and e-mail (see Chapter 6): all of them include some form of "call of attention" that is oriented towards the identification of the user's communicative intention and leads to the satisfaction of the informative intention. Instant messaging alerts the addressee user (with an emergent window that invades the screen and also with a sound) to the fact that another user has a communicative intention, and invites the user to engage in a relevance-seeking processing of this user's informative intention (often through a typed dialogue). Similarly, e-mail programs alert the addressee user with an icon (of an envelope, of a mailbox, etc. and with sounds) to the arrival of a message that might provide relevant information (i.e. carries a presumption of its eventual relevance). And the same applies to SNSs. The system sends e-mails to the users alerting them that a relevant comment has been typed on the profile, that some reply to one's comments has been posted, etc. (see Alandete 2009b).

From a cognitive pragmatics and relevance-theoretic point of view, SNSs are interesting because both profiles and the information made manifest therein are evidences of underlying communicative intentions and, ultimately, indices of the attributes of the user's identity. This information on the profile and on the site entries is interpreted by the readers with the aid of context, so that they can recover the information explicitly communicated and derive implications, that is, so that the intended interpretation(s) are correctly selected and inferred.

At the same time, the different options for interactivity that these sites offer (direct comments on entries, on pictures and videos, instant messaging, e-mail, *Twitter* messages...) favour a certain level of mutuality among the users. This mutuality also presupposes a command of the techniques for oralizing text that will be analysed in Chapter 5 and which signal user membership through their correct use and interpretation. Indeed, as we can see in (1) (from *Facebook*, January 2011) interactions inside SNSs exhibit a high oral quality that is coded by means of repetitions of letters, capitalization and creative use of punctuation marks, as well as the use of emoticons:

(1) [*comments on a photo depicting User 1 and User 2*].
User 1. YES!! Created by myself. im such a professional when it comes to hair!! WOOPWOOP
User 2. look how happy I am with the result! hehehe
I asked for Demi Moore! Thanks User 1 !
User 3. heehee! loving the friends quote reference there User 2 ! ;)
User 2. Thank User 3! glad u got it ;) hehehe
User 1. see User 2...... defo not in User 3's room..... :p xx
User 2. ah yes, how could I forget ahahaha :D

Ultimately, all of these forms of interaction provide users with an invaluable feedback for their identity shaping in a kind of circular process with a number of phases. Figure 4.2 reproduces my proposal of the steps of interaction, mutuality and information transmission on SNSs that play a part in the user's adjustments of identity (see also Georgalou 2010: 41–42).

1. The figure starts with what, in Chapter 2, I proposed as the sources of identity in physical contexts, which were represented as an inverted triangle with three layers: macro-social aspects such as race, sex, etc. (wide top part of the triangle), groups to which the person chooses to belong (middle area) and the person's self-identity as idiolect (narrow bottom part of triangle). I also argued that on the Internet this inverted triangle is *re-inverted*, as it were, since on the Net the wide top area is minimized due to the cues-filtered quality of cyberspace (attributes such as race or sex are no longer essential), the middle area is maintained but replaced with virtual groups, and the narrow bottom part is fragmented or amplified due to the possibilities of playing with multiple personalities and identities that Internet allows for, many of which fit the quality of *disembodied identities* suggested by Baym (2010: 105).

2. This picture of discursive sources of identity as triangles clashes with the picture, also described in Chapter 2, of today's tendency to hybridization of personal networks in physical and virtual settings and with the user as a node in intersecting networks. Therefore, it would not be a picture of "either" physical "or" virtual sources of identity, as can be deduced from two triangles that do not touch, but a picture of several sources of identity that get mixed and imbricated in a time when the dividing line between physical and virtual realms is increasingly blurred.

3. This blurring also happens on SNSs, which often sustain relationships that are created and developed exclusively on the Net, but which also help users to maintain ties that were created offline and connections that are created virtually and "jump" to physical scenarios at a subsequent stage. In all of these cases, the user's identity is shaped and adapted to the different networks and to

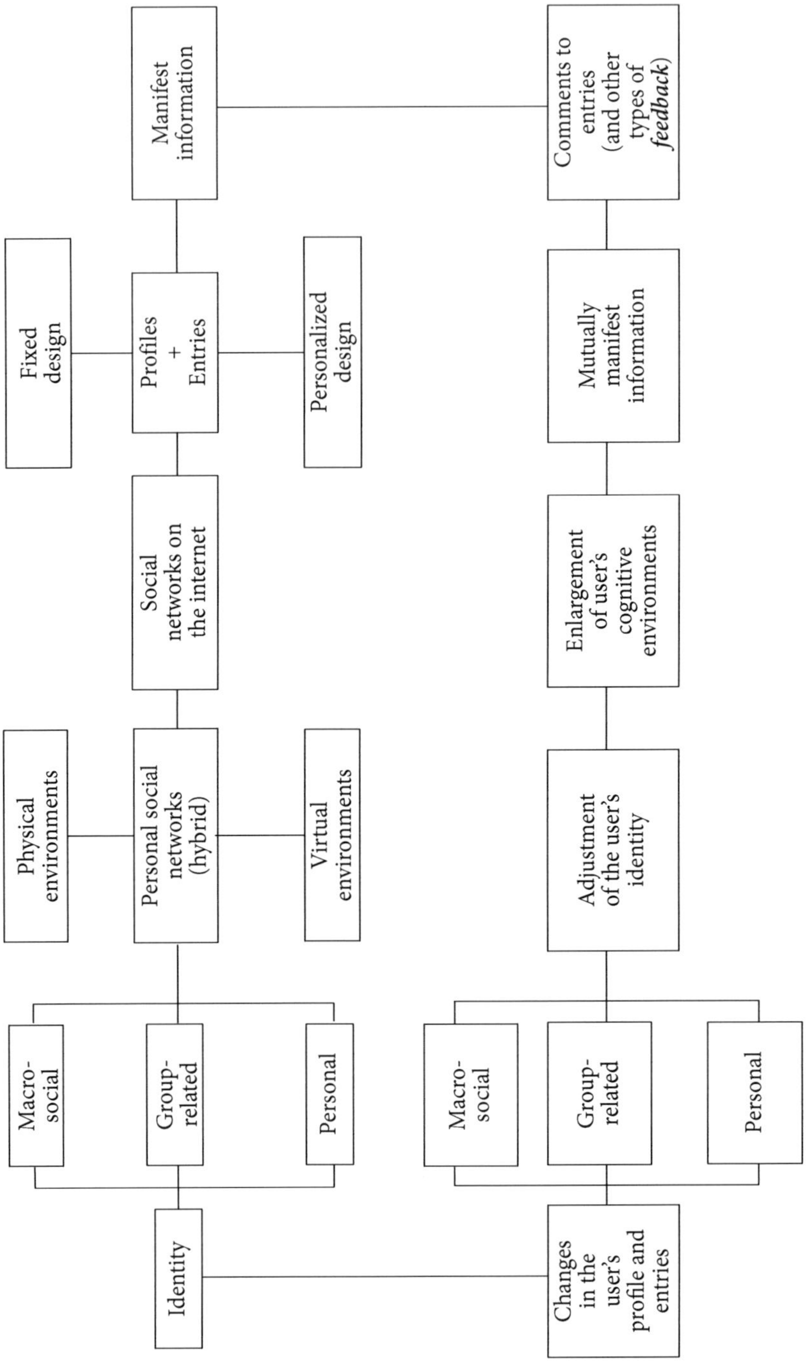

Figure 4.2 Profiles and (mutually) manifest information on SNSs and their influence on identity

the different intensities of ties and relationships with other users. And, of course, interactions with other users and the information made manifest in these interactions (through comments, etc.) is also an important source for the adjustment of identity. Notice, for example, how the interaction quoted in (2) (from *Facebook*, January 2011), where all the users make mutually manifest their opinions, helps User 1 to be more confident about her physical shape:

(2) [*replies to an initial comment by User 1 on her profile*].
User 1. I am still pissed! Tomorrow on the wagon, but am 7 pounds heavier! Why does it take months to lose half a stone but only a week to put it on??
User 2. You are a Piss Pot :) xxx
User 3. he he lol! xx
User 4. Bless ya honey.... Managed to keep it to 2lb but only because I was down the gym most days over xmas..... Don't 4get the planner on tues x x x
User 5. Keep drinking and find yourself a nice young man to work off the calories :) it works for me x
User 6. I lost 4 due to being ill, but reckon I've put 7 on in the last 2 days
User 1. Its shit isn't it! But have had a good time, well I think I have!!
User 7. Don't go on the wagon Hun, just stick with the workouts and it will fall off :-) xx
User 8. that's all that gets me through my workouts, the thought of having a beer as soon as I'm done :-)
User 1. I think if I have a month off the wine it will come off quicker! Feel like a little teletubby
User 7. Good luck x

4. Inside SNSs, identity is shaped basically at the users' profile, which contains the content that they upload plus comments by other users, the list of friends, etc. Although the picture or opinion that we can obtain from these profiles is always partial and, to a certain extent, schematic (boyd 2004b), at the same time it provides us with valuable clues about the identity of the user who owns the profile.

5. In the process of profile creation, there are several levels of personalization (*customization*) within a range between the extreme "zero personalization," when users have to follow strictly the rules and the default interfaces for profile creation by entering personal data on successive forms, to the extent that everybody on the SNS has identical designs of profiles; to the extreme "full personalization," for example providing freedom to integrate personal applications inside the profile.

As happens with blogs, owning a fixed standard profile or a highly personalized one has advantages and disadvantages. On the one hand, the repetition of a unique interface for profiles generates a conventionalization of the SNS genre that reduces, at least initially, the reader's effort (to locate information, etc.). Besides, these conventionalized profiles are a source for user identity shaping, because the user mimics other users' profiles and feels part of the community. The homogeneity of profiles produces a "group mark" of identity; or, as van Doorn (2010: 585) calls it, "a shared social reality":

> Instead of deriving social norms from other people's embodied presence, users have to create and interpret the semiotic resources (i.e. text, images, videos) that make up their profiles, which effectively constitute a digital infrastructure [...] These interactions dialogically produce a shared social reality through the distribution and interpretation of these artefacts.

This quality is also applicable to the kind of information that is uploaded on the profile. As boyd (2007b) correctly concludes from an analysis of SNSs for adolescents, when browsing and checking what information other adolescents upload on their profiles, they obtain a general idea of what they can or cannot provide about themselves on their own profiles. In such a way, a homogeneity is also generated in the content of these SNSs, besides homogeneity of design.

On the other hand, being given the chance to personalize the profile offers the user an alternative source of identity shaping based on individuation against the group. The readers will value, as additional cognitive effects, this personalization. But these effects should offset the additional effort that is involved in locating sources and types of information in non-conventionalized areas of the profile and which cannot be expected to be found in the same way as in fixed profiles. Nowadays, most SNSs offer users the possibility of personalizing, to a certain extent, their profiles by adding applications, changing colours, etc.

6. The next stage in Figure 4.2 indicates that one of the main reasons why profiles and entries are created is to provide other users (normally the ones labelled as "friends," see 2.4 below) with certain information about the users' lives, events, etc. That is, they want to make manifest information, in relevance-theoretic terms. Immediately, we ask ourselves the reason for this choice of content, the underlying intentionality. For Lin & Lu (2011: 1159), "enjoyment is the most important factor affecting the behavior of SNS users [...] [and] the number of peers and perceived complementarity effectively reinforce SNS usefulness and enjoyment." Making information manifest to other users aims at getting comments by other users, which makes them aware of the size and quality of their networks, while producing enjoyable effects.

Needless to say, by uploading information on the SNS, the user *exudes* several attitudes, beliefs, values, etc. that readers can infer without the user making them explicitly manifest on the profile, that is, beyond the user's intention. In fact, a mere tagging of information or choosing which words are going to be turned into links that bind chunks of text together influences the way in which this information will be subsequently located and processed, and regardless of whether the author had consciously assessed the interpretive consequences of this tagging or link design (see Lampe et al. 2007).

Tagging is usually referred to as *folksonomy* (see Ribes 2007), a term that reflects the intuition of many users making an effort to label and link discourses and influencing the quantity and quality of information that other users obtain and process, and hence of eventual relevance. Something similar happens to "RSS feeds," designed to satisfy specific informational needs of users. But they also influence the kind of information that users access and process, and therefore they also affect the eventual relevance (Daugherty et al. 2008).

The information made manifest on SNSs is distributed over the different areas that a profile is made of. A prototypical profile such as the one in *Facebook* contains typical areas of information such as the self-introductory text, personal interests, user's picture, and a wide area for entries and comments. Although these areas constitute valid sources of information, they do not provide the whole image of what the user is really like but, rather, the reader is faced with partial chunks of information and has to undertake the task of inferring implicit and implicated information by means of repeated interactions with the profile. These interactions lead to adjustments in the information that the reader has about the owner of the profile, and this information is a valuable preliminary context upon which subsequent interactions can be sustained.

However, although users can upload huge quantities of information on their profiles, the processing of this information as preliminary context for interactions differs substantially from the cumulative process which, in oral interactions, produces a gradual mutuality of assumptions (the "getting to know each other"). On SNSs information is often "simply there" and offered to the reader as a whole archive of the owner's life (see Lenhart & Madden 2007), including intimate details. A term has even been coined for this exposition of personal details: *extimacy*, the public *ex*hibition of *intimacy* that often abounds on SNSs (Pérez-Lanzac & Rincón 2009). Besides, on SNSs this "foundation" for future interactions does not necessarily have to be verbal. Pictures, for example have an important role on SNSs: "they establish communication paths between nodes, producing and reproducing social networks. More than being mere promotionalist self-advertisements, they are conversation pieces, necessary starters for the exchange of compliments-qua-gifts, which enable not only the formation of relations, but also their maintenance" (Schwarz 2010: 174).

Moreover, there is no unique pattern of information revelation or presentation. It is the readers who have to infer the information from the different options available on the profile.[17] As Gross & Acquisti (2005) argue, not all SNSs share the same options or content, but differ in significant aspects: (a) in the user's identification. On some SNSs users are encouraged to use their real names and pictures,[18] while on others the use of *nicks* is expected, especially in "love portals" such as *Meetic* or *Match*); (b) in the kind of information provided by the users. Frequently, it is about hobbies and interests, but it can be very different on other SNSs, for example on those centred upon a shared specialized topic that entails the use of jargons. (c) Finally, the visibility of information on these SNSs also varies, this time depending on whether all the users can access other users' uploaded information or there are filters so that only the intended audience interprets it.

There are multiple reasons for uploading (and making manifest) information on SNS profiles and these go beyond the basic desire to publicize the user's life, and there is an expectation of (relevant) reward in the effort of uploading information. In the case of adolescents the reward is obvious: the information on the profile can lead to a positive judgement by friends and contacts, as illustrated in this comment by an adolescent user (from boyd 2007b):

(3) I'm not the most popular girl in my class. I'm just a kid. I'm a little shy. And it's really hard in this school to impress people enough to be your friend... But I go on these really great vacations with my parents... And I take pictures of places we go. And I write about those places. And I post this on my Xanga. Because I think if kids in school read what I have to say and how I say it, they'll want to be my friend.

Self-disclosure is, perhaps, one of the most important reasons for uploading information on SNSs and part of the overall human tendency to obtain social benefits from interactions with other members of the site community, even if some users are obsessed with controlling the information that others can obtain from their profiles (see Peterson & Siek 2009). The eventual benefit will affect the

---

17. In this sense, Zhao et al. (2008: 1824–1826), in a study of *Facebook*, conclude that users resort to three basic forms of identity disclosure on the profile: (a) *visual identity claims* (the user as a social actor in the sense of "look at me and see how I am"), basically the publication of photos and videos with comments (one's or other users') on the SNS wall; (b) *enumerative identity claims* (the user as provider of hobbies, interests, etc.); and (c) *narrative identity claims*, when the user self-describes and claims an individual identity to the other users.

18. The user's main picture displayed on the profile plays a part in the initiation of relationships within the SNS. Wang et al. (2010) demonstrated that it had "a significant main effect on willingness to initiate friendships with the profile owners. Physical attractiveness was most salient as a visual cue when choosing whom to befriend when other verbal or non-verbal cues were limited. [...] The results suggest that both male and female subjects were more willing to initiate friendships with opposite-sex profile owners with attractive photos."

user's self-identity and his/her role and placement on a scale of popularity on the SNS. These adjustments of the self may also lead to multiple selves with different intensities in the user's life, as was suggested in Chapter 2. As Turkle (2011: 160) acknowledges, "we use social networking to be 'ourselves', but our online performances take on lives of their own. Our online selves develop distinct personalities. Sometimes we see them as our 'better selves'. As we invest in them, we want to take credit for them."

One drawback of this accessibility to other users' information is that it reminds users not only of their place and identity on the SNS, but also of their adaptability to an inherently social environment:

> *Facebook* is not a good place for a lonely person, and not just because of how precisely it quantifies your isolation. The news feed, the default point of entry to the site, is a constantly updated stream of your every friend's every activity, opinion and photograph [...] you know exactly how much more popular everyone else is [...] It can be, to say the least, disheartening. Without a real-world social network with which to interact, social networking sites act as proof of the old cliché: you're never so alone as when you're in a crowd. (Meltzer 2010: 26)

Other reasons for participating on SNSs are listed by Gangadharbatla (2008): (a) *need for cognition* (already cited in Chapter 3, it is the individual's tendency to get involved in certain tasks, even if they entail much mental effort); (b) *need to belong* (that is, of meaningful and positive interpersonal interactions); and (c) *collective self-esteem* (arising from feelings associated with belonging to a social group and its attributes). More predictable reasons are listed in Brandtzæg & Heim (2009): to establish new relationships, maintain contacts, socialize, get information on topics, chat with friends, kill time, surf profiles, etc. These reasons also vary depending on the "culture" of the users who interact on the SNS (i.e. prototypical habits, beliefs, etc. shared and taken for granted within a community). For example, a study that compared the broad cultures of the USA and Korea (cited in Kim et al. 2011: 367) concluded that Internet users from Hong Kong, a *collectivistic culture*, tended to view the Internet primarily for social interaction, whereas Americans, a typically *individualistic culture*, were more likely to use the Internet as a means of seeking and gaining information. "Individualistic" and "collectivistic" are taken here as the classic terms proposed by Hall: in a collectivistic culture, people value group identity, and tend to foster lifetime relationships, whereas in individualistic cultures independence is highly valued, which results in fragmented and short-term relationships with one another.

7. However, and following the chart in Figure 4.2, although it is important to make information about oneself manifest to other users, what is very significant is to get some level of mutuality of this information, to reach a *mutual manifestness*

and enlarge one's *mutual cognitive environment*. This is an optimal level at which users can extend and overlap their personal cognitive environments and fine-grain the specificity and extent of information that is really mutual and can be used as a preliminary context in subsequent interactions. Ellison et al. (2011: 133) claim that mutuality lies in the heart of all the SNS activity: users seek cues about each other to create common ground, and profile fields reduce the cost of finding these commonalities among users. This suggests that *Facebook* users may be more likely to use online information to find others with whom they share some kind of offline connection, as opposed to finding others whose connection is based on common interests like music or movies.

This mutually manifest information is essential for effective communication, since the presumption of this mutuality leads users to leave much information implicit, non-coded, taken for granted, and addressee users have to fill in the blanks by resorting to this overlapping space of their cognitive environments, as in (4a), which demands from readers the answer to questions such as the ones provided in (4b):

(4) a. Hey! The idea of a party in the dunes sounds great. We'll have to get everyone to go to the offie to get some booze and we can meet at the usual place on Saturday to pick up Tom's car.

b. Hey! The idea of a party [*organizing it? go to it? what party? what kind of party? whose idea was it?*] in the dunes [*which dunes?*] sounds great! We'll have to get everyone [*get whom?*] to go to the offie [*which offie?*] to get some booze [*which kind of drink? which brand?*] and we can meet [*who? all the mates?*] at the usual place [*which place?*] on Saturday [*which Saturday?*] to pick up Tom's car [*which Tom?*].

In everyday interactions, very often the only communicative purpose of utterances is to reveal areas of mutuality between the interlocutors' cognitive environments. This is the case of Ann's irony in (5) about a pub, whose objective is mainly to determine if the information in (6) about her preferences is or not shared by her partner, and it is used as an interpretive premise for the derivation of the implication (*implicature*) in (7). The successful outcome of dialogue (5) makes Ann and Peter aware, at that precise moment, that the information in (6) is mutually manifest to both of them, which produces an enlargement of their mutual cognitive environment (see Yus 2009a):

(5) [*Peter and Ann enter a pub. It is filled with people singing and dancing*].
Ann: [*smiling ostensively*] There's nothing like a lively pub!
Peter: Indeed! Shall we go to another pub?
Ann: Please!

(6) Ann hates overcrowded places. She prefers a quiet atmosphere where she can have a chat without loud background noise.

(7) Ann is being ironic and, in fact, she does not like the atmosphere of the pub we have just entered. She'd rather leave the pub.

In a similar fashion, part of the information made manifest on SNSs can reach an adequate level of mutuality and serve as a preliminary context for future interactions. This interactive mutuality would be complemented by some "*ex post facto* mutuality," achieved when the interactions do not presuppose mutuality of information, but *reveal* aspects that overlap in the users' cognitive environments. This happens, for instance, if during an interaction on a SNS two users discover that they have been on holiday at the same place and they can, from then on, assume that certain information about this place will be mutually manifest (even if they are not sure of that). And again, this is a useful preliminary context from which to build up subsequent interactions.[19] It is not surprising, then, that many searches on SNSs are intended to find people with similar interests, beliefs, or hobbies that can serve as foundations for future interactions.

Mutuality of information is possible because SNSs exhibit many forms of interaction (comments on posts, dialogues on a picture, instant messaging facilities...) and users code their messages assuming the existence of this mutuality, unlike traditional web pages. Furthermore, mutuality takes place in an environment that, according to Miller (2008: 393–395), values interactivity over information:

> we see a shift in emphasis from blogging technology which encouraged the creation of substantive text along with networking, to social networking profiles which emphasize networking over substantive text [...] communication that retains a general sociability without the exchange of real information [...] towards what are being called 'phatic technologies': technologies which build relationships and sustain social interaction through pervasive (but non-informational) contact and intimacy.

It is undeniable that interactions on SNSs differ from the ones in physical contexts. However, although in the past it was easy to dismiss Internet interactions for being plain-text-based, nowadays, the increase of bandwidth and the use of pictures, videos and applications play a part in these interactions, creating a

19. Feld (1981) proposed the term *focus* to describe all the situations, hobbies, interests, etc. that reveal connections among people and shape or allow for the formation of social networks. These *foci* vary from those which favour frequent interactions and tight links (such as belonging to a family, for example) to more relaxed ones that generate more occasional interactions and weaker ties. This scale would also be found on SNSs.

communicative platform rather than a unique source of interaction.[20] Besides, on these SNSs the users often do not know the extent of their readers' cognitive environments and how much information they can assume to belong to their mutual cognitive environments, and interpretations may differ from the intended ones. Finally, interactions on SNSs can be performed in different ways and formats with varying degrees of "visibility" for the readers, as can be seen in Table 4.1 (slightly adapted from Joly et al. 2009: 55). To this table, it would be necessary to add some information about the intensity of these possibilities of interaction. As van Dijck (2009: 45) comments, the concept of "participation" on a SNS is vague, ambiguous and imprecise. There are degrees both in the production of content, in the quality of users' feedback and in the intensity of interactions.

Also related to this issue is the fact that mutual manifestness may be intended for a specific user, for example when posting a comment on the user's "wall," but in fact this comment may also be read by other users who share the label of "friends," and who may also aim at mutual manifestness. Walther et al. (2011: 33) correctly stress the importance of this issue when they state that a comment on the user's profile "is, by definition, a public message, bordering on being broadcasted (or at least, narrowcasted within the social network) for others to see. Facebook users have noted that one of the main uses for social networking technology is relational maintenance [...] Are such wall posts 'mass' messages or 'interpersonal' messages?" And from our perspective, which mutual manifestness is intended? Is this "collectively achieved mutuality" beyond the intention of the author of that comment on the user's profile? And, incidentally, is this mutuality possible in the first place?

8. This mutuality of information favoured by interactions on the SNS leads to a number of adjustments affecting the identity of the users, who will obtain from other users' comments and dialogues a valuable source for their positioning in the group or network and personal introspection.[21] These comments are also

20. The most paradigmatic case is the re-design of *Facebook* profiles, which now include spaces for personal applications and has integrated an instant messaging service, a wide set of options for interaction within the same interface (see Keenan & Shiri 2009: 444).

21. For Jones et al. (2008), a problem is that these SNS users not only have to adjust the revelation of their identities on the profile and its entries, but are often forced to make a coherent identity display for multiple potential readers. For example, the same profile can be read by intimate friends and occasional acquaintances, workmates and relatives, a reality that has been labelled "context collapse" (see boyd 2008, Marwick & boyd 2010), and the identity that the user has shaped online may be adequate for intimate friends but not for workmates, for instance. One of the problems that is arousing media attention is, precisely, that many adolescents own profiles that are adequate for their peers but these profiles might be problematic and even dangerous if accessed by unknown readers.

Table 4.1 Different possibilities of interaction on a SNS (adapted from Joly et al. 2009: 55)

Interaction	Recipient(s)	Visibility	Intention
Profile message	Contact / own profile	Public (all contacts)	– To introduce a newly added user. – To show publicly one's opinions, hobbies, etc. or recommend something to other users. – To let the recipient's contacts know what is going on between them.
Bulletin, posted item	Contacts / own profile	Public (all contacts)	– To share interesting content with contacts. – To announce an important event to contacts. – To ask contacts for their feedback.
Gift	Contacts	Public (all contacts)	– Public display of interests, hobbies, etc with more impact on the profile than a message, because gifts are usually not free.
Events (invitation)	Contacts	Public or private	– To invite (some) contacts to an event. – To facilitate communication between those who intend to attend an event (e.g. for arranging a common gift, adding contacts). – To share information related to the event (e.g. photos, videos, links).
Groups (invitation)	Contacts	Public or private	– To gather users around a common interest or facilitate a dialogue about it. – Opportunity to add contacts.
Poke	Any person	Private	– To say "hello, check out my profile" to someone that the user has probably just met offline. – To include the recipient in the sender's contacts temporarily, allowing visibility of his/her profile and rich communication.
Private message	Any person	Private	– To have private interpersonal discussions (no particular interest for social networking)

visible for other users, who can derive conclusions and a more accurate picture of the owner of the profile. For example, one of the conclusions drawn by Walther et al. (2008) referred to the comments that friends make on the user's profile, which have an impact on the impression that other users form about the user whose profile contains these comments, specifically on the assessment of social attractiveness and credibility. And Toma (2010) adds that the most prominent feature is that SNSs allow users to 'collect' information from friends and their contributions to the SNS. Since friendships and personal relationships are the most widely used sources of self-affirmation, she predictably concludes that "SNS profiles appear to restore users' sense of selfworth by reminding them of the important aspects of their lives: their connections with friends, their identities and group membership. As such, a surreptitious effect of the selective self-presentation and social connectedness afforded by SNS profiles can be a boost in morale and feelings of self-worth" (ibid.: 1752).

9. Therefore, as Figure 4.2 reflects, the information uploaded on the profile, the comments on entries and pictures and other users' access to this information may generate important adjustments in the user's identity, not only as an individual, but also in other sources of identity shaping such as group identity and macro-social identity. These three sources (individual, group and society) would be related to the three modes of social influence proposed by Kelman (1974, quoted in Cheung & Lee 2010: 25): *compliance* (subjective norm), *internalization* (group norm), and *identification* (social identity).[22] Although "subjective norm" is important at the moment of choosing to belong to a SNS, once the user is part of the collectivity of users the "social" sources of identity are essential in an environment such as the SNS, where the "feeling of belonging" is prominent. As Cheung & Lee (ibid.) stress,

> unlike the traditional individual-based approach (personal intention to perform an individual act), social interaction and connection is the objective in Web 2.0 technologies, including online social networks. Associated with these new phenomena in human communication and interaction patterns, we believe that We-Intention, encapsulating social behaviors by the collectivity, is a more appropriate approach to study user participation in online social networks.

22. Social identity has three major components, all of which are clearly applicable to identity shaping on SNSs: (a) *cognitive social identity* (the self-categorization process blends the users into the group but, at the same time, differentiates them from other groups); (b) *evaluative social identity* (the evaluation of self-worth on the basis of belonging to a particular group); and (c) *affective social identity* (a sense of emotional involvement with the group, which is characterized by identification with, involvement in, and emotional attachment to the group) (Cheung & Lee ibid.: 26).

Similarly, boyd (2011: 43) acknowledges this social side of identity shaping when she stresses that, in fact, the design of SNS profiles creates an environment in which self-presentation is often beyond the user's control: "profiles are a place where people gather to converse and share. Conversations happen on profiles and a person's profile reflects their engagement with the site. As a result, participants do not have complete control over their self-representation."

These adjustments of identity and self-presentation on SNSs will eventually affect what (and the type of) information is uploaded on the profile and its entries, forming a new preliminary context from which the whole process depicted in Figure 4.2 would start all over again.

2.4 Adjusting the concepts of "friend" and "friendship" on SNSs

When we interpret an utterance, we have to answer three basic questions: (a) What does the speaker intend to communicate explicitly?, (b) what does the speaker intend to communicate implicitly (i.e. what does he/she intend to implicate)?, and (c) what contextual information does the speaker expect us to access in order to obtain (a) and (b) correctly? To illustrate these questions, we can analyse Ann's reply in the dialogue (8) below (Yus 2010c):

(8) Tom: By the way... Did you buy that table I told you about?
Ann: It's too wide and uneven.

If Tom wants to interpret Ann correctly, he has to turn what she has said (the logical form, what she has literally coded) into a contextualized, meaningful interpretation. He will use his inferential ability to obtain the propositional form that Ann intends to communicate explicitly (the *explicature*, question (a)) and will use it plus contextual information (question (c)) to derive an implicated conclusion (question (b)), and all that in a mutual parallel adjustment of explicit proposition, context and implications. Among the inferential procedures that Tom has to apply to answer question (a), the following can be listed: *disambiguation* (a table can be "uneven" because its surface is not smooth or because its legs are not properly levelled), and the *free enrichment* of the content that Ann's utterance lacks in order to really make sense ("too wide and uneven [*for what?*]"). The inferred outcome as an answer to question (a) would be a proposition similar to the one in (9):

(9) Explicature: The table that Tom told Ann about is too wide to go through the bedroom door and its surface is irregular.

Since (9) is not really the answer to Tom's question, he will have to combine (9) with contextual information (*implicated premises*) to yield the intended interpretation as an implicature (*implicated conclusion*). In this example, the contextual information would be some encyclopaedic commonsense assumptions about how unlikely it is that someone would buy a table that is too wide and uneven (in the senses already inferred), and this contextual information allows for the derivation of the implicature in (10):

(10) Implicature: Ann didn't buy the table I told her about.

Among the inferential operations required to turn the zero-context schematic logical form into a fully contextualized proposition and also important for the derivation of implicatures is the so-called *adjustment* of the prototypical concepts that underlie the words uttered by the speaker. In other words, to turn the conventional meaning of the words (coded concepts), as one would find in dictionaries, for instance, into more specific, contextualized *ad hoc concepts* adjusted to the speaker's intentions and the hearer's interpretive needs in a specific communicative situation (i.e. these *ad hoc concepts* may not be valid in a different situation).[23] For example, someone who interprets the concept coded by the word *tired* in (11) will have to adjust it inferentially so that the resulting *ad hoc concepts* fit the type of tiredness that the speaker intends to communicate in each case:

(11) a. I've been running for three hours. I'm very *tired*.
(*ad hoc concept*: physical exhaustion).

b. When a person is *tired* of London, he is tired of life.
(*ad hoc concept*: vital boredom).

c. I want to split up. I'm *tired* of this relationship.
(*ad hoc concept*: dissatisfaction with a relationship).

d. I can't type anymore. I'm *tired*.
(*ad hoc concept*: mental exhaustion).

The analysis of SNSs reveals that something similar happens to the coded concepts of "friend" and "friendship," which have a prototypical meaning that has to be adjusted correctly by users to make them fit the specific qualities of relationships, ties and contacts that they establish and foster within these SNSs. As Cambra González (2009) correctly points out,

23. Similarly, in Example (8) above Tom will have to adjust the coded concept "uneven" (once disambiguated) in order to obtain a more relevant *ad hoc concept* that fits Ann's intended interpretation.

> different types of interpersonal relationship are reduced to the ambiguous label (raised to the status of category) of "friend," reducing part of the distinctive character of each particular case (one can simply be an acquaintance, and we even know that there are different types of friendships related to different types or degrees of ties). This biassed effect that is created when everybody linked [to a user] in *Facebook* is reduced to the same category is the first step towards the upcoming of certain distorting effects that socialization within this medium introduces in interpersonal relationships.

This adjustment operates on two concepts that are difficult to define and delimit. These are inherently ambiguous terms both *intra-* and *inter-*culturally and involve variations that depend on how people feel the intensity of their ties with others. As Adams & Allan (1999, quoted in boyd 2006b), point out, friendship must be analysed in context because context influences the forms that these friendships adopt. Even in the bibliography there is little consensus on what model of friendship should be applied to SNSs. In this book I have claimed that friendships on SNSs, even those that are created and sustained online, can reach a surprising level of intensity, whereas for other authors such as Stefanone et al. (2011),

> while a subset of these online networks may be composed of traditional close friends, the majority are likely characterized by much lower levels of emotional closeness and intensity placing them on the far end of the weak tie spectrum. In other words, weak tie relationships on sites like Facebook.com may not represent meaningful connections because generally people invest comparatively little in these relationships.

Moreover, each SNS seems to have its own conceptualization of the kind of "friend" and "friendship" that can be developed inside it, with a danger of misunderstanding if several users do not share the same *ad hoc concepts* from the processing of the coded "friend" and "friendship." One of the SNSs, *Friendster*, encourages users to attract people and even coined the term *friending* for that task. But there are great differences among users when it comes to weighing the reasons for *friending*. On another SNS, *LiveJournal*, the concept of friendship is very unspecific and does not require reciprocity. Anybody can add others to the contact list without the presumption of a minimal relationship. Since messages in *LiveJournal* have to be labelled as public, private, only for friends or available for sub-group of friends, misunderstandings abound due to dissimilar adjustments of the concept "friend," to the extent that analysts such as Fono & Raynes-Goldie (2006) suggest the term *hyperfriending* to describe the variability of concepts that underlie what users conceptualize as "friendship."

This lack of agreement about the meaning of this term is also transferred to the realm of types of comment on the SNS. To choose between a mere comment and a private message entails parallel decisions about the level of friendship that binds users together. In boyd (2008: 126) there is an interesting quote by an adolescent user that indicates differences in the conceptualization of message types and message purposes: "a message is like if you want to like to a person like talk and talk and the comment is just like to just drop by and say 'how are you' and stuff." Hence, the type of message is connoted according to shared implicit rules. The same applies to the channel chosen for communication. For adolescents, e-mail is too formal and only useful for student-teacher interactions and homework assignments (see Chapter 6). The SNS is used for more informal, humorous and "flirtatious" goals. And instant messaging is for intimate conversations with peers. Not all *cyber-media* possess the same function and the fact that most users exhibit similar choices of channels and for similar purposes indicates the existence of collective negotiations about these qualities and communicative goals, and a prediction of mutual manifestness of this information.

Fono & Raynes-Goldie (2006) also suggest a typology of senses of "friendship" in *LiveJournal* that can be re-interpreted in terms of "conceptual adjustment" and "ad hoc concept," as proposed by relevance theory:

1. *Friendship as content.* Some contacts appear especially as mere lists of users and other users access this list as part of a search for content.

2. *Friendship as offline facilitator.* The use of this SNS can facilitate the creation of ties outside the Net, and part of the friendship on this SNS entails listing offline friends who also have profiles there.

3. *Friendship as online community.* For others, friendship is mainly focussed on users that will never meet face-to-face, but only inside the SNS.

4. *Friendship as trust.* For some users, to be on a restricted-access list of friends is the foundation of true friendship.

5. *Friendship as courtesy.* Although *LiveJournal* does not require reciprocity, the lack of it is often considered rude, asymmetrical and devoid of the true qualities of friendship.

6. *Friendship as declaration.* Most of the users of this SNS consider friendship to be a kind of "declaration of intentions." By publicly listing a user as friend in the profile, it is understood that a relationship is publicized.

7. *Friendship as nothing.* For some users, *friending* is nothing more than putting a person on a list, without further connotations or implications, that is, it is not a signal of an underlying friendship. This also happens on *MySpace*, where adding someone to the contact list does not indicate any kind of feelings for this person (see Jones et al. 2008).

Apart from misunderstandings due to differing adjustments of the concepts of "friend" and "friendship," problems may also arise if these concepts are mixed with the concepts that we usually employ in physical scenarios and we try to transfer them to the Net (Thelwall 2009a, 2009b). Although I do not claim that offline interactions are necessarily better than their virtual counterparts (see Chapter 2), it is true that there is an idealized concept of friendship for physical interactions in specific spaces (bars, streets, parks…) and this idealization is not directly applicable to the "friendship" that arises on the Internet or in mixed physical-virtual interactions. Therefore, the users will continuously have to make adjustments to determine the intensity and labels of all the relationships that they foster on SNSs, which range from intimate friendship, to occasional contacts and ties with people that will never be met offline, and all that with an increasing number of interactions in hybrid personal networks, as proposed in Chapter 2 (see Antheunis et al. 2008). Perhaps a solution would be to dismiss the notion of "friend" altogether, as Isidro Maya (quoted in Gosálvez 2010: 28) suggests: "Friend" is "an inadequate use of the term. Contacts is more adequate. Facebook is a peculiar context for socialization, like a square or a bar, and in this context these contacts from the past simply turn up because the software facilitates and promotes the re-activation of latent relationships."

3. The microblog *Twitter*

3.1 Introduction

At present, a number of *microblogging* technologies are being developed, with *Twitter* as the most popular example. This is a short-message service (normally of less than 200 characters, *Twitter* messages are up to 140) that allows users to post "in real time" what they are doing, either through the Net or through mobile phones or PC tablets. Although some analysts claim that this is an ephemeral means of virtual communication,[24] in my opinion it is an interesting option for interactions and for transferring everyday information to other users. As Johnson (2009) corroborates, *Twitter* may have disappeared in a few years' time, but what will be perpetuated are the structure and communicative essence of *microblogs* (live micro-messages, access from multiple devices, communication centred upon ordinary life, etc.).

Definitions of *Twitter* include the following:

24. Beckett (quoted in Hughes 2009) says that "*Twitter* is definitely an important tool but it's also important to note how fast this technology can change. In five years' time, sites such as *Twitter* or *Facebook* may not exist at all -something else will have replaced them." For Jones (2007), the most evident danger for *Twitter* is its imbrication within SNSs, which would make its existence as an independent technology unnecessary.

> online service you can use to send out short (140 characters or less) notes to the world via the Web, IM and text-messaging. People use it to issue updates about what they're doing, eating, seeing, feeling, etc., to their family, friends and whoever else might be following them ("following" is *Twitterspeak* for signing up to receive somebody's "tweets," which is what the individual updates are, adorably, called). (Grossman 2009)

> A service that connects you with your friends, that answers questions about what you are doing through a mobile phone, a web page, instant messaging and e-mail. It allows you to keep in touch with people in real time.
> (Jack Dorsey, creator and president of *Twitter*, in Reventós 2008)

Although *Twitter* can be accessed from multiple devices and, in fact, more users do it from mobile devices than from computers (Lenhart 2009), *Twitter* also includes a personal web page for the user which resembles SNS profiles. This page is divided into two main areas, a wide one with the user's picture (a photo or an icon) and the list of messages or *tweets* that are published in real time. On the right there is a second area, a frame with personal information about the user, and a list of contacts with micro-icons, as on SNSs.

There are various reasons for using this *microblog* service, but most of them are related to a human need for "permanent connection" with other users and the desire to be constantly updated about what others are doing. As Chen (2011:760) concluded, "people who actively seek out *Twitter* are doing so out of a basic human need to connect with others that they can then gratify by using this computer medium." Specifically, Zhao & Rosson (2009) list the following reasons for using *Twitter*: (1) you can inform people about your most ordinary activities at the same time as they are taking place; (2) it is easy to provide information in real time; (3) messages are short; (4) you can send updates on your activities very easily; and (5) messages can be sent from many kinds of devices. On their part, Java et al. (2007:62) list the following reasons: (1) to comment on ordinary topics; (2) to engage in interactions (answer each other's *tweets*): (3) to share information and Internet addresses; (4) to comment on news; (5) to use it as a source of interesting information; (6) to meet new people; and (7) to search for information.

Reasons (2), (3) and (4) in Java et al. (ibid.) are indicative of a typical phenomenon of information and communication technologies (ICTs): that they are re-designed by users according to patterns and needs that were not predicted by the original designers of these technologies.[25] This happened with mobile phone

25. The theory of *The Social Construction of Technology*, proposed in the 80s by Bijker & Pinch (quoted in Mischaud 2007:10) affirms that technologies possess a certain degree of *interpretative flexibility*, since different social groups may have non-predicted ideas and generate different interpretations of the technology, beyond its initial design.

texting, a marginal service by design, but whose massive usage surprised mobile manufacturers. In the case of *Twitter*, it was initially designed to answer the question "what are you doing?" (reason (1) in Java et al. ibid.). But users have redesigned this service in order to be able to engage in short-message conversations by using the *ad hoc* nomenclature "@username"[26] at the beginning of the typed message. Actually, 480 messages out of my corpus of 1.000 *tweets* are conversational.[27] Some of the *tweets* are also intended to provide interesting information (85 in the corpus). Both uses are exemplified in (12a–b) and (12c–d), respectively:

(12) a. @usuario Ohhhh… me acabas de destrozar… Melendi no, porfa! (T1).
[*@username Ohhhh… you've just destroyed me… Melendi no, please!*].

b. @usuario desde siempre desde siempre, pero las fotos que ha subido ahora ya no dejan lugar a dudas eeeh? xDDDDDDD (T4).
[*@username Always, always, but the photos he has uploaded now leave little room for doubt huh?*].

c. Simplemente maravilloso: las transmisiones de radio de las misiones Apollo sobre música ambiental/electrónica/chillout http://bit.ly/bBU9r (T1).
[*Simply marvellous: radio transmissions from Apollo Missions about ambient / electronic / chillout music*].

d. Nuevo Post: Un breve resumen del podcasting en España (o como lo recuerdo yo) http://is.gd/20tR (T10).
[*New post: A short summary of podcasting in Spain (or as I remember it)*].

3.2 Cognitive effects vs. processing effort

From a pragmatic relevance-theoretic perspective, it is necessary to analyse the communicative and interactive qualities of *Twitter* and check the extent to which these qualities, together with the design of *Twitter* web interface, influence positively or negatively the user's estimation of the eventual positive or negative relevance of *tweets*.

If we focus on the initial purpose of *Twitter*, namely to answer the question "what are you doing?," the immediate intuition is that most of these messages

26. From now on, I will use the neutral nomenclature "@username" when quoting *tweets*, instead of the real name or *nick* of the user, so as to preserve anonymity.

27. Collected in August 2009 from 10 users that will be generically labelled "T" (*Twitterers*). Five of the users are male (T1, T2, T3, T6 and T10) and five are female (T4, T5, T7, T8 and T9). The *tweets* are in Spanish but, where necessary, a translation will be provided.

will tend to be, on most occasions, utterly irrelevant. Indeed, it is very unlikely that users will find any relevance in interpreting *tweets* whose content deals with "interesting" issues such as "making a sandwich" or "switching on the TV." This intuition explains some negative comments that *Twitter* has raised:

> Twitter is a massive waste of time [...] Twitter has turned distraction into an art form. (S. Karp, in Edemariam 2009)

> Twitter breeds a false sense of intimacy. Much of the communication that occurs on Twitter is the type of thing you normally say only to people you're very close to. (Govella 2008)

> *Twitter* [...] is too much... I think, because in my community there are all of those exhibitionists and I can't manage hundreds of messages each day. If they wrote less and more directly to a reduced group of friends I would feel more in contact with them. (M. Hodder, in Pisani 2007)

This intuition of irrelevance seems to be corroborated in the corpus of *tweets*, in which *tweets* abound that provide few or no cognitive effects that offset the mental effort required to process them:

(13) a. Hora de comer... hay que pensar en ir preparando algo (T3).
[*Time to eat... I'd better think about preparing something*].

b. Voy a cenar, ahora vuelvo (T4).
[*Off for supper, back soon*].

c. Intentando recuperarme de una resaca importante (T9).
[*Trying to recover from a heavy hangover*].

d. Mi mama me ha hecho cocretas... [*sic*] xD Desde que me fui de casa no las probaba (T10).
[*My mother has cooked croquettes. I haven eaten any since I left home*].

This feeling of irrelevance is accentuated in the case of sequences of *tweets*, very close to one another, that provide redundant information about an event that is, in itself, of little relevance, as in Example (14) from the corpus:

(14) a. Voy a hacer un wallpaper-collage chachi de los mios (T4, 10:43 am).
[*I'm going to make one of my nice wallpaper-collages*].

b. Vale, no, me voy to the shower y ahora vengo a hacer el wall (T4, 10:47 am).
[*ok, no, I'm going to the shower and I'll come back in a while to make the wallpaper*].

c. Horas después... me pongo a hacer el wallpaper (T4, 12:34 pm).
[*Several hours later, I start making the wallpaper*].

d. Vale pues que le den al collage ò.ó me estreso con tanta foto. Pero ahora que me pongo yo de wallpaper?! QUÉ!? (T4, 1:49 pm).
[*ok, fuck the collage... I get stressed out with so many photos. But what shall I put in the wallpaper now? WHAT!?*].

However, beyond an analysis centred upon cognitive effects and mental effort, there may be other sources of user satisfaction in *Twitter* interactions and messages that provide a certain cognitive reward which is not constrained by excess effort, as will be commented upon below.[28]

Firstly, a possible interest in these trivial *tweets* may lie in what Thompson (2008) called *ambient awareness*, a term that has already been mentioned in this book. It refers to a non-stop updating on other users' daily activities, which provides a feeling of closeness. Indeed, instead of inferring certain information and deriving conclusions about other people from the information they *exude*, with *Twitter* it is the users that intentionally inform their followers about these activities. This creates a different kind of "proximity in the virtual" or at least an awareness of its existence. For Richmond (2009), *Twitter* brings us the ordinariness of life in all its fascinating, beautiful and often tedious details. For Zhao & Rosson (2009), *Twitter* users obtain a high level of *cyberspatial presence*, a feeling of "being there" and they can get an additional level of connection with other users.

Secondly, knowing all these ordinary details, even if trivial, generates a cumulative background knowledge that can be recovered later as part of the (supposedly) mutual cognitive environment between users, and as a preliminary context for building up subsequent interactions. As a *Twitter* user acknowledges (in Zhao & Rosson 2009: 246):

> By reading someone's updates, you get more present understanding of what's on that person's mind, what he or she has been interested, so that it's more easily to get a conversation started and flow.

Thompson (2008) finds it a paradox, considering the irrelevant and trivial components of messages as an update on the individual; but if we take all the messages globally and in a time span, they turn into a detailed portrait of the user's life, like many dots that, together, form an identifiable image.

So far, I have addressed the initial intuition that *tweets* are trivial and produce few cognitive effects. Let us now consider that other condition for relevance: mental effort. On paper, these short trivial messages should posit no challenge for their

28. Interest from the point of view of the reader. But from the sender's point of view there is also a cognitive gratification, this time in the constant desire to send trivial messages about ordinary life as part of a general human tendency to feel close to other individuals and share information about the most immediate (physical/virtual) environment (Muñóz & Riveiro 2009).

interpretation. However, users code very little text (there is a 140-character limit) and leave implicit as much information as they can. As a consequence, the reader faces short texts, often sub-sentential ones, for whose interpretation they have to fill in many implicit blanks in order to turn the schematic *tweets* into meaningful contextualized interpretations. Besides, the design of the *Twitter* interface may also increase processing effort, in a similar way to the chat rooms interfaces that will be analysed in Chapter 5.

An additional source of mental effort is the fact that *tweets* can either be open (for all users) or conversational (for a specific user). Both options co-exist on the list of messages that arrive at the user's personal *Twitter* page and the system publishes them in strict order of arrival (as in chat rooms and instant messaging, see Chapter 5) and it is often difficult to follow the conversational threads. Moreover, on the user's page we can only see half of the turns in the conversation, which entails additional processing effort, since most *tweets* are meaningless without the contextualized information provided by the other turns, as can be seen in (15):

(15) a. lo reconozco… pero que nos quiten lo "comido" ;-) (T3).
[*I admit it, but I enjoyed what I "ate"*].

b. y eso que es? xD (T4).
[*and what is that?*].

c. pero yo pensaba que tu ibas a seguir! xD (T4).
[*But I thought you were going to continue!*].

d. Cómo se llama? Yo a veces tengo suerte para esas cosas (T5).
[*What is it called? Sometimes I'm lucky with these things*].

Twitter users have devised two strategies in order to reduce the processing effort of their *tweets*. One of them has already been mentioned: to type "@username"[29] at the beginning of the *tweet*, so that only the intended user reads it and replies to it, as in (16):

(16) a. @usuario A mí me pasó eso con Michael Jackson ;) (T1).
[*@username The same happened to me with Michael Jackson*].

b. @usuario la noche es sexi y peligrosa y bueno me siento acompañado con tantos twittfriends jejeje (T2).
[*@username the night is sexy and dangerous and well I feel that I'm in the company of so many twittfriends hehehe*].

29. When the *Twitter* company discovered the massive use of @username, it re-designed the program and now, every time a user types this nomenclature, this stretch of text turns automatically into a link that leads to the main page of the user whose name or *nick* appears after the @ sign.

c. @usuario igual para ti amigo mio! (T6).
[*@username same to you my friend!*].

d. @usuario pal calor no sé, pero igual pa la resaca sí, no? (T9).
[*@username for the heat I don't know, but maybe it works for hangovers, doesn't it?*].

The problem is, as Honeycutt & Herring (2009) qualify, that not all of these "@username" are typed with a conversational purpose, nor are they a requisite for dialogues between users, as can be observed in (17), where T4 prefers to type the name of the addressee in the *tweet* without the @ sign. Therefore, without a higher conventionalization of this nomenclature, it will not be truly effective for reducing processing effort.

(17) a. Irene: "el msn causa estragos en la autografia" HOSTIA, NI QUE LO DIGAS xDDDDDDDDDDD (T4).
[*Irene: "MSN devastates autobiography" SHIT, DON'T TELL ME ABOUT IT*].

b. Nat: "ays q calores me estan entrando" Ro: "eso es x apellidarte infernal" LOLAZO xDDDDD (T4).
[*Nat: "ays I'm getting really hot" Ro: "that is cos you're named infernal" BIG LOLA*].

c. Lu: "preveo que me voy a cargar un vaso proximamente" (T4).
[*Lu: "I predict that I am going to smash a glass soon"*].

The second strategy to alleviate mental effort is to insert a message that the server automatically copies under each *tweet* indicating which user is being replied to, as in (18):

(18) a. @usuario Nah, lo tuyo es mal de vacaciones. Seguro. Debes dormir de lado. Izquierdo. Seguramente ese es el lado de la cama del Tweet-Deck. (T1). 10:47 AM Aug 18th from TweetDeck in reply to user.[30]
[*Nah, you suffer from holiday disease. Sure. You must sleep on your side. The left side. Maybe this is the side of the bed of TweetDeck*].

b. @usuario Siento cortarte el rollo, pero la vieja de los Goonies lleva tiempo muerta (T3).
about 7 hours ago from web in reply to user.
[*I am sorry to disappoint you, but that old woman of the Goonies has been dead for some time*].

---

30. The real name or *nick* of the user has been deleted and instead the word "user" is quoted in order to preserve anonymity.

c. @usuario jajajajaja xDD nah mas vale tarde que nunca (T4).
3:27 AM Aug 20th from TwitterFox in reply to user.
[*hahahaha xDD nah better late than never*].

On the other hand, *Twitter* users have devised another nomenclature to forward *tweets* by other users, the so-called *re-tweeting*. It also entails challenges for a correct comprehension. The nomenclature is to write "RT" before the text of the *tweet*, as in (19):

(19) RT @usuario: El DNI electrónico en manos de 11.5 millones de españoles y la mayoría no sabe usarlo http://bit.ly/3l7oo (T3).
[*RT @username: 11.5 million Spaniards have the electronic ID card and most of them do not know how to use it*].

From a pragmatic perspective, this "RT" nomenclature is particularly interesting because it entails alterations both in the way *tweets* are coded and in their interpretation (see 3.3 below). In the first case, there are coding alterations because the user who *re-tweets* tends to summarize the message so as not to exceed the 140-character limit and the user inevitably alters the content of the initial *tweet* that is being forwarded. Therefore, its propositional form and eventual interpretation are also affected. This alteration may arise because the users change the initial text of the *tweet* or because the users simply erase words that they consider unnecessary. These alterations may also increase due to the multiple platforms and services from which *tweets* can be forwarded. As Marwick & boyd (2010: 117) summarize, "it is not uncommon for people to forward tweets via email or by copying and pasting them into new communication channels. Furthermore, various tools allow users to repost tweets to Facebook, MySpace, and blogs."

The comprehension of this "RT" nomenclature is also altered by the lack of agreement on how this *re-tweeting* has to be typed, that is, by the lack of a proper conventionalization of this strategy. Although the "RT" is very frequent, there are alternative ways of showing *re-tweeting*. In boyd et al. (2010) some possibilities are quoted:

(20)	RT: @	retweeting @	retweet @	(via @)
	RT (via @)	thx @	HT @	r @

An additional challenge for readers of *re-tweeted* messages is that very often these messages form a chain of forwarded messages and this makes the *tweets* difficult to interpret correctly and it is also difficult to locate the initial author of the message, as can be seen in (21):

(21) a. RT @usuario: RT @usuario: Desde Cuando Hay Que pedir permiso para Circular Libremente por la Ciudad? #venezuela #freemediave (T6).
[*RT @username: RT @username: Since when do we have to ask for permission to circulate freely in the town?*].

b. RT @usuario @usuario @usuario El Geek Errante tiene que volver! Mañana traigo el equipo de grabación a @usuario (T10).
[*RT @username @username @username The Wandering Geek has to return! Tomorrow I'll bring the recording equipment for @username*].

It is interesting to comment on the use of "#" in (21a). The "#" sign is another nomenclature in *Twitter* that refers to a specific topic of interest that the user labels as such, so that other users can locate the *tweets* that deal with this topic. It is, therefore, a kind of *thematic labelling*. It is also a sign that favours collective action on the Net: "users may be very widely dispersed and usually unknown to each other. Twitter provides a structure for them to act together as if in an organised way, for example through the use of hashtags – the # symbol – and keywords that signpost topics and issues. This provides a mechanism to aggregate, archive and analyse the individual tweets as a whole" (Hermida 2010). De Moor (2010) adds: "Tracking the tweets involved in conversations is relatively easy through searching on both replies and hash tagged-topics. However, the resulting linear list of contributions is sometimes difficult to interpret due to the immediacy, sheer number, and lack of thread structure."

3.3 Interpreting *tweets*

In this book, I have previously commented on the intuition that *tweets* should be easy to process due to their short length. But a more exhaustive analysis reveals that these short messages demand the whole range of inferential steps that are applied to the schematic logical form of the message in order to turn them into relevant interpretations. One of the most interesting contributions of relevance theory has been to demonstrate that obtaining the explicit interpretation of utterances demands as much contextualization and inferential activity as deriving implicated conclusions (implicatures). And *tweets* demand inferential activity similar to the one we apply to the interpretation of other utterances, with the additional task of turning schematic 140-character messages into meaningful interpretations.

Among the range of inferential strategies that we normally apply to the interpretation of utterances, *reference assignment* is pervasive because it has to be performed in the processing of almost any *tweet*, either because the reader has to find a referent for the name or *nick* of the author or because the *tweet* contains a number of indexicals (pronouns, time adverbs, etc.) for which a referent has to be found. In (22), for example, the readers have to find referents for the pronouns in the *tweets* or they will reach no interpretation:

(22) a. Wow, yo de eso no sé nada. Pero eso es lo tuyo. Te irá muy bien. :) (T5).
[*Wow, I know nothing about that. But that's your stuff. You'll be alright*].

b. a mi me gustó, debes verla ^^ (T6).
[*I liked it. You must see it*].

Other inferential strategies include disambiguation, conceptual adjustment, free enrichment, the compensation of elided content in sub-sentential utterances and the ascription of propositional attitude in the author. These are briefly commented upon below.

1. *Disambiguation.* Sometimes *tweets* contain polysemous words whose intended sense has to be inferred, as in "banco" (that can either mean "financial institution" or "bench" in Spanish) in (23) below:

(23) #FAIL veo la pagina del banco (T6).
[*#FAIL I see the page of the bank/bench*].

2. *Conceptual adjustment.* As has already been mentioned in this chapter, very often the prototypical concepts coded by the words, as we would find in a dictionary, for instance, are inadequate in the specific context in which these words are uttered and have to be adjusted inferentially to meet the speaker's intended *ad hoc concept*. Sometimes the speaker intends a broader, less exact *ad hoc* concept than the one coded by the word, as in (24a–b). On other occasions, though, the speaker intends a narrower, more exact *ad hoc concept* than the one coded by the word, as in (24c–d):

(24) a. We entered a pub, but we left because it was *empty*.
[not literally empty; there was surely a waiter, a few non-interesting people, etc.].

b. I've got *a thousand things* to do this morning.
[not literally a thousand; rather, a lot of things].

c. Tony *drinks* too much.
[specifically, he drinks too much alcohol].

d. I've got *nothing* to wear for the party.
[specifically, nothing nice, nothing classy].

In the same way, the readers of *tweets* have to adjust the concepts coded in these messages and infer the (broader or narrower) *ad hoc concepts* that the author intends to communicate. Some examples are quoted in (25), where the concepts coded by the words in italics have to be adjusted for a relevant interpretation:

(25) a. Senderos de Traición es el mejor... pero El Espíritu del Vino es *bestial* y Avalancha el más "*pesado*"... son 3 *joyas* en realidad (T1).
[*Senderos de Traicion is the best... but El Espíritu del vino is huge and Avalancha is the "heaviest"... They are three jewels in reality*].

b. Listo bañado vestido y ready pal party jejeje ya *activado* esperando que se vistan pa salir a buscar la *gasolina* ;) (T2).
[*Ready, had a bath, got dressed and ready for the party hehehe already activated waiting for them to get dressed and go for petrol*].

c. jejeje, q honor, gracias por el #FollowFriday y #TwitterAdicto total, muchas gracias por considerme [sic] *amigo*, es mutuo, un abrazo! (T6).
[*hehehe what an honour thanks for #FollowFriday and #TwitterAdicto great, many thanks for considering me a friend, it's mutual, hugs!*].

d. Hala, pues ya estoy más *tranquila*. Qué mar más *malo* había hoy! (T9).
[*Well, I am more relaxed now. What a bad sea there was today!*].

In (25a) the reader has to infer the *ad hoc concepts* that underlie the concepts coded by the words "bestial" (huge), "pesado" (heavy) and "joya" (jewel), which are used metaphorically. The reader will have to select features associated with these concepts and adjust metaphorically the ones that might be applicable to the referents in question. Something similar happens in (25b), where "activado" (activated) and "gasolina" (petrol) are again used metaphorically (meaning "ready" and "alcoholic drink" respectively). (25c) demands the adjustment of the coded concept "friend" in a similar way to the one already commented upon for SNSs in this chapter. Finally, the reader of (25d) has to adjust the concept coded by "malo" (bad) to fit the context in which it is used. "Tranquila" (relaxed, calm) also demands adjustment since the coded concept covers a whole range of states of mind, most of which are not intended.

3. *Free enrichment*. It takes place when the utterance demands from the hearer the "inferential filling" of some elided part. Despite being a grammatical utterance, it makes no sense unless this non-coded part is inferred correctly, as in (26), where the square brackets suggest this inferential compensation:

(26) a. This girl is too small [*for what?*].
b. The other medicine is better [*than what? for what?*].

Similarly, readers of *tweets* often complete their non-coded parts inferentially:

(27) a. Pobrecica Pero ya te queda menos, ya te queda menos!! (T4).
[*Poor girl But there is not much left (for what?)*].

b. Jajajajaja! Te hacía falta un babero? (T5).
[*hahaha Did you need a bib? (for what?)*].

c. jejeje, si, es cierto, el dolor es grande, el primer año me la pasaba a punto de motrin, q fino q ya estás cerca de terminar (T6).
[*hehehe yes, it's true, the pain is huge, the first year I was on Motrin all the time, it's nice you are about to finish (what?)*].

4. *Sub-sentential utterances*. Utterances can be arranged on a scale from the most explicit to the least explicit depending on how much information is actually coded by the speaker. On paper, the more information is left implicit, non-coded, the higher the interpretive challenge for the interlocutor, who has to fill these information gaps inferentially, as happens in the strategy of "free enrichment." For example, utterance (28a) is the most explicit one, while (28b-d) are increasingly less explicit, thus demanding more inferential activity by the interlocutor with the aid of context:[31]

(28) a. John has left the book by Larsson on the dining room table.
b. John has left the book on the table.
c. He has left the book there.
d. On the table.

Tweets are, in essence, prone to being sub-sentential utterances due to the 140-character limit, and users tend to suppress all the coded content that they expect their readers will be able to recover by themselves, as in (29):

(29) a. Gran juego ;-) (T3).
[*Great game*].

b. Bueno, va (T4).
[*Well, ok then*].

c. Cuándo vuelves, para hacerte uno? :P (T5).
[*When are you coming back, to make you one?*].

d. no se, y no creo, pero de que vuelan vuelan (T6).
[*I don't know, and I don't think so, but they do fly*].

5. *Propositional attitude ascription*. To enrich the coded message with the aforementioned inferential strategies is not enough to yield a fully relevant interpretation. It is also necessary to ascribe the user's propositional attitude (or the speech

31. This does not mean that the hearer invariably expects the most explicit utterance on every occasion. As a matter of fact, leaving information implicit (non-coded) is the norm, rather than the exception. For example, a person would sound strange if, to the question "where has John left the book?" the hearer replied (28a). Rather, (28d) would be more appropriate. But if where the table is located is not mutually manifest to both interlocutors, the speaker will have to be more specific (i.e. explicit) and rephrase the utterance by adding "on the dining room table."

act schema of the *tweet*) that underlies its production plus an estimation of the feelings and emotions that the user holds when typing the message. Propositional attitude is essential in human communication, because the same utterance can communicate a wide range of attitudinal intentions. Certainly, it is not enough to interpret utterance (30a) as the neutral (30b), but hearers normally aim at obtaining a correct underlying attitude, as in (30c–e):

(30) a. Boss to employee: "you're leaving this project."
b. My boss is informing me that I am leaving the project.
c. My boss is *asking* whether I am leaving this project or not.
d. My boss is *ordering* me to leave this project.
e. My boss is *advising* me to leave this project.

The corpus of *tweets* provides us with examples in which propositional attitude ascription has to be inferred with the aid of context. Occasionally, it is the users that make this attitude explicit, as in (31):

(31) Quiero, necesito ir al FNAC… qué digo, EXIJO ir al FNAC Ò.Ó (T4).
[*I want, I need to go to FNAC… I mean, I DEMAND to go to FNAC*].

As far as the ascription of feelings and emotions is concerned, users tend to colour their *tweets* with words that indicate their feelings or emotions. An even more interesting strategy is to resort to the techniques for oralization of text, which will be analysed in Chapter 5 for chat rooms and instant messaging. Certainly, these techniques for oralization (repetition of letters, playing with capitalization, creative use of punctuation marks…) offer a good repertoire of written means for the communication of feelings and emotions, as in (32):

(32) a. ayyyyyy dolorrrrrrrrrr!!!!!!!!!!!!!!!!!!!! (T7).
[*pain*].
b. QUÉ HA PASADO?!?!?!?! (T4).
[*what's happened?*].
c. TE ODIOOOOOOOOOOOO!!!!!!!!!!!!!!!!!! (T7).
[*I hate you*].
d. Por fin juevesssssssssssssss!!!!!!!!!!!!!!!!! (T9).
[*Thanks God it's Thursday*].
e. Quéeeeeeeeeee?????? En Lisboa????? qué fuerte me parece (T9).
[*What? In Lisboa? This looks heavy to me*].

Besides, visual nonverbal behaviour is communicated with the aid of emoticons (again, as in chat rooms and instant messaging and even SNSs), combinations of punctuation marks to yield iconic compositions. Among them, the ones most frequently used are the emoticon of happiness [:-) or :-D], of sadness [:-(] and of winking [;-)]. The corpus of *tweets* contains many of these emoticons, some of them with innovative combinations of punctuation marks and letters [Ò.Ó]:

(33) a. Ya ves Ò.Ó Es que encima sigue a Brendon, Gerard y Danny. TOCATE LOS ********!!! ¬¬ (T4).

b. como que ugh!??!?!? xD lo que ha dicho! ò.ó (T4).

c. RT @usuario: la cancion mas hermosa de este mundo =) <3 (T6).

There are also instances of what Poyatos (1975, 2002) called *alternants*, sounds that, as the name indicates, may "alternate" with speech, facilitating their transcription in the *tweet*.[32] In (34) there are some examples of laughter (34a–b), surprise (34c) and admiration (34d):

(34) a. jajajajaja xDD nah mas vale tarde que nunca (T4).
[*better late than never*].

b. M VOY A FLICKR MWAHAHAAHAHAHHAAH (T4).
[*I am off to Flickr*].

c. uuuffff, ok ok, considero seriamente irrumpir en tu casa para jugarlo! (T6).
[*I am seriously considering popping round to your place to play it!*].

d. wow pana, muchas gracias, que honor, un abrazo y feliz viernes! (T6).
[*pana thanks, what an honour, hugs and happy Friday!*].

All of these strategies for oralizing typed text will be analysed in more detail in the next chapter, devoted to virtual conversations.

3.4 *Twitter* conversations

The introduction of new interactive capabilities with nomenclatures such as "@username" , "RT" and "#topic" allow for authentic micro-blogging 140-character conversations among disperse users all over the world.

The hashtag (#) is convenient in its ability to sustain dense interactions under the same label or tag and with an explicit wide audience. De Moor (2010) compares this capability with SNS conversations and concludes that in *Twitter*

32. Defined as "nonverbal, marginal and nonspeech sounds or clusters of sounds, articulated or not [...] which do not affect the verbal utterance [...] Alternants occur either isolated or alternating with the verbal utterance and with the kinesic behaviour" (Poyatos 1975: 294).

> it is very easy to join a conversation with complete strangers purely based on interest, instead of being limited to talking to people currently in one's circle of friends. As the effort of reading and replying is minimal, over time a deep conversation web with strands to a large group of relevant people can develop, both on an ad hoc basis (joining conversations based on a search) and permanently (by being their followers and joining in when an interesting topic passes by.

Besides, what de Moor (ibid.) calls "tangential conversations" are also inherent in this *microblogging* service. *Tweets* are short, provide little information, and therefore the thread of the conversation builds up on the micro accumulation of many *tweets* that are incorporated to the discussion.

Finally, as pointed out above, *tweets* can either be open (for all users) or conversational (for a specific user). But they can also be "self-oriented" and "other-oriented," the latter being more typical, because the system invites users to follow *twitterers*. It seems that *Twitter* might mesh all of these possibilities into an effort-producing mixture of messages with different intended audiences, but Marwick & boyd (2010: 120) claim that this is not really the case: "users write different tweets to target different people (e.g. audiences). This approach acknowledges multiplicity, but rather than creating entirely separate, discrete audiences through the use of multiple identities or accounts, users address multiple audiences through a single account, conscious of potential overlapping among their audiences."

CHAPTER 5

The virtual conversation

1. Introduction

On the Internet there are multiple options for engaging in synchronous conversations: chat rooms, instant messaging, Internet-mediated phone calls, videoconferencing, 3D virtual worlds, etc. Some of them have been incorporated into other more typically asynchronous forms of Internet-mediated communication, as happens with the instant messaging application on social networking sites.

In this chapter I will review most of these options for virtual conversations. One of the central issues in this chapter will be to analyse how users compensate for the lack of oral features that their typed texts exhibit, compared to the contextual richness of face-to-face interactions.

2. Chat rooms

The chat room is one of the most popular forms of virtual conversation. Although it has evolved into an enhanced medium with the incorporation of web cams and sound, many users still prefer the traditional text-based utterances sent to a chat portal on the Net. These portals for synchronous conversations contain a number of design features that influence the quality of interpretations and eventual relevance, as will be analysed below.[1] This section deals with the pragmatic implications of chat rooms and of the synchronous oralization of the text typed by users.

Chat rooms are not only interfaces for virtual conversations, but also encourage community bonding. Many of the attributes that were commented upon in Chapter 2 as indicative of a feeling of community among users (Yus 2007b, Baym 2010: 71–98) are also reproduced in these synchronous interactions. Paolillo

---

1. As Baron (2003a) comments, chat rooms, as they are used nowadays, were not created until 1988, when Jarkko Oikarinen, a student at the University of Oulu (Finland), created software that was later known as *Internet Relay Chat* (IRC). At the beginning of the 90s, it turned into an open-access program and started being offered by Internet providers such as *America Online*. For a short history of chat rooms see Mariottini (2004: 29–30).

(2001: 185) asserts that chat rooms can foster the formation and consolidation of social networks that are similar to the ones found in physical settings. Their users spend much time chatting there, which can be compared to "spending some time together" in physical scenarios and these casual conversations foster the formation of networks in urban spaces. Besides, chat room users exhibit linguistic strategies that indicate community membership. This is why non-members are very easy to spot. These "outsiders" will have to "train themselves" in the use of the linguistic strategies that abound in this type of virtual conversation and mark boundaries for group membership. Regular interactions in chat rooms will turn these neophyte users into experts who discursively blend into the community (the *familiarity principle* that Peris et al. 2002: 44 suggest).

2.1 Utterance, propositional attitude and audio-visual context

The steps of interpretation, according to relevance theory (Sperber & Wilson 1986, 1995) start with the identification of the logical form of the utterance, which is enriched inferentially to yield the proposition expressed by the utterance. This proposition can be selected as the explicit interpretation of the utterance (*explicature*) or it can be used as one of the premises that, together with contextual information, enable the derivation of implicated conclusions (*implicatures*). The inference of explicit and implicated interpretations is guided by an inherently human search for relevance (see Wilson 1999, 2000). And the speaker has to predict that the addressee will be able to access the necessary contextual information that allows for the inference of these interpretations. As will be analysed below, in the case of Internet-mediated communication one of these predictions includes the addressee user's command of a number of typical discursive techniques that are inherent in text-based interactions (abbreviations, oralization of text, and emoticons, among others). As Fuller (1994) asserts, if the models that users construct differ greatly, communication may break down: "the models of other people's expectations and prior knowledge that people bring into communication can influence not only the tone of the discussion, but also the expectations of one person regarding someone else's personality." In other words, users need what Simpson (2005) considers as electronic communicative competence for managing and agreeing on conversational strategies and discursive rules that differ so much from oral interactions, and which I label *cyberliteracy*. This explains why chat room users tend to agree on conventions "on the fly" that guarantee mutuality in how certain communicative strategies have to be expressed and interpreted, as in the following example quoted in Campbell & Wickman (2000):

(1)	<Wickmansa>	I was thinking of a coded way to quickly indicate "busy"
	<Wickmansa>	that the other person wouldn't respond to
	<DRCSC>	ok, what?
	<Wickmansa>	but would know to stop sending messages
	<Wickmansa>	it could be anything we decide that is one keystroke
	<Wickmansa>	like *
	<Wickmansa>	or /
	<DRCSC>	ok
	<Wickmansa>	you pick one
	<DRCSC>	doesn't matter
	<Wickmansa>	whichever is hit first I guess
		[...]
	<DRCSC>	/ may be the easiest
	<Wickmansa>	Yes
	<DRCSC>	/
	<DRCSC>	back
	<DRCSC>	that worked

Besides, in the interpretation of the intended interpretation it is important to identify the speaker's propositional attitude upon coding the utterance (or in a more general sense, the relationship between the speaker and the thought expressed by the utterance), because the eventual interpretation will be different depending on whether the speaker is regretting, ordering, asking, advising, etc. with the utterance. This attitude can be communicated in different ways: (a) syntactically (with verbal mood, for example); (b) lexically (using assumption schemas – as in speech acts – that include attitudinal markers such as "I regret that...," "I suppose that..." or "I wish that...," and also with adverbs such as "unfortunately," "probably," etc.); and (c) nonverbally (a smile can reveal an underlying ironic intention). In ordinary interactions, interlocutors devote much effort to identifying the speakers' attitude towards what they are saying and metarepresent their intentions. Some of this cognitive reasoning may be really complex, even though the hearers are normally unaware that they are performing it.

In (2) there is an example of three levels of metarepresentation (summarized in (3)), generated in the hearer's search for the speaker's underlying intention (Sperber 2000):

(2) Mary is picking berries. Peter happens to be watching Mary. Mary intends that Peter should be aware of her intention to inform him that the berries are edible.

(3) Mary intends... that Peter should believe... that Mary intends... that he should believe... that these berries are edible.

These metarepresentational inferences are universal and, like the human search for relevance, they are biologically rooted in human psychology.[2] It is not surprising, then, that chat rooms should reproduce the same inferences of attitudinal attribution, as in this conversation quoted in December (1993):

(4) [*<wabbit> is surprised at a message by <KMOORE>*].

<KMOORE>	wabbit well i thought that you thought that i meant something else!!!
<KMOORE>	wabbit that was a confusing line i just wrote.
<wabbit>	i think what i meant and what you meant didn't mean the same thing and we're all confused now.
<KMOORE>	wabbit yes.. exactly. and now i know what u meant and you know what I meant!
<wabbit>	kmoore: that's what i meant!

As I have already suggested, a textual way of communicating attitudes is to use speech-act expressions such as "I'd like to ask you if…" or "I recommend that…." Curiously, the commands that are used in most chat rooms include keystrokes that automatically turn a sequence of characters into a complete speech act (see Hassell 1998, Cherny 1995b, and Goutsos 2005).

2.2 "What is important is to be able to talk"

2.2.1 *Introduction*

The title of this heading corresponds to a famous slogan of a Spanish phone company in 2001 ("lo importante es poder hablar"), showing the main reason why users spend hours engaged in virtual conversations, namely, to be able to chat with people regardless of their physical location. To achieve this goal, users sometimes have to make the most of typed texts with similar goals to the ones that underlie conversations in physical environments: to attract the interlocutor's attention, direct this attention towards the user's intention and, finally, reveal this intention

2. Metarepresentations are essential cognitive operations of humans, according to which, when a person is faced with a mental representation, this person is capable of making a representation of this representation. There are several possible metarepresentations: A thought about another thought, as in (a); an utterance about a thought, as in (b); a thought about an utterance, as in (c); and an utterance about another utterance, as in (d) (see Wilson 1999, 2000):

a.	John thinks:	Tom *wants me to leave.*
b.	Mary says:	Tom *thinks that he is intelligent.*
c.	John thinks:	Mary says that *she ate all the chocolates.*
d.	Mary says:	John says *that it rains a lot in England.*

(S&W 1986: 153–154). For this triple task, users resort to a number of conversational strategies and alter their texts as much as necessary to convey their intentions, but if these are not properly understood, there may be a significant increase in mental effort or even room for a wrong interpretation.

Therefore, users of chat rooms are expected to master several norms and conventions about how to interact in these environments (Araujo & Melo 2003, Savas 2011: 308). Sometimes users also *exude* information about their command beyond their intention to communicate it explicitly. For example, a user who is sent a message that is full of abbreviations, acronyms, emoticons, etc. may get the impression that the sender is an expert in chat room communication, although the sender may have no intention to communicate that information ostensively (S&W 1986: 58). In fact, there is usually a parallelism between how short and altered the message is and the impression of command that the user exudes.[3] For example, in Chapter 1 the lack of familiarity of <mariabisb> in (5), correlated to a lot of typed text, was compared to the expert user <Bisbaal> in (6a), whose short message communicates much information with just a few keystrokes, as can be seen in (6b), the type of information that the (familiarized) users of this chat room, *Operación Triunfo* (the Spanish equivalent of the British *Fame Academy* programme), can recover without much processing effort:

(5) <mariabisb> rosa tiene una voz bonita pero le falta mucha autoridad en el escenario en eso le dan 100 vueltas chenoa y bisbal, y manu. [*Rosa has a nice voice but she lacks authority on the stage and in that chenoa and bisbal and manu beat her hands down*].

(6) a. <Bisbaal> y creo q n tienen dntro d la academia.

b. <Bisbaal> y[o] creo q[ue] [los concursantes de Operación Triunfo] n[o] tienen [un ordenador conectado a Internet] d[e]ntro d[e] la academia [de O.T.]. [*I think that the contestants at Operación Triunfo don't have a computer logged onto the Net inside the Operación Triunfo Academy*].

Initially, one might think that users interact in text-based chat rooms *despite* the limitations of text-based communication, that is, these users type what they would like to be saying and they read what they would like to be listening to. However, this assertion has to be qualified. In fact, many users interact in these

3. There is a parallel relationship between the level of coherence between messages sent to the chat room and the users' command of the technology to sustain conversations there (see Cornelius & Boos 2003).

environments not "despite" but *precisely because* chat rooms possess this textual quality (Yus 2001b). For example, a user in Savas (2011: 308) explains why he prefers chat rooms: "I am more comfortable with chat because I can overcome my pronunciation problems. When I speak I have to be careful with my pronunciation. In chat, I can express myself better."

2.2.2 *Limitation or advantage*

Anonymity and textual quality are, on first sight, limitations of chat rooms if compared to the richness of face-to-face interactions. Undoubtedly, the vocal and visual information that accompanies speech in situations of physical co-presence are essential for a correct interpretation. "Proximity allows participants to gauge whether or not they are being understood and take appropriate action if they are not" (Freiermuth 2011: 129). Therefore, users are likely to be dissatisfied with some reduction in the possibilities for expressing their thoughts properly on the keyboard and also with the problems involved in checking the extent of each other's cognitive environments. But in reality, many chat room users are satisfied; they hide behind the security of the *nick*, and free themselves from the pressure that being face-to-face with another person exerts. In this environment they express themselves more accurately, freely, spontaneously, and even play with the multiplicity of identities that the system fosters.

According to Caldwell & Taha (1993), it may be concluded that many chat room users shy away from direct face-to-face contact because of the challenge of controlling the interrelation of verbal and nonverbal information. The control that users have over other users' impressions and interpretations may lead to a preference for virtual conversations. But the new developments in this type of interactions on the Net, for example the introduction of web cams and the microphone, add a new dimension because users have to assess to what extent they are willing to let other users perceive their vocal and visual nonverbal behaviour (intentional or exuded) and which impressions they want to convey (see Becker & Stamp 2005, Peter et al. 2007). Nowadays, despite these developments in chat room interfaces, many users still resort to (more secure) plain-text-based communication, except when interacting with close friends and relatives. In that case they are not so reluctant to reveal information.

Altogether, text-based communication is an interesting feature of chat rooms that deserves pragmatic analysis, especially in its oralized quality and in the way it affects processing. Suler (1997a) points out that users find it attractive to see how others express themselves through text in spite of its limitations: "they love to immerse themselves in the quiet flow of words that feels like a more direct, intimate connection between one's mind and the minds of others [...] without the distracting sights and sounds of the ftf world" (see also Belson 1994).

2.2.3 *Conversational interaction*

Both in physical and virtual conversations there is a similar goal: to engage in interactions and make information mutually manifest, information that alters each interlocutor's cognitive environment.[4] Chat rooms are "transit places" in which users exhibit their predisposition to interact with one another in a predominantly casual way.[5] Thanks to this interactive environment, users can feel the presence of others and mind-read (i.e. metarepresent) their thoughts, select intended interpretations and agree on the direction that conversations are going to take (Bellamy & Hanewitz 1999). Nevertheless, the differences between face-to-face and chat room interactions are also notorious, as will be commented upon in the next few paragraphs (see Suler 2000).

1. *Synchronous vs. asynchronous.* Face-to-face conversations are centred upon the co-presence of interlocutors and there is a possibility of overlappings and interruptions. Chat rooms are synchronous, but they depend on a rigid succession of messages as they arrive at the computer system (Giese 1998). Although both users who are having a chat-room conversation are online simultaneously, chat rooms suffer from succession in time and space. This fact reduces communicative options and naturalness. And the scenario gets even more complicated if more than one user is involved in a chat room conversation. As a user in Savas (2011:307) explains, "flow and continuity is difficult to establish with more than two people at a time. Continuity can be established with two people if there is substantial "waiting time" while the other person waits for the response from the other. But still, disjointed discussions are common in chat and poor typists suffer." Nevertheless, the evolution in virtual environments for text-based interactions permits nowadays a true feeling of synchronicity. For example, some systems allow users to read, word by word, what the other user is typing on his/her screen, drawing understanding closer to the inferences in face-to-face interactions. In the latter, the processing of previous stretches of the utterance generate anticipatory inferential hypotheses that are (dis)confirmed with the processing of subsequent stretches of discourse. Frequently, listeners have to backtrack and re-interpret previous discourse in the light of new interpretive evidence just processed. This is now possible with modern systems for synchronous text-based interactions (apart from the obvious capacity of voice-enabled chat rooms). As a consequence, classifications of synchronous virtual interactions such as Anderson et al.'s (2010) are welcome:

---

4. Especially now, when both types of interactions tend to be amalgamated in personal networks of a hybrid nature. See Suler (1997c) for a terminological proposal of in-person relationships and cyberspace relationships for physical and virtual interactions, respectively.

5. See, among others, Mayans (2000a, 2000b, 2002a: Chapter 2), Baldwin (1996: Chapter 3), Rafaeli & Sudweeks (1997) and Nilsen (1999).

Systems can be said to predispose communication to *non-simultaneity* (i.e. asynchronous, one-way systems, such as email, discussion boards, blogs), in which conversation is isolated spatially, chronologically, and contextually; *near simultaneity* (i.e. synchronous, one-way systems, as in instant messaging, multiparticipant chat rooms, and text chat in multiplayer games), in which users typically respond to others' comments as soon as they are received; and *high simultaneity* (i.e. synchronous, two-way systems, such as the VAX "phone" and the contemporaneous UNIX "talk" programs), in which not only immediate responses are possible, but also communications overlap.

2. *Contextual cues in the unfolding of conversations*. The lack of vocal (see 2.3 below) and visual (see 2.4 below) nonverbal information (as part of accessible contextual information) are two challenges for users of text-based chat rooms (Ten Have 2000). To these challenges we can add more technological issues (the *scroll factor*, for instance, the speed at which threads of messages disappear from the top of the screen) and physical ones (there is no shared space, apart from the screen, that anchorages the identification of indexicals), which posit a burden for chat room effectiveness.

3. *Turn taking*. Terms such as move, act, sequence, and opening are typical of conversation analysis. But the most famous term is the speaker's "turn." In chat rooms these turns are subject to the sequencing imposed by the software that manages interactions.[6] Until more "natural" chat room software becomes popular, for instance software (already available) that conveys the user's voice, as in *Second Life* (see 4 below) or one's visual image, the sequencing of utterances will be the norm in chat room interactions, unlike face-to-face interactions. Besides, in chat rooms all utterances are sequenced, even if simultaneously produced. As Nilsen & Mäkitalo (2010: 92) comment, "chat systems are designed so that several persons can post messages simultaneously, which means that there is no competition for the floor since all messages sent off will be posted."

4. *Multiple interactions*. One of the most typical attributes of chat rooms is the juxtaposition of conversations in the same (main) area of the screen, also typical of instant messaging (see 3 below) and *Twitter*. Indeed, unless the user is engaged in a private conversation with another user (in a different window), the norm is that all the messages arrive at the "central area" of the chat room, together with messages that the system creates automatically. This quality might produce increased effort when following threads of conversation that are mixed up without a clear arrangement, and hence affect users' eventual estimations of relevance (Werry 1996: 51).

---

6. See Zitzen & Stein (2004: 991–993). The convergence of interactive aspects of humans and attributes of computer software for Internet communication is usually called *groupware* (Feenberg 1989: 28). The term is also used to describe the creation, among multiple users, of a single text (Greller & Barnes 1993).

Furthermore, while an "addresser user" is waiting for another user's reply, he/she may initiate other conversations in the central area of the chat room or privately with other users in other rooms of the site, or engage in several one-to-one private conversations, and therefore when this user receives the initial addressee user's reply, this might be totally irrelevant for someone who is already carrying out other interactions. This is why Serpentelly (1992) concludes that a "serious" conversation in a chat room is impossible, because we cannot control the multiplicity of simultaneous dialogues that can co-occur in the same virtual space.

A possible strategy of mitigation that reduces increased mental effort when the user tries to follow the conversational threads in the chat room is to type the *nick* of the "addressee user" at the beginning of the message, similar to the convention "@username" in *Twitter*. In this way, it is easier to obtain an adequate level of coherence among the threads. Herring (1999) studied, precisely, the coherence between turns in chat room conversations. Her conclusions were that these conversations suffer from fragmentation, and this provokes a growing lack of interest in users, who have to pay attention to multiple dialogues simultaneously. Herring (ibid.) wonders why chat rooms are so popular if they are so communicatively limited. Two explanations are plausible: (a) the users' ability to adapt themselves to the peculiar idiosyncrasy of chat rooms; and (b) the advantages of losing coherence in exchange for higher interactivity and textual playfulness. In her study, Herring analyses the short conversation in (7), which exhibits an overlapping of conversational threads, as summarized in Table 5.1.[7] There are up to three levels of juxtaposition, and the "jumps" among threads are constant:

(7)	1.	<ashna>	hi jatt
	2.	<Dave-G>	kally i was only joking around
	3.	<Jatt>	ashna: hello?
	4.	<kally>	dave-g it was funny
	5.	<ashna>	how are u jatt
	6.	<LUCKMAN>	ssa all
	7.	<Dave-G>	kally you da woman!
	8.	<Jatt>	ashna: do we know eachother?. I'm ok how are you
	9.	<kally>	dave-g good stuff:)
	10.	<Jatt>	kally: so hows school life, life in geneal, love life, family life?
	11.	<ashna>	jatt no we don't know each other, i fine
	12.	<Jatt>	ashna: where r ya from?

7. See O'Neill & Martin (2003) and Panyametheekul & Herring (2003) for similar analyses of overlapping threads of conversations in chat rooms.

Table 5.1 Overlapping of conversations in a chat room (adapted from Herring 1999)

User (initials)	[1] A	[2] D	[3] J	[4] K	[5] A	[6] L	[7] D	[8] J	[9] K	[10] J	[11] A	[12] J
Dialogue 1	A		A/B		B/C			C/D			D/E	E
Dialogue 2		A		A/B			B/C		C			
Dialogue 3									A	A		

5. *Ephemeral conversations.* One of the drawbacks of the chat room interface is that the messages keep arriving at the main area and, if there are many users in the same channel and are participating actively, their messages immediately disappear from the top of the screen and it is sometimes impossible to process the information that the users intended to communicate (the so-called *scroll factor*). This affects expectations of relevance and the effort that needs to be devoted to a fast processing of these ephemeral messages, and it even constrains a correct contextualization of these messages.

In general, this fast scroll of messages on the screen has consequences, for example regarding their brevity. It is indeed difficult to find messages of over three lines. Werry (1996: 53) and Rintel & Pittam (1997) suggest another important reason for these short messages: the users are fighting for other users' attention. A potential reader of one's messages may focus his/her attention on another conversation if one spends too long typing the message. In this case, the eventual relevance of one's messages has to be predicted very quickly. In chat rooms, cognitive effects must offset the mental effort required to process the messages that do not stay on the screen for very long. In this sense, Ruane's (1999) words are illustrative:

> It's fast: Try talking to six people at once. It's brief: three or four words per exchange. It takes wit, concentration and nimble fingers. And it requires tremendous linguistic economy. There's neither time nor space for exposition. The solution has been to abbreviate, contract and condense. On a huge scale. Why, for example, consume precious keystrokes telling six friends you have to go smack your little brother when BRB (be right back) will do?

But constant advances in chat room communication and interface allow for an increasingly contextualized and "natural" interaction in this virtual environment.[8] Advances have brought voice and image to the chat room. But if one likes to preserve anonymity, there are also 2D graphic chat rooms such as The Palace (www.thepalace.com), where users can choose a virtual image (which, in Section 4, will be labelled *graphic avatar*) that complements their *nicks*. This is a kind of ritual in which we can wear masks for performing interactive tasks (Goffman 1987).

8. See, among others, Suler (1997b, 1999), Smith et al. (2000), and Smith, Cadiz & Burkhalter (2000).

6. *Clipped messages.* Another aspect of chat rooms that may influence message comprehension is that very often a message turns up on the screen divided into chunks that are allocated by the server as different turns and in different positions on the list of messages. For example, in (8), taken from a corpus of chat room conversations (Yus 2003b), the answer to utterance 262 is divided into two messages (268/271) and between them there are messages by other users. The counter-reply (285) can only be found after several messages. Finally the initial user sends another two messages (292/293), this time luckily allocated by the server contiguously:

(8) 262 <fryski> ves soys unos pardillos
[*You see? You are novices*].
268 <zen80> eeeeeeeeeeeeeeen fryski no te pases
[*fryski don't push us too far*].
271 <zen80> que de pardillos aki nadie
[*there's no novice here*].
285 <fryski> q pasa zen no te pongas nervioso
[*what's up zen, don't get nervous*].
292 <zen80> no fryski nervioso nooooooooooooo
[*no frysky not a male nervous*].
293 <zen80> nerviosa en todo caso
[*a female nervous, if any*].

7. *Conversations in progress.* One always enters a chat room and finds threads of conversations that are already "in progress" (Mayans 2002a: 34). The user is faced with many messages, some addressed to a general audience and some to specific users, and has to try to make sense of them and decide in which thread he/she is willing to participate (Yus 2009e). In this sense, Miura & Shinohara (2005) propose a model of interpretation of chat room messages that starts, precisely, at the moment when the user accesses the main area of the site. At a first stage, the *information-acquisition phase*, the users face a screen filled with conversations that have already been initiated. They have to structure these messages in coherent threads, understand the context in which they were typed, and devise their own messages. Since messages tend to disappear very quickly from the screen, this task has to be performed very rapidly.[9] Some of these messages and

9. Therefore, although in theory typing on the keyboard gives users time to devise their messages, in reality these are as spontaneous and unplanned as oral utterances in face-to-face conversations. I do not agree with Pano (2008a: 89) when she states that the action of typing characters one by one and sending them to the server to appear on the screen reduces spontaneity and gives users a longer time to think about what they actually mean to communicate, compared to what one really says in a prototypical oral conversation, full of slips.

conversations are held in the user's short-memory storage, including the topics that these conversations are about. This leads to a *situational awareness*. Then the users have to decide in which conversations they are going to participate. At the second stage, called *information output*, the users decide what message they are going to type. To do so, the users devote more processing effort to interpreting some of the messages, and the information they communicate will probably be remembered longer (and stored in the long-term memory). At the same time, this memory is accessed in the phase of composing the messages, together with information on how to use the chat room interface and its typical discursive techniques of oralization.

2.3 Compensating for the loss of the audio channel in chat rooms

2.3.1 *Introduction*

Virtual conversations do not differ so much from oral conversations in terms of what steps interlocutors have to follow in order to reach a relevant interpretation, but they do differ in how these steps are performed. As in any conversation, chat room users choose, among a range of possible options for coding their messages, the one that adequately serves as evidence of their underlying intentions and makes interlocutors grasp the intended interpretation without increased mental effort. Among the predictions, the "addresser users" will expect interlocutors to know about typing conventions and how to oralize text (abbreviations, acronyms, emoticons, among others) and about the use of chat room software (how to send a private message, how to change *nicks*, etc.). This is the type of information that users expect to belong to their mutual cognitive environment. In (9), the author of message 1 presupposes that the readers know the kind of message it is and how to type it, and the same happens in message 2, where it is presupposed that the user knows how to access a channel of the chat room and that the nomenclature "#" is necessary. Finally, in 3–4, we can see an adjacency pair in which a user asks how to type a private message and another user replies:

(9) 1 <ESIGUAL> Porfavor no mas privados
[*please no more private messages*].
2 <ESIGUAL> el chat con naim es en #naim_thomas
[*the chat with naim is in #naim_thomas*].
3 <Naiara> como se hace un privado???
[*how is a private message typed?*].
4 <ESIGUAL> Naiara- /q NICK

Therefore, chat rooms demand some kind of *cyber-literacy* as to how to oralize text and how to use the different commands for engaging in virtual conversations. The former is not only typical of chat rooms, but also of instant messaging and even dialogues in social networking sites. As will be analysed in 2.5, chat room messages are *oralized written texts*, hybrids of typed texts and the users' willingness to communicate their thoughts orally. It is even possible that users "hear" their own voices while they are typing their messages. This "written voice" leads to a textual deformation that aims at transcribing on the screen the message that the speaker would have said orally in a face-to-face conversation. This idea is corroborated in Savas (2011), where a number of informants who exhibited forms of text oralization in their messages were asked if they perceived chat as a written or spoken form of language, and most of them considered their chat room discourse to be "talking." For these informants, "chat was a different way of having a conversation. The only difference between online chat and spoken language was the typed responses in chat. Their choice of vocabulary, grammar, and style during the online discussion reflected how they spoke rather than how they wrote" (ibid.: 309).

This strategy of oralization as a hybrid form of Internet-mediated communication involves a number of techniques that will be studied in 2.5 below. These may strike the reader as unusual or, if they have become conventionalized among the regular chat room users, they might even go unnoticed as a taken-for-granted code. Indeed, textual deformation in chat rooms is continuously contrasted with well-established, normalised forms of written communication against which the users of chat rooms rebel. In Kataoka's (2003: 125) words,

> graphemic features may serve as a means of the writer's affiliation with particular groups, community, contexts, and cultures. Affective signs, exploited by young writers with a certain emotional drive, can index facets of the encoder's self through the ways s/he reveals and responds to affective events. We could take affective signs and punctuation to serve as a means of connecting emotion and youth identities [...] Youth identities are closely tied to the community-sanctioned ways of representing emotions that are shared between senders and addressees and appropriate to the epistolary context.

2.3.2 *Typographic innovation*

The keyboard offers users some possibilities for connoting typed text (Reid 1994: 31–32, Ruedenberg et al. 1994, Mariottini 2004). Jaffe et al. (1995) group these under the generic label of *emotexts*, a wide-ranging term that includes intentional variations in orthography (e.g. visual dialect), strategic use of capitalization, lexical substitutions (metalinguistic cues of paraverbal quality, for example to type "hmmmmm"), grammatical markers (e.g. reiterative use of exclamations), and iconic compositions of characters (emoticons). These possibilities qualify chat rooms as purposely playful.

In chat rooms, a frequent typographic resource for vocal connotation of text (e.g. to give emphasis) is phonematic repetition, as in (10), while other vocal qualities such as "shouting" are communicated with the aid of capital letters, as in (11):

(10) <tardío> tenia gaaaaaaaaaaaaaaaaaaaaanas de verte! dooooooonde estabas?
[*I was looking forward to seeing you! where were you?*].

(11) <tardío> NO LO SE!!!! YA TE LO HE DICHO!!!!! alguna otra pregunta?
[*I don't know! I told you! Any other questions?*].

It should be borne in mind that there are techniques of typographical compensation that are inherent in each language. Since Internet allows for multi-geographic synchronous conversations, a process of mutual awareness of (and eventual agreement on) conventions often has to take place to avoid misunderstandings (Fouser et al. 2000, Launspach 2000). This risk of misunderstanding depends on the existence of mutuality and whether users are strangers or already know one another offline. As Kavanaugh et al. (2005) correctly stress,

> While there is still a need for such verbal cues online in networked geographic communities, the fact that most people already know each other at least as acquaintances, provides a lot of background information about personality and manner from prior face-to-face interactions [...] The fact that members of groups interacting online typically already know each other in networked communities mitigates against some of the problems of social presence online.

2.4 Compensating for the loss of the visual channel in chat rooms

2.4.1 *Introduction*

In text-based chat rooms there is no simultaneity between the user's verbal and nonverbal behaviour. The hypothetical nonverbal behaviour that would take place while the user is typing also has to be typed (Reid 1994: 21–32). The user who reads the message has to imagine what nonverbal behaviour accompanies it. Instead of "reading" the nonverbal contextual cues that the speaker *exudes*, the reader has to make do with the *spectrality* (Feenberg 1989: 25) of their mere textual manifestation.

Of course, many authors find a positive side to this lack of nonverbal information, the most obvious being the suppression of stereotypes and social prejudices associated with the user's personal image (Lameiro & Sánchez 1998, Lee 1996). Walker & Bakopoulos (2005) also value positively this "visual silence" that suppresses the social obsession with visual appearance and liberates us from the limitations that it imposes on us. The reader, then, has to hypothesize this absent nonverbal information (Suler 1997c), which often does not match the interpretation that it would have produced in a face-to-face scenario. A quote by a user (in Baldwin 1996: Chapter 3) is illustrative about this point:

> I was challenged in the sense that I was having to interact with people who had no faces, including no smiles or frowns, people without hands and bodies so that I might read their body language [...] I did not know how they looked, how they dressed, nor was I able to hear their distinct southern drawls or haughty yankee accents; I was forced in this manner to be totally objective. They were merely words on the screen.

2.4.2 *Nonverbal communication and simultaneity with verbal language*

An important aspect of text-based chat room language is that verbal and nonverbal communication do not overlap or occur simultaneously, but can only be typed in succession. In face-to-face interactions, the intersection of (non)-intentional nonverbal behaviour and verbal utterances can generate four main prototypical cases (Yus 1997b: 334–335):

1. A non-intentional nonverbal behaviour strengthens the information provided by the verbal utterance.
   [*a person shivers and his hands sweat while saying "I'm nervous"*].
2. A non-intentional nonverbal behaviour contradicts the information provided by the verbal utterance.
   [*a person shivers and his hands sweat while saying "I'm not nervous"*].
3. An intentional nonverbal behaviour strengthens the information provided by the verbal utterance.
   [*a person puts his index finger on his mouth while saying "Be quiet!"*].
4. An intentional nonverbal behaviour contradicts the information provided by the verbal utterance.
   [*a person winks ostensively while saying an ironic remark*].

These possibilities of overlapping are absent in text-based chat rooms because of the succession of typed characters on the screen, a problem shared by all written texts. Hence, it is not correct to claim, as Danet et al. (1998) or Ruedenberg et al. (1994) do, that the interpretation of both sources of information is simultaneous (a kind of *gestalt* following the terminology of Danet et al. ibid.).

2.4.3 *Intentionality in verbal-nonverbal information*

In text-based chat rooms there is no unintentional nonverbal behaviour, that is, information conveyed nonverbally that the users *exude* without consciously intending to communicate it, because all written texts involve a willingness to produce them (Masterson 1996, Lang 1995). By contrast, it is possible to misunderstand the intentional nonverbal behaviour typed by the user. The two main strategies of intentional transference of nonverbal information are the *emoticon* (2.4.4 below) and the *stage direction* (2.4.5 below).

2.4.4 *The emoticon*

This is one of the most famous ways of connoting typed text with visual nonverbal information (also called *smiley*) (see Quinion 1996, Witmer & Katzman 1997). The most typical ones are "happiness" [:-)], "sadness" [:-( ] and "wink" [;-)]. They were initially inherent in chat room communication but have been exported to instant messaging, mobile phone *texting*, *Twitter* and social networking sites, among others. The source of the term, a combination of the words *emotion* and *icon*, is misleading. As Dresner & Herring (2010: 252) correctly clarify, the purpose of emoticons is not only to express the users' emotions, since

> many facial emoticons do not seem to express a single emotion, or indeed any emotion at all. Is a face with the tongue sticking out – for example ;-p – a sign of a specific emotion? Various sources attribute to it the meanings of teasing, flirting, and sarcasm, all of which may be associated with emotional states, but are not emotions per se. Or consider the familiar winking face ;-): Conventionally, it indicates that the writer is joking, but surely jokes are not associated with a single emotive state. People may joke when they are happy or sad.

In some contexts (institutional ones, for instance) the use of emoticons is considered inappropriate, but they are common in many *cyber-media* and signal that the user is aware of the typical strategies for oralization.[10] Nowadays, the production of emoticons has been automatized by the software and even 3D and Flash-animated emoticons populate chat room conversations turning them into a kind of *pastiche* of colourful verbal-visual information. Besides, the current software for chat room management offers the option to generate emoticons by typing a sequence of characters. For example, by typing ":," "-" and ")" the system immediately generates a 3D emoticon of smile.[11]

10. Emoticons are typically informal, so it is not surprising that Derks et al. (2007) concluded that the use of emoticons is much more frequent in environments of camaraderie, friendship or tight-knit groups than in more neutral scenarios such as the workplace.

11. The emoticon was created in 1982 by Scott Fahlman, from Carnegie Mellon University. At that time, Fahlman wrote: "I propose the following character sequence for joke markers: :-). Read it sideways. Actually, it is probably more economical to mark things that are NOT jokes, given current trends. For this use :-(." (quoted in Baron 2003a).

In this book, the term *emoticon* will refer to textual combinations of characters to create iconic images. However, other authors propose a wider semantic field for this term. It is the case of Metz (1994: 41), who proposes four types of emoticon: (1) verbalizing nonverbal behaviours (I will call this *commented stage direction* in 2.4.5 below); (2) writing nonverbal behaviours between asterisks (I will call this *autonomous stage direction* in 2.4.5 below); (3) marking the text with capitalization; and (4) composing images with punctuation marks (the only type that, in my opinion, should be labelled "emoticon").

Furthermore, emoticons are intentional and do not cover the whole range of nonverbal behaviours that people give off or *exude* without intention. In an ordinary situation such as (12), B's nonverbal behaviour – a yawn – can lead to many interpretive possibilities (13a–f) depending on the (in)adequate identification of the underlying intentionality (or lack of it) and on the correct interpretation of this nonverbal behaviour. But in text-based chat rooms only (13a) and (13c) are possible:

(12) [*During A's visit, the host – B – yawns*].

(13) a. B yawns intentionally so that A infers that B wants him to leave; A interprets the ostensive act correctly and decides to leave.
b. B yawns intentionally so that A infers that B wants him to leave; A interprets the ostensive act as a signal that B is tired, that is, A thinks that there is no underlying intention in the yawning.
c. B yawns intentionally so that A infers that B wants him to leave; A correctly interprets the act as ostensive, but chooses a different interpretation, for example that B is trying to communicate that A is very boring.
d. B yawns because she is tired, without any intention in its production; A interprets correctly both the behaviour as unintentional and its interpretation (tiredness).
e. B yawns because she is tired, without any intention in its production; A interprets correctly the behaviour as unintentional but chooses a different interpretation, for example boredom instead of tiredness.
f. B yawns because she is tired, without any intention in its production; A incorrectly interprets the act as ostensive and connotes it with an intentionality that B does not hold.

The origin of these iconic compositions of characters lies in the limitations of traditional punctuation marks that prevent authors from expressing vocal and visual qualities of conversations (Yus 1998g: 119 ff). On the Internet, the plain ASCII text, developed for the *global village* so that all computers with different operating systems could exchange messages, is also a hindrance to users' expressiveness, so they devised new conventions for connoting texts and bringing them closer to their communicative goals.

However, the analysis of emoticons reveals that they are frequently redundant with respect to the typed text that they precede, follow or are inserted within:

(14) a. I am very happy today :-)
b. Yesterday I had a terrible day :-(

Derks et al. (2008) concluded that using an emoticon produces positive judgements among users because it adds a supplementary emphasis compared to the neutral interpretation that the message would convey without it. For instance, the author of a positive message accompanied by the typical "smile emoticon" obtained an interpretation of more happiness with the message.

Additionally, the use of emoticons is reminiscent of the strategy of *lettering* in comics (Gasca & Gubern 1988, Gubern 1992, Yus 1997b: III.1, 2008f). This is a technique for adding emphasis to text in comics balloons by changing, for example, the size and shape of letters, and thus providing readers with a more accurate picture of what the character is feeling.

Although the main function of emoticons is redundancy, sometimes they can alter the meaning of the message, for example reducing its force or even invalidating its propositional content altogether, and on these occasions they play a similar part to nonverbal behaviour in face-to-face interactions. In these oral exchanges, a wink by the speaker can neutralize the effect of the propositional content of the utterance and turn it into an ironic remark, or connote it with politeness (Wilson 1993, Menges 1996). Similarly, an emoticon can soften the meaning of a message and even make it mean the opposite of its literal meaning, as happens in the "wink" that connotes the message (15) with irony:

(15) Only an idiot like you would have done something like that ;-)

Sarcasm would also be conveyed in a similar fashion. The previous example illustrates how an emoticon can aid interlocutors in finding an interpretation that differs from the content of the message completely, which contrasts with Derks et al.'s (2008) opinion that emoticons do not have the strength to turn around the valence of a verbal message. According to Dresner & Herring (2010: 253), these emoticons "seem to have no self-standing content on their own, but rather contribute to – indeed, provide a vital cue as to how to interpret – the linguistic content of messages. When used this way, emoticons seem to be a part of the text, as much as punctuation marks, which can also signal sarcasm." Besides, these authors propose that emoticons have a role beyond redundancy, namely,

> indications of the illocutionary force of the textual utterances that they accompany. As such, they help convey the speech act performed through the production of the utterance. These uses of emoticons neither contribute to the propositional

> content (the locution) of the language used nor are they just an extralinguistic communication channel indicating emotion. Rather, they help convey an important aspect of the linguistic utterance they are attached to: What the user intends by what he or she types. (ibid.: 255–256)

The problem with a non-redundant use of emoticons is that these combinations of characters may become really complex compositions and the reader may not grasp their meaning correctly. Indeed, even in the use of emoticons there is a continuum from the purely arbitrary to the explicitly iconic based on cultural rules of making sense of visual signs. These rules are shared by the community, that is, mutually manifest to all its members. In the case of the emoticon, it demands *emoticonic literacy* from the users beyond the simplest well-known compositions (Reid 1994: 31–32, Watson 1996). Therefore, it is very likely that the authors of the emoticons in (16) will not obtain the desired effect without the aid of the information of the message that precedes them, which *anchors* their meanings, in Barthes' (1977) sense, reducing the range of possible interpretations that these emoticons can convey:

(16) a. You know I don't agree :-e
b. I've recently been ill with a cold :-'I
c. Did you miss me? >:-> (malicious comment)
d. I send you lots of kisses :-X

More examples are found in the following messages (del-Teso-Craviotto 2008: 260): in (17a), the text that follows the two emoticons *anchors* their meanings as "cup" and "rose." In (17b), the user's *nick* helps us to deduce that the emoticon portrays a cowboy winking and his corresponding hat:

(17) a. TNCharmer: c(_) @}}~~~ coffee and roses for the ladies and hello room
b. BGHEARTEDCOWBOY: single m with pic on profile c);o)

Both in redundant emoticons and in cases where emoticons play a more important role, users seem to infer that the emoticon influences the complete message. According to a study by Provine et al. (2007), users tend to process the message as a whole and *then* add the meaning of the emoticon as additional or complementary information. After a classification of emoticons into (a) those which constitute the only content of the message or *naked emoticons*, (b) emoticons that are placed at the beginning or the end of the message, and (c) emoticons that are inside the message, they concluded that the second type is much more frequent. This corroborates a tendency to use emoticons as qualifiers of the whole message.

A possible solution to this inevitable shift of the emoticon to the more arbitrary side of the aforementioned continuum is to find some form of conventionalization, that is, to get all users to know on which occasions all types of emoticon can be used and with what meaning, independently of the message that they qualify. In this sense, there have been multiple attempts to establish a code among Internet users, especially from glossaries on websites, but these were doomed to failure. Faced with the lack of semiotic code for the use of emoticons (beyond the universal "smile" or "wink"), users are forced to introduce their most original iconic compositions only through redundancy towards the verbal content of the message, for fear of being misunderstood:

(18) a. I like chocolate a lot, :-P
b. I have recently changed my hairstyle &:-)
c. I went for a drink last night #-)
d. Now I have my hair parted in the middle {:-)[12]

The lack of a shared code for emoticon use is corroborated by the fact that the same emoticon is defined differently even in the dictionaries and glossaries that compile them. In Belda Medina (2000: 573, 2003b) there is a comparison between the definitions suggested by Silverlink (a), Milner & Burrows (b) and A. Fernández (c):[13]

(19) :-P (a) stick out one's tongue; (b) to talk in a joking mood; (c) to howl.
#-) (a) to be dead; (c) to go partying all night.
:-7 (b) to smile in an unnatural way; (c) to smoke a pipe.
:-3 (a) to eat a lemon; (c) smile of a man with a moustache.

Sometimes, the emoticon may even be the protagonist of all the interaction, and the knowledge and transfer of information through these emoticons provide users with a playful atmosphere and awareness of mutual manifestness of emoticon conventions. The following conversation in (20), quoted in Merchant (2001: 301), is an example in which most of the information communicated is emoticon-centred:

12. Most emoticons are interpreted by turning one's head 90° to the left. However, there are also horizontal emoticons, especially in Asian cultures. In the corpus of *tweets* analysed in Chapter 4, some of these emoticons can also be found, such as [Ò.Ó] and [ò.ó]. Similarly, Baron (2009) comments that American emoticons are read sideways and emphasize the mouth, whereas the Japanese *kaomoji* are read horizontally and focus on the eyes. For example, the typical emoticon for smile, :-), is the *kaomoji* ^--^. Several studies indicate that cultural differences between Japan and the US are reflected in the ways the Japanese interpret these two forms of nonverbal online expression.

13. (a) Silverlink: "Acronym list (including smilies)," (b) A. Milner & T. Burrows (1997): *Internet*. London: Dorling Kindersley; and (c) A. Fernández Conde-Cuadra: "Expresiones en el chat."

(20)

adz46:	hows you
pintsize:	fine thanx u?
adz46:	great
pintsize:	cool wot u up2?
adz46:	not A LOT
pintsize:	wot av u bin up2?
adz46:	Writeing a Macbeth Essay
pintsize:	o gr8 fun!
adz46:	mmmmmmm
adz46:	:-(
pintsize:	:)
adz46:	:-(
pintsize:	cheer up!
adz46:	:-|
pintsize:	Stop it!

2.4.5 *The stage direction*

With this term, I refer to texts that users type in order to describe nonverbal behaviours. It is taken from theatrical terminology, due to the similarities with stage directions that are typical in plays. I distinguish two varieties of stage direction:

1. *Commented stage direction*. In this variety, the text describes, as a whole sentence, the user's nonverbal behaviour:

(21) Tom is laughing out loud.

This is also labelled *emote* in some studies (e.g. Herring forthcoming). Commented stage directions are used not only as verbalizations of nonverbal behaviour, but also with the aim of creating an atmosphere or contextual environment that colours interactions with a higher feeling of realism. For example, in Gelléri (1998) some of these stage directions can be found:

(22)
a. Deadcow waves to everyone
b. neichy1 waves goodnight to jazzzz
c. Ik4u laughs at dinorex
d. frankay is happy now :)

The third person alleviates the monotony of the first person in chat rooms, turning conversations into a kind of theatrical script. This explains why it is not uncommon to find examples of stage directions that refer to the (unseen) environment (Gelléri ibid.):

(23) a. SteveC turns his central heating up, mmmmmm nice
b. DaProphet clears his throat
c. ^Prophet^ is pissed off coz his computer hanged
d. Merlyn was on the phone

These stage directions may comment on events, places or situations that have little to do with the on-going conversation. Sometimes, though, these do reflect moods and feelings that play a part in the interaction (Gelléri ibid.):

(24) a. Kali yawns
b. Graeling sits in the corner and hopes someone will talk to her

In this sense, Cherny (1995b) proposes five types of commented stage direction, exemplified in (25a–e):

(25) a. lynn waves.
b. lynn nods.
c. Mike pastes Tom's lips...
d. lynn packs for the trip.
e. lynn hated the film.

(25a) is a *conventional action* typical of chat rooms, waving upon entering the room. (25b) is a *back channel*, that is, a discursive strategy to make manifest the user's interest in the current conversation. (25c) is a *meta-discourse comment* of a humorous kind about another user's message. (25d) is a prototypical commented stage direction, specifically a *narration* of an ordinary nonverbal activity transcribed as text. Finally, (25e) is an *exposition* of a specific mood or opinion.

As an example of how commented stage directions are used, the following virtual conversation, quoted in (26) (from Menges 1996), combines most of the strategies described in this heading:

(26) [Joyce_] everyone agree the list is good enough to vote on?
[lizzie] no
[Joyce_] please feel free to discuss further guys :)
* Joyce_ wait and watches with interest
[JmpMstr] voting to find the top six???
[dori2] lizzie describing actions is more than just making actions - it is a more detailed thing
[Joyce_] JM: yes, that's the next step
* OldBear wishes there were a chalk board or place we could write each item on yellow stickies and then cluster them into groups. This medium is not very good for developing true group hubris. ;)

[lizzie] dori, convince me
* JmpMstr lines up with OB… I agree
[dori2] well I can say dori2 sits
[lizzie] dori…. you mean… dori sits on the chair hapazardly
[dori2] that is simply using /me command
[dori2] yes - that is closer lizzie
[lizzie] oooooooooooooooo, this group is creating a mind meld… we both knew you were sitting
* lizzie gets spooked
* Joyce_ smiles

It is also convenient to remark that some programs for chat room management, for example the ones analysed in Johnova (2004), include a command called "action" (or "think," depending on the program itself) that one can click on and the message that the user has typed is immediately turned into a commented stage direction. For example, if a user called <stunt> types "throws ice all over the place to freshen up his friends" and then clicks on "action," the program automatically yields the stage direction "*<stunt> throws ice all over the place to freshen up his friends*." In this way, the messages typed by users get mixed up with commented stage directions which, although they have also been typed by these users, exhibit a different format automatically created by the software, as in this example from Herring (2001):

(27) <Dobbs> come on, Danielle!!
<Danielle> No.
<Danielle> You have to SEDUCE me…
*** Action: jazzman reaches out for Danielle's soft hand.
*** Danielle has left channel #netsex
*** Action: Dobbs whispers sweet nothings in Danielle's ear
*** Action: Butthead moves closer to Danielle
<jazzman> danielle's gone dumbass

2. *Autonomous stage direction.* It occurs when the user's nonverbal behaviour is expressed with its closest translation, normally in one or two words, and framed by asterisks that separate it for the verbal content that they accompany. It is defined by Herring (forthcoming) as "predications that can function alone as complete performative utterances." An example is quoted below:

(28) <bull> What you're saying is funny *laugh*

In general, users resort to typical terms that one can find in a dictionary as prototypical of the equivalent nonverbal behaviour. Therefore, they vary inter-culturally depending on the lexical repertoire that languages offer for the description of nonverbal behaviour.

Again, these words between asterisks remind us of theatrical stage directions, and are typically used for saying hello and good-bye in chat rooms (Werry 1996: 60):

(29) a. <ariadnne> A N N E M A R I E!!!! *hugs*
<amya> *hugs* :)
b. <Untio> *kisses*
c. <Lola> I was joking *smiles*

2.5 Oralized written text

2.5.1 *Text-based chat rooms in the oral/written continuum*

The previous analysis has demonstrated that chat room discourse is a hybrid, somewhere between the stability (and often formality) of typed text, on the one hand, and the ephemeral (and often informal) quality of speech, on the other. As Baron (2009: 107) summarizes, although chat room discourse and other forms of computer-mediated text "are technically forms of writing, most varieties of online communication have often been thought of as forms of speech, with creative punctuation and typography substituting for paralinguistic cues (such as volume, proxemics, and facial expression) for expressing emotion." Therefore, many prototypical aspects of oral conversations can also be found in chat room interactions. Voiskounsky (1997) has summarized the oral and written qualities of computer-mediated communication (see Table 5.2) (see also English 1999, Kling 1996b and Kolko & Reid 1998: 213, 220).

Table 5.2 Oral and written aspects of computer-mediated communication

Oral	Written
Informal conversational style (e.g. use of first person, colloquial forms of address, recurrent idioms).	The user controls the composing process and the interlocutor cannot intervene in this process.
Search for textual equivalents to features of conversational interactions.	More complicated syntax than in oral conversations.
Short messages.	In the text, the "addresser user" makes the reason of the message explicit.
Intimate and emotional topic of conversation, which provokes the use of non-standard spellings and the use of symbols that are so typical of virtual conversations.	Possibility of revising the text and correcting errors before sending the message.

The anomalous situation of chat room discourse because of the simultaneity of oral and written features has led some authors to propose a new status for this discourse, a *third element* to be added to the traditional oral/written dichotomy, a hybrid that oscillates between the two extremes. In a similar fashion, I propose the label *oralized written text* for chat room discourse.[14]

Furthermore, Shank (1993) calls chat room discourse *multilogue*. As can be seen in Table 5.3 (Patterson 1996: Chapter 3), this *multilogue* differs from other types of interaction such as monologue, dialogue or discussion.

Table 5.3 *Multilogue* compared to other forms of virtual communication

	Monologue	Dialogue	Discussion	Multilogue
sender	one	one	initially one, maintains control	initially one, no control
receiver	one or more, passive	one, active	one or more, active	one or more, active
channel	FtF, mass media, other mediation	FtF or mediated	FtF or mediated	computer-mediated
examples	lecture, TV, books, radio, mailing list	FtF, letters, telephone, email	FtF, moderated groups	chat rooms, MUDS, news-groups, listserv

2.5.2 *Language games*

In headings 2.3 and 2.4 above I commented upon different strategies for the compensation of the loss of nonverbal features, vocal and visual, in text-based chat rooms. These strategies alter the typed text, and hence there is a presupposition that the interlocutors will choose the intended interpretation despite the increased effort that these strategies might posit. In text-based chat rooms, users make hypotheses about other users' ability to access a context in which the interpretation of the utterance will be relevant in the balance of cognitive effects and mental effort. Sometimes, the deviations from neutral text will demand extra processing effort, but this may be offset with additional interest (effects) from impressions (weak implicatures) such as the feeling of sharing the conventions of oralization, of an increase in sociability, or in humorous effects and playful atmospheres that are often generated in these text-based conversations.

14. Werry (1996) calls it *interactive written text* following Ferrara et al. (1991). In Elmer-Dewitt (1994) it is labelled *written speech*. Young (1994) calls it *writing conversation*. Merchant (2001) and Blanco Rodríguez (2002) prefer the term *written conversation*. Stein (2006) proposes *typewritten conversation*. Borreguero (2002, in López Quero 2010: 174) calls it *written simulations of oral conversations*. Finally, Fraca de Barrera (2007: 30) proposes the alternative term *pluridialogue*.

Among the most significant strategies of text deformation (and signals of oralized written text), the following can be listed:

1. *Orthographic mistakes.* Chat room discourse exhibits an informal style and, due to the pressure to type and send the messages as fast as possible, it abounds in orthographic mistakes. Some are involuntary, produced by errors in pressing the keys, but others are intentionally produced, as part of the language games that this medium favours (Mayans 2002b). Chat room users usually accept these mistakes as an inherent element of text-based interactions, although a user may occasionally complain if another user constantly makes too many mistakes.

2. *Phonetic orthography and eye dialect.* There are several strategies for the oralization of chat room messages. Following the terminological proposal by Androutsopoulos (2000: 521–522), some of them can also be found in chat rooms:

a. *Phonetic spellings.* It is the strategy of reproducing textually the text as it would be pronounced orally (for example, writing "imeil" instead of "e-mail" in Spanish). Chat rooms are prone to this strategy due to their hybrid oral-written quality.

b. *Colloquial spellings.* It refers to a colloquial reduction of words due to their pronunciation in the flow of speech. An example would be to transcribe strong and weak forms, as in (30):

(30) What d'yu wanna do?

c. *Regiolectal spellings.* These are transcriptions of regional variations of a language. It is also called *eye dialect*, an attempt to represent textually the phonetic qualities of a specific dialect, which involves phonetic elisions and text deformation. In (31a) there is an example to convey the typical accent of people living in the countryside, compared to a standard equivalent in (31b):

(31) a. You raaamblers don' realoyse the daaamage 'ee derz, traamplin' everywhoyre, leavin' gates open, with yourn dawgs runnin' oout o'control.
("Farmer Palmer," from *Viz*, quoted in Yus 1995: 56)

b. You ramblers don't realise the damage you do, trampling everywhere, leaving gates open, with your dogs running out of control.

d. *Prosodic spellings.* It refers to the textual transcription of prosodic contours of the voice by resorting to repetition of letters, capitalization, and the creative use of punctuation marks. This strategy is also frequent in instant messaging and *Twitter*, as can be observed from these examples taken from the corpus of *tweets* mentioned in Chapter 4:

(32) a. hasta la nocheeeeee (T4).
[*see you tonight*].

b. @usuario NOOOOO! Me rehuso a creerlo! (T5).
[*@username NO I can't believe it*].

c. Por fin juevesssssssssssssss!!!!!!!!!!!!!!!!! (T9).
[*Glad it's Thursday!*].

e. *Interlingual spellings.* This is a strategy that consists in transferring the phonetic attributes of a word from a foreign language but making it fit the orthographic conventions of the importing language. An example is the placement in Spanish of "e-" before several imported words from English (standard = estándar).

f. *Homophone spellings.* It refers to two kinds of textual alterations that do not correspond to parallel phonetic alterations that justify them:

f.1 *Lexical substitutions.* The strategy consists in writing a word (or part of a word) whose pronunciation is similar or equivalent to the initial word, but is shorter and complies better with speed-obsessed chat room users:

Word used	**Standard equivalent**
every1	everyone
18er	later
c u 18er	see you later

f.2 *Grapheme substitutions.* This strategy aims to replace one grapheme with another. In Spanish it is typical to replace "qu" with "k" (Mayans 2002a):

(33) <DUDU> jorrrrrr?
<DUDU> ke es esto? [*¿qué es esto? what is this?*]
<^XcyOnE^> es la cabeza de KaOs
<karin> diossssssssssss

It is quite clear that, for these chat room users, language is a flexible tool for transcribing speech, and they seem to obtain special pleasure from playing with the possibilities that the keyboard offers, a pleasure that certainly offsets any effort demanded in exchange.

3. *Abbreviations, acronyms, clippings.* Chat rooms are also full of these, turning paragraphs into a kind of hieroglyphic that only those users who master the conventions of these textual strategies can decipher. An example of text deformation is quoted in Belda Medina (2000: 562), which is now compared to its standard version:

Chat room conversation	Standard equivalent
– Hiya, r u new?	– Hello, are you new?
– Hi, no dewd	– Hi, no dude
– OK, a/s/l plz?	– OK, age/sex/location please?
– Lisa 20yo fm LA, u?	– Lisa 20 years old woman Los Angeles, you?
– Brad 22 m NY	– Brad 22 male New York
– kewl, gotta a selfpic for trade?	– Cool, have you got a photo of yourself to exchange?
– Yeap, sure, but what kinda format is urs?	– Yes, sure, but what kind of format is yours?
– Mine is a JPEG but I can't DCC, doesz't work. I can e-mail it though	– Mine is a JPEG but I can't send it via DCC, it doesn't work. I can e-mail it though
– Kewl, send it dewd, you first	– Cool, send it dude, you first
– Okay, g or x?	– Okay, general or x-rated?
– What? sorry, not into that stuff, bye	– What? sorry, not into that stuff, bye
– no prob, C U	– no problem, see you

The fact that only experts in these conventions are capable of understanding this conversation can be, in itself, utterly relevant to the users. The communicative success enlarges the portion of their environments that is mutual. Besides, chat rooms provide users with many impressions, in the shape of weak implicatures, that build up a particular source of satisfaction in these text-based interactions.

Abbreviations are also frequent in interactions. Again, some of them are raised to the status of conventions and enter the code of chat room discourse:

(34) msg (message)
thx (thanks)
tlk (talk)
bs (bullshit)

The problems of comprehension of *oralized written texts* increase with the use of acronyms that require *cyber-literacy*. These are usually integrated inside the messages, as in (35), re-written in (36) as standard English (Bunting 1999):

(35) BTW IMHO you deserve a :-* i'm sure you get plenty ITRW ;-). From my POV, you sound luscious (:-*;-*) and I'd like to spend time with you 24/7.

(36) By the way, in my humble opinion you deserve a kiss. I'm sure you get plenty in the real world. Flirtatious wink. From my point of view, you sound luscious (kiss, kiss) and I'd like to spend time with you, twenty four hours a day, seven days a week.

Another example is found in (37), published in Knight Ridder Newspapers (2002, quoted in Squires 2010: 467):

(37) RU der? GR8. Let's TLK bout all d abbrz & othr shrtcts poppin up mo&mo n MSGS… 'Are you there? great. let's talk about all the abbreviations and other shortcuts popping up more and more in messages. This might look like a word jumble -unless you're younger than 25, in which case you know it's the lingo used by kids to communicate with buddies.

4. *Ellipsis*. When users want to save time when typing their messages within such a dynamic environment, ellipsis is a useful tool. In everyday conversations, people usually leave implicit, non-coded, all the information that they expect that their interlocutors will be able to retrieve from context by themselves (an assumption of mutual manifestness). Needless to say, the more information is left implicit, the greater the role of the addressee and his/her responsibility in the eventual communicative success.

Ellipsis in chat rooms is used with a similar purpose, but time constraints must be added to the reasons for using it. Since chat room messages always contain the *nick* of the user at the beginning, the most frequently elided element is the first-person pronoun, more connoted in English than in Spanish because in the latter it is acceptable to omit this pronoun.

2.6 Attitudes and emotions in chat rooms

Oralized written texts can be a useful means to convey attitudes, feelings and emotions that are hard to code in neutral typed text. In this case, it is important to see to what extent this *text oralization* or *text deformation* (e.g. creative use of punctuation, capitalization and use of emoticons), as will be generically labelled in this chapter, allows "addressee users" to infer the intended interpretation of these attitudes and emotions correctly and also to measure their intensity (that is, to engage in what I call an *ad hoc measurement* of this intensity, see Yus 2005a).

Four hypotheses can be considered:

Hypothesis 1. Text deformation helps readers to identify the propositional attitude that underlies the composition of a message, that is, it has a *procedural*[15] role

15. The term *procedural* comes from the *conceptual / procedural* dichotomy. According to relevance theory, most words encode concepts and possess, therefore, a *conceptual* status. By contrast, some words such as connectives only code inferential "instructions" that aid the hearer in the relevant interpretation of the utterance, assuming a role of facilitator, a *procedural* role (W&S 1993, Blakemore 1987, 1992). For example, the connective "but" codes the instruction that the subsequent text should be interpreted as a contrast to the preceding one. See Yus (1998a: 328–329, 2010a) for a summary of this terminological dichotomy. See also Yus (2000a: thematic section 5.2) for a list of studies that have addressed this relevance-theoretic dichotomy.

facilitating the identification of this attitude. Besides, different amounts of text should be correlated to the intensity of these attitudes.

Hypothesis 2. Text deformation is useful to communicate propositional attitudes, especially when the content of the message is not sufficiently explicit to communicate them adequately. Again, it is also hypothesized that the more textual deformation there is, the higher the intensity of the attitude adopted by the chat room user.

Hypotheses 3 and 4. Users who want to communicate feelings (hypothesis 3) and emotions (hypothesis 4) resort to text deformation to assure a more accurate interpretation, and will engage in as much text deformation as the intensity of the feeling or emotion demands.

To test these hypotheses, 1.700 chat room messages were compiled, mainly from the Spanish chat room portal *Terra* (www.terra.es/chat). Besides, a questionnaire was handed out to students in order to corroborate or refute these hypotheses.[16]

2.6.1 *Hypothesis 1: Ad hoc measurement of procedural content*

This hypothesis tests whether text deformation works as a procedural element that guides readers in their inferential steps towards the interpretation of the subsequent part of the message that follows this deformed text. By doing that, the "addresser users" make sure that their attitudes are correctly identified and, at the same time, in an effort-relieving way in terms of mental effort. Kneepkens & Zwaan (1994: 129) add that "the emotional impression directs the attention of readers and helps them to decide which information is relevant for the situation and must be activated. This role of emotions is especially important when there are few textual and contextual cues, for example, at the beginning of a text." If textual deformation manages to convey a more precise account of the users' underlying attitudes in communicating the adjacent stretch of discourse, then the procedural role of some verbal elements will probably be enhanced.

From the analysis of the corpus, it can be concluded that most users resort to interjections for this procedural role of text deformation, in a similar way to their use in face-to-face interactions. Indeed, Wharton (2000: 194, 2009) concluded

16. The students lived in Alicante (Spain), studied at a High School (58.6%) or at university (41.4%). Most of them knew how to surf the Net and use the computer (77.52%) and how to use the chat room software and engage in text-based interactions (19.52% of them entered chat rooms on a daily basis). The students' ages ranged from 14 to over 21: 14–15 years old (34.32%), 16–17 (24.26%), 18–19 (4.73%), 20–21 (15.97%), and over 21 (20.72%).

that interjections can guide the addressee procedurally in the identification of the attitude regarding the subsequent text and activate several attitudinal concepts. From this point of view, saying "*wow!*" does not code a concept that the hearer interprets as "X is delighted"; rather, "*wow!*" activates a number of attitudinal descriptions that include, for example, happiness, surprise, or excitement. Something similar happens with the interjections found in the corpus of chat room messages, as in these examples:

(38) a. <stefany> jooooooo pronto empieza el curso.
[*blimey! school starts soon*].
b. <Jun-> aggg el brecol ta mu malo [i.e. el brécol está muy malo].
[*aggg broccoli tastes awful*].

In (38), "jooooooo" and "aggg" activate in the reader some attitudinal schemas, into which the text that is typed afterwards is inserted. The reader of (38a) is warned that he/she has to interpret the message under the schema "U regrets that *p*," "U" being "user" and "*p*" being "school starts soon." Similarly, the reader of (38b) is warned that the correct attitudinal schema for interpretation is "U is disgusted by *p*," "U" being again "user" and "*p*" being "broccoli."

Regarding the relationship between the amount of text typed and the intensity in the attitude, the hypothesis was that the informants would find differences between the neutral "jo" and "ag" and the text-deformation connoted interjections in (38). That is, it was predicted that the informants would interpret a supplementary layer of meaning relating intensity and amount of typed text, so that in (38) they would interpret a more intense "U regrets *a lot* that *p*" and "U is *utterly* disgusted by *p*," respectively. That hypothesis was confirmed in the questionnaire. The informants noticed differences between the neutral and the connoted versions, the latter revealing a more intense attitude. However, they found no differences between the version quoted in (38) (jooooooo / aggg) and a version with greater text deformation (jooooooooooo / agggggg). In other words, they identified a difference between the neutral and text-deformed versions but not between two text-deformed versions, as if text deformation worked as an "either-or" strategy: if text deformation is applied, a supplementary layer of intensity is added during interpretation, but no levels of intensity are paired to quantity of text deformation.

2.6.2 *Hypothesis 2: Ad hoc measurement of propositional attitude*

Users of text-based chat rooms (and hearers in face-to-face interactions) insert the proposition expressed by the utterance into a schema that indicates the user's (or speaker's) attitude underlying this utterance. There are several ways in which this attitude can be communicated. These include that-clauses introduced by an attitudinal verb, as in (39a), parenthetical clauses (39b), verbal moods (39c), illocutionary adverbials (39d), and evidentials (39e), among others:

(39) a. I regret that you failed your exam.
b. It's time to go, I guess.
c. Come here right now!
d. Frankly, I am not surprised.
e. No doubt, he is the best candidate for the job.

Most of them are also available to chat room users, as can be observed in (40b–d) for the attitude described in (40a):

(40) a. [*User B is delighted with what user A has just typed*].
b. <B> :-D (an emoticon that replaces a smile).
c. <B> Fantastic!!!!! (writing about the attitude).
d. <B> Wow!!!!! (an interjection halfway between the mere expression of feelings and coded communication)

On other occasions, though, as pointed out in Yus (2005a: 151), the underlying attitude (and also the feelings or emotions attached to the utterance) can only be derived from non-linguistic evidence, by focussing on vocal aspects (e.g. intonation) and visual aspects (e.g. facial gestures) of the specific context in which the act of communication is taking place. For instance, Carston's (2002: 156) subsentential example "Out!" in (41a) can be interpreted as an order, that is, as having the higher-order explicatures in (41b–c), depending on non-linguistically-coded paralinguistic information (and also contextual information about power relations between the interlocutors):

(41) a. Out!
b. The speaker is telling the addressee to get out of the room.
c. It is desirable to the speaker that the addressee get out of the room.

However, in my opinion (41b–c) do not fully reflect the kind of information that the speaker really intends to communicate with (41a). The speaker is probably also interested in conveying his/her feeling(s) when uttering "Out!" (e.g. anger, indignation, irritation), which may be highly relevant for the interlocutors' background knowledge on the kind of relationship which they hold. In face-to-face contexts, the combination of vocal cues (e.g. shouting) and visual cues (e.g. facial gestures) is a good resource for speakers to convey the intended extent of their feelings and emotions while uttering a command, but in virtual environments such as text-based chat rooms, letters and punctuation marks are the only resources available for this task.

In a similar way, the chat room user's propositional attitude is sometimes not "visible" (i.e. coded) and text deformation is useful to compensate for this non-coded information. In the corpus, this use of text deformation is frequently found:

(42) a. <Jun-> no chillesssssssssssssssssss.
[*don't shout!*].

b. <ZePeLiNa> NO XHILLE OTIAAAAAAAAAAA [¡no chilles, hostia!].
[*don't shout, damn it!*].

c. <^LoBe^> maaaaaan echao!!!
[*I have been kicked out (of the chat room)*].

d. <elia> malhe nooooooooo te vayas.
[*malhe don't go*].

e. <rey-swin> [Hazlo] pero yaaaaaaaaaaaaaaaa.
[*(do it) right now*].

Examples like these indicate that capital letters, repetition of letters and punctuation marks can aid in communicating propositional attitudes more accurately. These attitudes have to be interpreted and measured "ad hoc" during the fast exchanges of chat room conversations. A common example is the repetition of the punctuation mark, because it helps to recover both the user's attitude and its intensity. For example, given the first part of message (43a), (43b) would be the proposition obtained after inferential enrichment of its schematic logical form, whereas (43c) would be a higher-level explicature that provides an attitudinal schema favoured by the use of the punctuation mark. Finally, (43d) would be the proposition after *ad hoc measurement* of its intensity has been applied to the repetition of this punctuation mark:

(43) a. <beckham15msn> alguna valenciana?????????????????
[*Any girl from Valencia?*].

b. [*Is there*] any girl from Valencia [*connected to this channel*] [*who would like to chat with me*]?

c. <beckham15msn> is asking if there is a girl from Valencia [*connected to this channel*] [*who would like to chat with him*].

d. <beckham15msn> is asking *with insistence* if there is a girl from Valencia [*connected to this channel*] [*who would like to chat with him*].

The results of the questionnaire corroborated that users did add a supplementary layer of intensity in the attitude associated with the repetition of the punctuation mark. However, they did not, again, relate the amount of typed text to the degree of intensity, and contrary to my predictions.[17] In general, it can be concluded that users find neutral text insufficient to communicate attitudes and emotions and text deformation is a useful discursive tool to ensure their correct interpretation. For example, in (44a) the text itself reveals the higher-level explicature that includes an attitude of begging. On the contrary, (44b) might be interpreted as an informative text if its text deformation did not indicate a negative attitude towards being fat. Finally, in (44c) capitalization is used (although capitalization is not a polite strategy in chat rooms) to communicate how desperate for an answer the user is:

(44) a. <Quesalid> bastaaaaaa… Bastaaaa pofavooo [por favor].
[*stop, stop please*].

b. <saratogo> estoy gorrrrrrrrdddooooooooooooooooooooo [gordo].
[*I am fat*].

c. <rubita69> DEJAMEEEEEEEEE DEJARMMEE VEROSSSSSSSSSS
[*let me see you*].

2.6.3 *Hypothesis 3: Ad hoc measurement of affective attitude (feelings)*

In this case, textual deformation is a consequence of the need to communicate feelings in a more accurate way, given the limited options of language for that task. Again, the prediction was that there would be a coupling between the intensity of the feeling and the amount of text typed. In the analysis of the corpus, it can be observed that some users simply type the utterance that best describes the feeling that they hold, as in (45a), while others use text deformation to connote the feeling with a greater intensity (e.g. 45b), or even use conventions such as "Z" to show boredom, as in (45c):

(45) a. <ferrari> me siento solo.
[*I feel lonely*].

b. <morena> me aburroooooooooooooooooooooooooooo.
[*I'm bored*].

c. <solito_casa> zzzZZZzzzz.

17. An explanation for this may be that my informants were reading the messages. Several studies have concluded that, in general, readers can only infer general categories of feelings and emotions from the texts that they read, and are rather inefficient when it comes to identifying subtle variations of intensity in the same feeling or emotion (see Gygax et al. 2003).

Interjections are also a recurrent means to connote utterances with an affective layer, as in the feeling of surprise communicated in (46) as "U is surprised while typing *p*":

(46) a. <sigma> uy!! ya se a parado esto.
[*hey this (computer) has frozen*].

b. <Malhe> uys esta celosita esta cambiando las buenas costumbres.
[*hey this celosita is changing good habits*].

Other examples are quoted in (47), together with the interpretation of the underlying affective attitude ("U" for user and "M" for message), connoted with higher intensity thanks to the repetition of characters:

(47) a. <xica_gogo> abridme un privado por favor!!!!!!
[*Send me a private message, please*].
[*U is feeling very eager while typing M*].

b. <yo> HOLAAAAA ALGUIEN QUIEREE HABLARR CONNMIGOO QUEEEE YO ACEPTOOOOO A CUALQUIERAAAAA.
[*Hi! Does anybody want to talk with me? I'll accept anyone*].
[*U is feeling incredibly anxious while typing M*].

c. <cufi> t kieroooooooooooooooooooooooooooooooooooo
[te quiero].
[*I love you*].
[*U is feeling absolutely in love while typing M*].

d. <BEFLY> vaya cuerdas vocaleeeeeeeeeeeeeeeeeeeeeeeeeeeeeeee
[*What vocal chords!*].
[*U is feeling utterly amazed while typing M*].

The analysis of the questionnaire also revealed that users interpret greater intensity in the feeling when text deformation is used. Informants were also asked about a possible relationship between the amount of typed text and the intensity of the feeling, that is, whether in (48) the intensity of the feeling when typing "hola" (*hello*) was higher in <sevillana14> (48f) than in <patricia> (48c):

(48) a. <Elitrix> holaa
b. <RuBiOWaPo> HoLaa!!!!
c. <patricia> hola¡¡¡¡¡¡¡¡¡¡¡¡
d. <tere_rubia> ola a todo el mundooooooo.
[*hello everybody*].
e. <chico_20> hhhhhhoooooooooooooolllllllllllllllllaaaaaaaaaaaa
f. <sevillana14> holaaaaaaaaaaaaaaaaaaaaaaaaaaaaaaaaaaaaaa

Again, and against my predictions, the informants found no difference in intensity regarding the amount of typed text. This corroborates the idea, already concluded after previous hypotheses, that users can only distinguish between neutral and textually deformed text. Either a supplementary level of intensity is communicated or no intensity is communicated at all.

2.6.4 *Hypothesis 4: Ad hoc measurement of emotions*

It is very likely that users will type text-deformed utterances in order to communicate their emotions more accurately and connote them with some intensity.

There has been an on-going debate on the differentiation of feelings, attitudes and emotions. On paper, attitudes seem to be cognitively more lasting than emotions, even though one can temporarily hold a certain attitude to what one is saying (propositional attitude) or feeling (affective attitude). As pointed out by Pilkington (2000: 152–153),

> an emotion, such as fear or anger, is a temporary state, a response to some perceived event or state of affairs in the world […] An attitude, such as love or hate, involves the storage of a belief and/or phenomenal state in long-term memory, attached to a conceptual address […] Whereas an emotion is a temporary response to a situation involving the creation of a new desire or the strengthening of an existing desire, an attitude is focussed upon a particular object.

In general, emotions can be defined as "acute, intense, and typically brief psychophysiological changes that result from a response to a meaningful situation in one's environment" (E. Rosenberg 1998, quoted in Kidron & Kuzar 2002: 130). Besides, it is interesting to stress that emotions can be divided into two broad categories: those that are produced unintentionally (as part of *emotional behaviour*) and those that are produced intentionally (belonging to *emotive communication*).[18] The former are usually inter-culturally valid, whereas the latter are influenced by the culture to which the individual belongs, which regulates their (in)adequate uses in interactions.

Generally, text-based chat room users can only communicate emotions intentionally (there is usually no unintended typing on the keyboard), but other users can infer emotions from text beyond the user's control. This fact introduces a danger of misinterpretation of emotions compared to the straightforward access to emotional cues in face-to-face interactions. As Riordan & Kreuz (2010: 167) point out,

18. Caffi & Janney (1994: 328) characterize these types as "a type of spontaneous, unintentional leakage or bursting out of emotion in speech" (*emotional*) and "the intentional strategic signaling of affective information in speech and writing in order to influence partners' interpretations of situations and reach different goals" (*emotive*).

> the ease of interpreting nonverbal cues in FTF [face-to-face] relative to CMC [computer-mediated communication] may be the result of socialization skills accumulated by FTF contact over a lifetime; as CMC is relatively new compared to FTF socializing, it is still a channel in which cue use and interpretation is negotiated between users and learning to encode and interpret emotions using these cues is an ongoing experience.

But emotions abound in this virtual environment where anonymity and lack of physical co-presence allow for emotional exaltation without the control that society usually exerts on individuals in physical settings, something that also applies to social networking sites (see Blincoe 2009, Derks et al. 2008). The analysis of the corpus reveals that text deformation is systematically used for communicating emotions and, perhaps, also their intensity. The most typical emotions are the easiest to type, those that often *alternate* with speech (Poyatos 1975), such as laughter (49) and shouting (50):

(49)	<xicagogo16>	os kiero muchisimo!! jeje
		[I love you a lot!! ha ha].
	<ZePeLiNa>	Quesaild: jajajajajajajja jajajaajajajajjajajaja.
(50)	<cristina_c_a>	aaaaaaaaaaaah.
	<ESIGUAL>	waaaaaaaaaaaaaaaaa.
	<zuMBaooo>	WweeeeeeeeeeeeeeeEEEEEEEEEEEEEEeeeeeeeeeeee.

Emoticons are also useful for conveying emotions in chat rooms, especially broad categories such as happiness [:-)] and sadness [:-( ].[19] It would be useful, in this sense, to test whether the repetition of an element of the emoticon communicates a higher intensity in the emotion or not. One of the questions in the questionnaire dealt with this hypothesis. The informants were asked whether they found differences in emotional intensity between (51a) and (51b) (prototypical emoticon vs. connoted emoticon with repetition of one of its elements). Then, they were asked whether they also found differences between (51b) and (51c) (intensity of the emotion related to the amount of typed characters):

19. Although there is a large variety of emoticons, only a handful of them are frequently used by Internet users, especially the ones communicating happiness [:-)], wink [;-)] and sadness [:-( ] (see Walther & D'Addario 2001). For instance, Schulze's (1999) quantitative analysis revealed that only nine emoticons are extensively used. Taking into account the process of stabilization that certain emoticons have undergone, especially with the online publication of various emoticon glossaries, it can be stated that any repetition of one or more of the typographic signs composing the emoticon should be taken as a connotation of the default type, and hence the other users reading these emoticons will tend to infer an additional layer of meaning associated with this repetition.

(51) a. <Quesalid> :-)
b. <Quesalid> :-)))))
c. <Quesalid> :-))))))))))))

The answers yielded results that are consistent with the informants' answers for previous hypotheses. Again, most of the users found high intensity in the emotion communicated by (51b) compared to the neutral (51a), but found no differences between (51b) and (51c). This is consistent with previous conclusions that corroborate that identifying attitudes, feelings and emotions in chat rooms is an "either-or inference," in the sense that either they are interpreted as neutral (no text deformation) or a supplementary level of intensity is inferred (via text deformation), but beyond this initial dichotomy, there are no fine-grained inferences of intensity related to parallel amounts of typed text.

3. See you on messenger

The title of this heading has turned into a typical phrase among adolescents, who have replaced the physical "see you" (in a bar, square, or park) with an electronic equivalent within *instant messaging* (henceforth IM). Ours is a highly technified society with a convergence (and clash) of increasingly virtualized physical settings for interactions and a growing *physicalization* of virtual environments for interactions. Since communication has evolved towards a total hybridization of physical-virtual scenarios in personal networks (Yus 2007b), IM allows for synchronic conversations that substitute physical interactions or complement them efficiently (see Baron 2008c).

3.1 Instant messaging compared to other forms of interaction on the Net

IM shares attributes with other forms of synchronous virtual interactions. Specifically, it shares with chat rooms many characteristics such as the fact that they are still typically text-based but have the possibility of visual and audio contextualization (web cam, microphone) (see Peter et al. 2007). Besides, both of them generate lists of messages in a central area of the screen while other private conversations are taking place in multiple windows, and they both exhibit similar strategies of oralization and text deformation. However, there are also differences between chat rooms and IM:

1. Chat rooms typically hold one-to-many interactions, even though private conversations can be held in multiple windows. IM is typically a one-to-one interaction, as can be seen in Table 5.4 (adapted from Baron 2008c). But, again, it

is also possible to engage in multiple conversational threads in the same area of the screen.

Table 5.4 Instant messaging compared to other forms of Internet-mediated communication

	Synchronous	Asynchronous
One to one	Instant messaging	E-mail, SMS
One to many	Chat rooms, videoconferencing, blogs	Newsgroups, e-mail distribution lists

2. On most occasions, chat room users simply access a website that contains the interface (for example http://terra.es/chat/). The user can have the window of the chat room minimized without interfering with the user's task at hand (although modern chat rooms produce noises such as telephone calls that do interfere). On the other hand, IM generates pop-up windows on the taskbar that warn users of an in-coming message while, at the same time, a sound is heard. From a pragmatic point of view, this "visual warning" is useful to stress the user's communicative intention underlying his/her informative intention. However, it also produces disturbances in the user's task at hand.

3. Chat rooms are available as soon as the user accesses the website that contains the interface, whereas IM demands the installation of one of specific software programmes that, at least for the time being, are not mutually compatible, in the sense that all users who maintain an IM conversation have to use the same software (but this is not a problem since most users install the same popular IM software).

4. The IM software warns the user when someone from his/her *buddy list* logs onto the system. In one of its most popular versions, a small window opens on the taskbar with a photo of the user and a message communicating that this user has logged onto the IM, and a sound is heard. By contrast, chat rooms do not offer information about contacts, but a text is offered in the main area of the chat room if a user with a *nick* has entered the room.

5. Chat rooms arouse users' interest mainly due to the possibility of meeting strangers that hide behind the *nick* (although it is also used for conversations with friends and acquaintances). IM is typically useful for dialogues with people that users already know and meet in physical settings on a regular basis (for example, users who go to the same college, high school or university, see Quan-Haase 2008). IM is therefore often a complement to face-to-face interactions. As a consequence, although IM interlocutors do use *nicks*, these do not differ substantially from their real names (see Dietrich 2004, Quan-Haase & Collins 2008). As Abril (2006: 39) qualifies, "unlike chat rooms and newsgroups, which

are macro-communities open to anybody who wants to leave a message, with IM each individual chooses how, when and with whom." A female user corroborates this feature when she comments that IM "is like being in a coffee shop to which friends and acquaintances enter to chat for a while [...] and you avoid all the weird people that inhabit chat rooms" (in Abril ibid.).

6. Chat rooms are open to everybody, whereas only IM users who belong to the list of contacts (buddy list) can interact with a specific user.

7. IM offers options for telling other users about the situation of a user, especially the reason for a temporary absence (Cameron & Webster 2005: 86), often with connotations of group membership and personal status within a network of friends and acquaintances.

8. Chat rooms and IM share the users' strategy of text deformation and emoticon use. However, they are more frequent in IM, where conversations are richly decorated with a visual display of coloured emoticons and text.

9. As far as the use of distinctive avatars[20] is concerned, Kang & Yang (2006: 1177) state that IM avatars are more directly related to the user's personality and looks, while the ones used in chat rooms are more imaginative and disconnected from the user. Since IM users meet face-to-face in physical settings, they do not mind using avatars that resemble their physical appearance. By contrast, chat rooms are places for strangers and it comes as no surprise that avatars are so different from their actual users.

IM has also been compared to e-mail in spite of the fact that IM is synchronous and e-mail is asynchronous. Besides, the latter has a more "formal" connotation for adolescents, who use it only for contacting teachers and parents, whereas IM is more informal and relaxed (Grinter & Palen 2002). But despite the obvious differences, there is a similarity between them in terms of how long users have to wait for a reply (i.e. the *chronemics* of IM communication). Indeed, although there can be intense IM interactions composed of chained messages and multiple conversational threads, part of the users' knowledge of IM conventions includes the expectation that the interlocutors may take some time to answer a message, just as in e-mail communication (Nardi et al. 2000). Nevertheless, within IM interactions misunderstandings are frequent due to the differing conceptualizations of how long it is advisable and acceptable to wait until one's interlocutor replies, with negative effects on the regulation of IM turn-taking (Voida et al. 2002: 191).

Finally, IM has been compared to SMS *texting* (Ling & Baron 2007, Baron forthcoming). In principle, they seem to be difficult to compare, because IM is

20. Normally bi-dimensional avatars that cannot be altered. Later in this chapter I will differentiate between these bi-dimensional unchangeable avatars, which I will call *graphic avatars*, and 3D avatars that can move, express feelings, etc., labelled *corporeal avatars*.

normally performed between people located at specific places, in a synchronous way and typed from computer keyboards (although there are increasingly popular IM applications for smart phones), and *texting* is often performed by people located anywhere, asynchronously and typed from the small mobile phone keyboard. From a linguistic point of view, SMS messages tend to be longer than their IM counterparts, probably due to the habit of IM users to fragment messages into several micro-texts so as to maintain their "presence" in the IM interactive environment (more on this below). Besides, SMS messages usually contain more abbreviations due to the 160-character limit that can be typed in one single message.

3.2 Why use instant messaging?

IM is widely used among adolescents as a "natural" form of interaction on the Net and its use decreases at an older age. Therefore, it seems pertinent to determine the benefit IM offers adolescents to justify the massive use of this *cyber-medium*. Within the framework of this book, it is important to determine what expectations of relevance are satisfied with the use of IM. And both the expectations and the use of IM are influenced by the constant evolution of *cyber-media*, which affects not only the kind of use that is made of this medium, but also the assessment and resulting relevance of IM exchanges in ordinary situations. In this sense, Baron (forthcoming) summarizes some of the current characteristics of IM that can be considered evolutions from its initial design and applicability: (1) many IM conversations are conducted asynchronously (rendering them not "instant"), (2) IM now supports multi-person chats, (3) users can now be logged on to IM but "lurk," rendering them invisible to members of their buddy list, (4) some IM software offers voice and video options besides typed messaging, (5) IM is incorporated into other computer platforms, especially commercial web sites offering clients opportunities for live "chatting" with customer-service representatives, and (6) IM can now be done on mobile phones.[21]

In my opinion, IM offers adolescents a wide range of relevance-generating attributes, several options for personal reward both from an individual point of view (achievement of communicative purposes) and from a collective point of view (feelings of group or network membership, and satisfaction at sharing a virtual setting). But several reasons can also be found beyond the age constraint. For example, the IM interface is a user-friendly environment that offers immediate

21. IM is also one of the synchronous options for avatar-mediated conversations in 3D virtual worlds (see 4 below). It has also been incorporated to SNSs such as *Facebook* and *Tuenti*. Although SNS software labels this service as "chat," in reality it is an IM application for one-to-one conversations, even if many of them can be carried out in parallel.

reward for users who seek synchronous interactions on the Internet, and the effort associated with using the interface (which in previous chapters of the book has been signalled as a potential source of alterations of relevance) is reduced significantly (Chung & Nam 2007: 227). According to Grinter & Palen (2002) users' needs for interaction can be summarized as follows: (a) *needs for socializing* (informal conversations to spend some time with friends, with no pre-arranged topic), (b) *for event planning* (to arrange meetings, for example to go to the cinema), and (c) *for schoolwork collaboration* (to clarify problems with homework, for instance). Quan-Haase (2008: 109) adds a fourth need: to be able to engage in multiple one-to-one conversations simultaneously, an aspect that I prefer to address under the individual / group interface in the next heading.

3.3 The individual versus the group

Cognitive pragmatics is especially interested in IM as a tool for interactions that satisfies communicative needs and allows the possibility to draw relevant conclusions through inferential strategies that turn the schematic logical forms of messages into fully contextualized and meaningful interpretations. These interpretations may be relevant both in a purely informative sense and also as a source of phatic socialization, together with the reinforcement of group membership (see Boneva et al. 2006, Flanagin 2005). The individual / group dichotomy that IM fosters is one of the key elements that, in my opinion, explains why IM is massively used by adolescents and youngsters as part of their process of individual and social identity-shaping, and is not so widely used by adults. Another reason might be that the benefit that adolescents obtain from IM offsets the drawbacks involved in using this *cyber-medium*. By contrast, adults find it too annoying and distracting as it interferes with other tasks (Birnholtz 2010).

From an individual point of view, IM is a tool for fast synchronous communication that satisfies specific expectations of relevance with little mental effort in exchange, and with greater emphasis on interactions between users who already know one another in physical settings.[22] Besides, the content exchanged through IM, about apparently irrelevant topics, favours phatic strategies. These "useless" topics have, nevertheless, an impact on users' identities and social awareness. Therefore, if we establish a scale of communicative intentions, we will find that on IM very few messages are intended to satisfy individual needs; most of them

22. This function of "complement" that IM plays with respect to physical interactions makes it difficult to draw a clear dividing line between physical and virtual IM interactions, and it is also difficult to conclude which kind of interaction is more important to users in terms of intensity and impact (see Bryant et al. 2006: 586).

possess a connotation of satisfaction of collective communicative[23] intentions (a sort of *we-intention*), whose fulfilment demands the participation and cooperation of all the users who are synchronously logged onto the IM system at a specific moment.

From a group or social point of view, we can find many instances of IM interactions of a phatic kind, filled with (apparently) irrelevant utterances in a purely informative sense. But they do provide relevance in making mutually manifest assumptions such as awareness of co-presence inside the group or network of friends who are synchronously inter-connected, as well as relevance in the mutual manifestness of being present in the conversation, even if not actively participating. In IM conversations among young users, there is an obsession with demonstrating that the user is part of the interaction, part of the synchronous collectivity. A female user's comment (in Lewis & Fabos 2005: 487) is illustrative of this: "if I don't get on, like if it's broken, like if the Internet's not working, I'll, I'm like ahhhh! So I'll call my friends, and I'll be like "Who's on? What are you talking about?!… I'll be like, Get on my name and pretend you're me for a little bit!" This obsession with group recognition partly explains the habit of sending utterances as several chunks of text in successive messages (Baron 2010a). With this strategy, the user maintains other users' attention and awareness of his/her presence in the on-going interaction. This is why Baron (2010b) proposes a specific terminology that I think is worth quoting and in which there is no obligatory pairing of the utterance and the typed message:

1. *Transmission Unit.* An instant message that has been sent:

   (52) Tom: how are you doing, mate?

2. *Utterance.* A sentence or sentence fragment in IM:

   (53) a. Susan: I've just returned from a restaurant! [sentence].
   b. John: Come to think of it… [sentence fragment].

3. *Sequence.* One or more IM transmissions sent sequentially by the same person:

   (54) Mark: hi man!
   Mark: what are you up to?
   [this sequence equals two IM transmission units].

---

23. From a philosophical point of view, collective intentions are considered intentions to participate in a group to carry out an activity in which the participants feel that they are members of the group (Cheung et al. 2007). IM clearly fits this kind of satisfaction of collective intentions from the involvement of the users in the positive and effective development of virtual interactions.

4. *Closing*. A series of transmissions (between IM partners) at the end of an IM conversation, beginning with one party initiating closure and ending with termination of the IM connection:

(55) Sam: Hey! gotta go [first symptom that Sam is willing to terminate the conversation].
[...] [subsequent exchanged messages].
Sam: I'm off to work! [last transmission in the conversation].

5. *Utterance chunking*. Breaking a single IM utterance ("sentence") into two or more transmissions:

(56) Joan: that must feel nice
Joan: to be in love
Joan: in Spring

6. *Utterance break pair*. Two sequential transmissions that are grammatically part of the same utterance:

(57) Ally: what are you bringing to the dorm party
Ally: on Saturday?

It should be noted that several aspects of IM that have been addressed as mutually independent share, in reality, some attributes concerning their social connotation. I am referring to (1) *multi-tasking*, which is more frequent in IM than in other forms of Internet-mediated interactions; (2) *multi-windows*; (3) *buddy list management*; and (4) strategic use of automatic *away messages*, which are often personalized. I shall briefly comment on them below.

1. *Multi-tasking* has already been mentioned in this book. It refers to the simultaneous engagement in different activities, normally with parallel computer applications. It is typical of adolescents and very frequent in IM interactions, since young users frequently have several windows open for one-to-one IM conversations, plus the Internet browser and the word processor for doing homework.[24] An explanation of this high frequency of IM-related multi-tasking may

24. Nowadays there seems to be a shift in multi-tasking. Instead of using different programs for parallel tasks (e.g. word processor, Internet browser, IM software...), the trend now is to use the same portal for all kinds of multi-tasking activities on the Net. For example, as commented in Aldama (2011:6), in the Chinese social networking site QQ, which competes with *Facebook* in the number of registered users, people can use e-mail (*QQMail*) and a virtual hard disk (*Wangluo Yingpan*), update a blog (*QQZone*) or a microblog (*Tencent Weibo*), download music and ring tones (*QQYinyue*), buy goods online (*Paipai*) and play online games (*QQYouxi*). Users can also buy plane tickets, look for romance (*QQTongchang*) and even look after a virtual pet (*QQChoungwu*).

lie in adolescents' need of mutual manifestness of virtual co-presence of friends during an IM session (a sort of *ambient awareness*, as Thompson 2008 would call it). This need of mutuality prevents users from leaving the IM session, and forces them to reply to messages even though they are engaged in a different task (these messages are, predictably, short and schematic, worthwhile only for maintaining mutuality of co-presence).

Multi-tasking (and parallel multi-windows) may retard the user's replies to IM messages, but this time a gap between transmission and reply is accepted by the users, who do not expect immediate feedback and assume that the other user is probably doing something else while logged onto an IM session. This is what Baron (2008c) calls *language under the radar*. In other words, instead of devoting most of our cognitive resources to following and participating in IM interactions (as face-to-face conversations demand), IM is becoming something that is followed as a background to other activities within multi-tasking, just as background music might be playing while we are engaged in a different task, and the IM "volume" can be lowered or increased depending on how involved the user is in the current IM session. This idea was corroborated in another piece of research by Baron (forthcoming), where 98% of 158 informants (half male, half female) were busy doing at least one different task (on the computer or away from it) while participating in an IM interaction.

2. *Multi-windows* is clearly related to multi-tasking. It takes place when the user has opened several windows for one-to-one IM conversations, and the user tries to monitor and follow in a relevant manner several conversational threads with different people and about different topics simultaneously, despite the effort-producing challenge that this involves. This challenging activity refers again to the social need to make it clear to other users (i.e. to obtain a mutual manifestness) that the user is connected and able to sustain several interactions as a signal of sociability and as a source of prestige for other users. IM can indeed be a source of social positioning within the network of friends and acquaintances. Schwarz (2011:77) quotes the following illustrative example:

> A 16-year-old blogger published an IM conversation with a boy she didn't know who tried to lead the conversation to sex, while she played dumb, wittily using puns and double-meanings she found in his formulations to respond to all of his questions with innocent, non-sexual answers. Sharing the evidence of her victory, the girl won her friends' esteem.

A third of the informants in Boneva et al. (2006) commented that one of the strong features of IM is, precisely, the possibility of carrying on multiple conversations simultaneously, what Garrett & Danziger (2007) call *polychronic communication*. This "strong feature" certainly demands extra cognitive resources and supplementary mental effort to maintain "interactive congruency" throughout all the conversations in these windows. It comes as no surprise that the adolescents interviewed in Grinter & Palen (2002) admitted that there is a limit to the number of opened windows beyond which it is impossible to maintain congruency or coherence. As a general rule, the average number of windows opened simultaneously is four, which users arrange in parallel on the computer screen as a mosaic of windows. The most typical strategy to maintain congruency and coherence in these multiple conversations is to follow with greater interest one or two of them and simply follow "in the background" the other conversations (see Lewis & Fabos 2005: 218–219 for an example).[25]

Hence, many of the IM exchanges are basically to say hello and carry a phatic connotation, i.e. they are short messages that stress a social connection and co-presence in the virtual environment of IM, and they are useful for social grooming beyond their informativeness. In this sense, the other users assume that the user is surely participating in several conversations simultaneously and that they will have to fight for his/her attention by offering a presumption of relevant inferential outcomes.

3. The *buddy list* is the exponent of the contacts on the screen that belong to the user's personal networks. Therefore it is vitally important for the adolescent user. These friends and acquaintances are grouped under categories that reflect, with greater or lesser accuracy, the mental picture that the user has of his/her personal networks.

4. Finally, *away messages* are often personalized by users, and they regulate the degree of social involvement that they desire within IM interactions, even when the user is not logged onto the system. It is a tool for the management of interactive availability in a type of interaction where, as Bays (2010: 43) puts it, "everything from the onscreen activity with its colours and mosaic of windows to the physical environment of the user who may be listening to music, talking on the phone or engaged in exogenous conversations are also important to forming the whole IM experience." When the user personalizes messages, very often they aim at relevance with them, trying to call other users' attention, for example with humorous quotes (Nastri et al. 2006: 1027), or explaining in detail the reasons for

25. Emoticons can also be used in isolation for the purpose of manifesting presence in other users' windows while focussing more on a particular window. As Bays (2010: 57) points out, emoticons can "show continual presence in front of the screen and in a particular conversation, the participant can simply send a smiley as his turn, as a kind of conversation filler. This accords him time to scroll back up the conversation window to see what the topic was and how it evolved while he was away in order to answer the adjacent pair or contribute a relevant message to the topic. The ambiguity of its semantic meaning allows the smiley to be relevant in many situations and to retain the general tone of the conversation."

being absent. For example, one of the away messages quoted in Grinter & Palen (2002) reads "I'm currently removing all dirt, grime and other dead biological matter from my body. I can be found in the nearest decontamination center."

3.4 Oralized written text in instant messaging

The texts typed on IM exhibit similar strategies of oralization and text deformation that have been studied in the section devoted to chat rooms (see Varnhagen et al. 2009, Baron 2010a, 2010b, Bays 2010, Herring forthcoming). The evolution of chat rooms into a more contextualized *cyber-medium* incorporating web cam and sound is also valid for IM. The messages exchanged through IM exhibit oral properties, with short utterances and dynamic interactions, typical of ephemeral oral dialogues. But users also value the typical properties of written communication such as the lack of a need for immediate feedback. The informants' opinions quoted in Voida et al. (2002: 188) are enlightening, because they describe IM as

> being nearly synchronous but able to be attended to when opportune. The former characteristic is shared with most verbal communication; the latter, with most written communication. Implied in the interviews of our participants is that instant messaging is valued because of the unique balance it holds in affordances between the conventions of verbal and written communication.

Curiously, users also make typical gestures of face-to-face interactions while typing their IM messages, even though their interlocutors cannot see them most of the time (see Marcoccia et al. 2008). Bays (2010: 46) draws similar conclusions from the analysis of IM users' behaviour:

> When the new message appears, there is an immediate often physical reaction. Among others, we found laughs, shaking heads, pointing fingers at the screen, having a dumbfounded expression and moving lips to read the new message. Generally, these are the same physical attitudes that can be found in other conversational settings from face-to-face interaction to talking on the telephone, whereas this behaviour is rare (at most) for email or reading a blog.

Emoticons seem to be more abundant in IM than in chat rooms. Xu et al. (2007) have made an exhaustive analysis of emoticon use in IM. Initially, they proposed three basic uses: to accentuate or emphasize the meaning of a message, to convey the user's mood or impressions, and to enrich a verbal utterance with visual information. They then concluded that the greater or lesser frequency of emoticons in IM seems to be influenced by three factors: the degree of (in)formality of the conversation, the relationship that both interlocutors hold, and the personality of the interlocutor. These are commented upon below.

1. The degree of formality is usually established according to two main types of interaction: *task-oriented* and *socio-emotional*. In the former there is a clear objective and the exchange of information is prominent. In the latter, by contrast, the interaction is informal, of a phatic quality, with no predetermined topic, and dominated by the expression of feelings and emotions. It is not surprising that using emoticons in task-oriented communication is not considered appropriate and is seen as a source of unnecessary distraction, whereas in socio-emotional conversations their use is not only predictable, but expected, as happens in most informal conversations among adolescents.

2. There are several levels of relationships between users within IM. In the case of intimate friends, it is more likely that the users will be willing to express feelings and emotions while typing their messages, and emoticons are one of the resources at hand to communicate them more efficiently. The opposite occurs when the conversation takes place between acquaintances.

3. As far as the user's personality is concerned, the use of emoticons is acceptable or inappropriate depending on the interlocutor's personality, since some users do not like finding emoticons in the messages that they read.

4. Chatting in 3D: Advances, avatars and *Second Life*

Chat rooms and IM have evolved enormously in the last few years, even though text-based interactions are still very frequent. In fact, the chat room and IM interfaces have not changed dramatically over the years. In chat room interfaces there is still a wide central area for open messages, a list of *nicks* that one can click on for private conversations and a space for typing one's own messages. When typing, the interface now gives users options for fonts such as bold, italics and colour, and also pre-designed 3D or animated emoticons that were not available in the 90s, when chat rooms first became popular. Furthermore, several authors have proposed interfaces for a better contextualization of conversations (see Yus 2003f). Some of them are briefly reviewed below.

1. Viegas and Donath (Donath 1996, Viegas & Donath 1999, Donath et al. 1999) proposed an interface for chat rooms called *Chat Circles*, in which there is only one room for interactions where several circles represent the users who are logged onto the system. The user can move the circle on the screen, and these circles change size to accommodate the text typed inside. There is also a possibility of engaging in private conversations, since users can only chat with other users whose circles are close enough to theirs.

2. *BodyChat* (Vilhjálmsson & Cassell 1998) is another proposal of an interface. It uses anthropomorphic figures, known as *avatars*, to construct a more

contextualized virtual conversation (avatars are essential in interactions within virtual worlds such as *Second Life*, see below). Among the options for expressiveness, these figures can visually manifest that they are paying attention and also blink their eyes. The problem is that the design of avatars is very rudimentary; these avatars cannot reproduce even a minimum of gestures to the user's satisfaction, and hence they distract rather than aid in virtual conversations.

3. *Chatscape* (Lee 2001) offers a low-quality graphical environment that attempts to provide more contextualization in virtual interactions. Users' messages appear on the screen as comic-style balloons that arise from polygons whose shape can be varied in certain ways.

4. *The Palace* is a 2D chat room where users have to choose the bi-dimensional avatar (which I will call *graphic avatar* below) that best suits their personality. After making a choice, the user enters a room where text-based conversations are taking place. There are some options for personalization of the avatar. Besides, users can make their avatars "jump" around the room and get closer to other avatars, even though closeness to an avatar is no pre-condition for starting a conversation. The biggest limitation of this chat room is the impossibility of conveying nonverbal information with the fixed avatar.

Besides *The Palace*, other graphic chat rooms offer similar options for image-supported text-based conversations: (a) *The Manor* (www.madwolfsw.com/) offers several advances, for example the possibility of changing clothes and the availability of 15 animations. (b) *TowerChat* (www.towerchat.com/) is also a 2D environment, but with an original "bird's-eye view" over the scene, giving users the feeling of a 3D scenario. It is divided into eight spaces or "towers" (politics, music, love, and others). One can choose a male or a female avatar. (c) *Voodoo Chat* (www.voodoochat.com/) is similar to *The Palace* but requires the installation of special software on the computer. (d) *VPchat* (www.vpchat.com/) is similar to *Voodoo Chat*. It contains rooms with avatars that interact with one another through text, gesture and voice. (e) Humphrey (2009) proposes a version of chat room mediated by what she calls "masks," that is, avatars that users create in order to interact without the danger of visual exposure and also to play with a multiplicity of online identities. In her study, she concludes that through these masks some aspects of the users' personalities are inevitably distinguishable. (f) *V-Chat* (Smith, Farnham & Drucker 2000) is a program in which rooms have a 3D appearance and users (up to 25 simultaneously) type their messages. As in *The Palace*, users choose a 2D avatar. These may be chosen from a gallery of pre-determined avatars or they can be created by the users themselves. There are seven possible nonverbal behaviours: anger, sexual innuendo, sadness, gesture of puzzlement, silly face, smile and wave. All participants can read what other users have typed, but there is an option that allows a user to send personal "whispers."

5. *Comic Chat* (Kurlander et al. 1996), as its name indicates, is a program that automatically generates panels that resemble the ones in comic books, and inside which the users are visually represented, as well as their messages in the form of comic-style balloons (the users have previously selected a character). The problem with this kind of chat room is that the characters suffer from a lack of expressiveness, just as in most comics.

6. *Coterie* (Spiegel 2001) is limited from a contextual point of view. It is basically an accumulation of oval figures that represent the users who are logged onto the system and interacting. Their colour and proximity to other figures indicate visually who is interacting with whom and with what intensity.

7. Smith, Cadiz & Burkhalter (2000) propose an interface called *Threaded Chats*. It is a visual arrangement of "trees" of conversational threads that help users identify and follow them. The software automatically links a message to the one it refers to and inside the overall thread of the conversation.

8. Ryu (2008) proposes the integration, in the same IM window, of both the text that the user is typing and a three-dimensional avatar with the ability to convey a range of emotions. The author starts with an acknowledgment that users get too distracted if the avatar is fully animated, and that distraction worsens if the avatar and the text share the same space in the window. Therefore, the avatar should be as "non-intrusive" as possible. But, on the other hand, placing the avatar outside the IM window would entail losing some of the user's attention towards this avatar. This is why the avatar appears as a faint image and the user's text appears superimposed on this image. The avatar can communicate a number of emotions, which can be personalized to a certain extent.

9. Fabri et al. (2005) propose a software for IM interactions called *Virtual Messenger*. It adds an animated face to the typical window for typing text. This face is capable of communicating several emotions (it contains animations for the eyes, the eyebrows, the cheeks, the mouth, and the whole head). As a result, six universal facial expressions can be conveyed: joy, surprise, anger, fear, sadness and worry. One of the conclusions when testing this interface was that the "addressee users" tended to imitate the gestures produced by the animated face, a kind of *avatar empathy* that reproduces a typical quality of human cognition, since there is a biological and cognitive component in imitation, to the extent that we learn to behave in the world by imitating what others do. And we possess a number of "mirror neurons" that have evolved specifically to check what others are doing and that are connected directly to other areas of the brain in charge of movement and comprehension of the outside world. One limitation of this interface was that users have to produce nonverbal behaviour intentionally, whereas daily face-to-face interactions are full of *exuded* information that "leaks" from the person beyond his/her conscious control.

But undoubtedly, the most impressive evolution in virtual conversations is the development of 3D virtual worlds such as *Second Life* (www.secondlife.com, henceforth SL), which possesses interesting attributes for a pragmatic analysis of Internet-mediated communication.[26]

SL is a 3D virtual world where users, by means of three-dimensional *alter egos* or *avatars*, interact with other users-avatars. It is much more than a simple environment for interactions, though. In SL many parameters of physical scenarios are reproduced, including bank transfers, and the purchase of land, buildings, or clothes, using virtual money (later turned into real money). Within SL, it is the creative, interactive and social activity of users that qualifies SL as a "world" where many real-life parameters are reproduced.[27] Therefore, these users are *prosumers*, devoting as much time to consuming information and virtual goods as to producing them (D. E. Jones 2005). Some examples of how blurred the distinction between physical and virtual lives has become in these virtual worlds are quoted below:

> A 17-year-old Dutch teenager was arrested this week on suspicion of stealing furniture worth £2,800 from a hotel room. Four other teenagers were also questioned about the offence. It is believed they moved the stolen furniture into their own hotel rooms. Such a minor incident might not have merited a paragraph in the local paper had it not been for one extraordinary detail of the case: the crime happened not in real life but in a "virtual" hotel in the three-dimensional world Habbo Hotel, a children's game that only exists on the internet. (Keegan 2007: 16)

26. Although the analysis in this section will focus on *Second Life* (SL), there are other similar virtual worlds with similar options for the personalization of avatars, interactions and socialization. This is the case of *Blue Mars* (www.bluemarsonline.com/), *OpenLife* (http://openlifegrid.com/), *Inwordlz* (http://inworldz.com/), *Kaneva* (www.kaneva.com/), *ActiveWorlds* (www.activeworlds.com/), *Entropia Universe* (www.entropiauniverse.com/), *Twinity* (www.twinity.com), *Evolver* (http://evolver.com/), *Worlds.com* (http://worlds.com/), the Habbo hotel (www.habbo.es/), *Cybertown* (www.cybertown.com), *Dubit* (www.dubitchat.com/), *Moove* (www.moove.com/), *The Sims* (http://thesims.ea.com/), *Sora City* (www.soracity.com/) and *There* (www.there.com, now closed).

27. An extreme case of his reproduction of real-life attributes can be found in the film *Avatar* (James Cameron 2009), in which the protagonist (disabled in the physical world) literally *fuses* with the avatar that allows him to interact and move around the incredible Pandora jungle (another physical world in a distant planet). While the protagonist is locked inside a chamber, he lives an autonomous life inside the body of the avatar, and only when this intimate connection is interrupted by someone pressing a button outside the chamber (the avatar then collapses) does the protagonist return to his normal physical life as disabled. It comes as no surprise that, at a certain stage during the film, he comments that "everything is upside down," that his life in Pandora is what seems "real" to him, and that his normal life is strange to him.

> A British couple who got married after meeting in Second Life are divorcing after the wife caught her husband chatting up another woman in the virtual world. (Keegan 2008: 28)

> A loft in New York City. The singer Regina Spektor is performing songs from her new album. People wander in, sit down and discuss the music. Everything seems normal. [...] The loft is a 3D computer animation [...] it exists only on the internet. The audience is made up of virtual representations of real people. The real people sit at their computer screens around the world, living their lives through avatars, the characters that appear on the screen. Regina Spektor and her music are real people selling themselves in a virtual world. (D. Smith 2006: 13)

In general, SL offers an interesting intersection, imbrication and hybridization between physical and virtual life. We are offered a chance to escape from our boring ordinary lives and explore new identities. Nevertheless, as Boellstorff (2008: 120–121) stresses, we cannot easily escape from our physical lives. Actually, SL interactions frequently mould and define accurately the attributes that individual users already have in physical scenarios, rather than providing alternative lives or identities that do not overlap with their physical identities.

4.1 Terminological explanation

Virtual world is a label that I apply to a three-dimensional space such as SL, in which human figures or *avatars* exhibit a great capacity for nonverbal behaviour and interact with one another. As such, it is different from other similar Internet-mediated environments and thus a terminological explanation is required.

According to Bainbridge (2007), a virtual world is a simulated environment on the Internet that emulates the real world and whose inhabitants interact with avatars." Bartle (2010: 24) defines it as "an automated, shared, persistent environment with and through which people can interact in real time by means of a virtual self." Book (2004) underlines attributes such as a shared space, a graphic interface, immediacy, interactivity, persistence, socialization and a tendency to communal bonding. Besides, Hua & Haughton (2009: 889) stress the visual experience that virtual worlds offer users. SL possesses these features, and hence *virtual world* is an appropriate label for it. Another term, *virtual environment*, is similar but broader, since it also includes interactions between 2D avatars. And the term *collaborative virtual environment* is even broader, defined as "a digital system that allows geographically separated individuals to interact via networking technology" (Yee et al. 2009: 286), where avatar-mediated interactions are only part of the opportunities for interaction that the system provides.

SL and other virtual worlds are sometimes included in the broad category of *online games* and, specifically in the category of *massively multiplayer online role-playing games*, typically abbreviated as MMORPG, where the famous *World of Warcraft* is a paradigmatic example. If we compare it to SL, there are many analogies: in both there are 3D avatars interacting with one another and controlled by users, who display different nonverbal behaviours. However, in SL there is no pre-determined goal that, once achieved, signals the end of the game. Besides, SL recreates ordinary activities that cannot be labelled "playful" even if they are sometimes highly creative. In addition, many activities within SL resemble the ones performed in physical scenarios (Fetscherin & Lattermann 2007: 4).

It is also important to distinguish between *avatars*, created by users for their interactive goals, and *computer agents*, anthropomorphic visualizations of computer applications that are, therefore, of little interest for a pragmatic study of user-to-user Internet-mediated communication.[28] In any case, in the same environment there is very often a mixture of avatars (controlled by users) and agents (controlled by the computer system), as happens with many online games (Fox & Bailenson 2009: 148). In this case, these agents are usually labelled *embodied agents*.

Lastly, it is necessary to make a differentiation between types of avatars. Some studies do not differentiate between bi-dimensional avatars, with no mobility or capacity to generate nonverbal behaviours, and three-dimensional avatars that are fully animated. In this book I propose a distinction between *graphic avatars*, bi-dimensional and fixed, and *corporeal avatars*, three-dimensional and fully animated (the ones that we can find in SL). The definition of avatar by Bailenson & Blascovich (2004: 64) fits the kind of avatar that is interesting for pragmatics, the corporeal one: "a perceptible digital representation whose behaviors reflect those executed, typically in real time, by a specific human being." By contrast, definitions such as "general graphic representations that are personified by means of computer technology" or "graphic icons representing users through various forms," quoted in Vicdan & Ulusoy (2008), refer to graphic avatars. Although there is research on users' reactions and inferences when faced with these graphic avatars, it is more interesting for pragmatics (and *cyberpragmatics*) to study fully animated corporeal avatars such as the ones that interact in SL.

28. A possible interest for pragmatics would be, perhaps, to analyse the inferential steps that lead to an interpretation of the avatar created and controlled by a user and compare them to the inferential steps for the interpretation of the avatar-shaped computer agent. Nowak (2004), for instance, concluded in her study that there are no great differences between users' reactions to agents and their reactions to avatars, even in terms of credibility. Similar conclusions were drawn by von der Pütten et al. (2010).

At this point it also convenient to distinguish, as Boellstorff (2008: 133) does, between *primary avatar* and *alternative avatars*. The former is the one that the user chooses when he/she first joins SL. This avatar usually reproduces, with greater or lesser fidelity, the physical shape of the user. By contrast, subsequent *alternative avatars* are much more creative and even less anthropomorphic. These complement the *primary avatar*. Of course, this is not a norm that is invariably followed. For instance, a teacher at the University of Leeds comments:

> My original avatar is a furry and as different from me as possible. I deliberately went in [SL] as a fantastic creature unlike myself. I use this avatar for creative and social activities. My alt [alternative avatar] was created more recently so that I can participate in educational activities as myself. She looks a bit like me, though she is pale green, younger, and more glamorous. (pers. comm., November 24th, 2009)

4.2 Identity

The analysis of virtual worlds such as SL entails, again, the analysis of identity on the Net. The kind of identity display and shaping that we can find in this virtual world fits the term *liquid identity* proposed by Bauman (2005) or *hyper-identity*, as Adrian (2008: 368) calls it,[29] subject to specific goals that are changeable and made to fit different situations. These avatars may be mere additions to a solid physical identity, or authentic protagonists and the main sources of identity shaping for users who are not satisfied with their offline identities. These users find on the Net an escape from physical interactions and a transit to more *ad hoc* and satisfactory identities, created for specific goals (see Morie 2008).

Perhaps the most interesting research for an analysis of physical identities vs virtual identities in SL would be to analyse to what extent the physical ones influence the virtual ones and, especially, whether the latter influence the former. For example, a user (quoted in Adams 2006: 9) acknowledges how much his SL identity differs from his physical identity:

> [SL] hasn't changed much… but I certainly have. You need to be an extrovert to thrive in here. In RL [real life] I'm the opposite. I have a wife and family, but I spend a lot of time not really interacting with the outside world.

To explore other aspects of identity and overcome the limitations that one's body imposes on us in physical settings seem to be objectives that underlie SL

29. He uses a comparison between hyper-links and identities. In the same way as one is offered several links to click on when one accesses a web page, a user is offered several identities in virtual worlds to choose from and to fit specific communicative goals.

participation. Of course, avatar-mediated corporeal virtual identities do not have to resemble their physical counterparts. In this sense, McKinnon (1995, quoted in Kang & Yang 2006: 1175) suggests three forms of self-description of the user's identity in virtual contexts: *transparent expression*, when virtual identity and physical identity coincide; *translucent expression*, when virtual identity resembles more or less physical identity; and *opaque expression*, when both identities differ completely.

4.3 Body

An essential and obligatory step when joining SL and other similar virtual worlds is to select a body for the avatar, shape it in its main features and choose clothes for it. These choices inevitably generate inferences in other users, some of which will match the user's intentions, while others will be constructed beyond these intentions. This intended-exuded duality is of particular interest for a pragmatics of avatar-mediated communication. In theory, only the information intended to be communicated should be analysed by pragmatics, but valuable conclusions may be drawn from the avatar's exuded information. In this case, sex and body (and also clothes) stereotypes are usually at work in the form of mental schemas stored as part of "the culture" of a collectivity and which are inevitably reproduced in virtual worlds as intensely as in physical settings.

It is undeniable that when two individuals interact, the information about their sex, physical appearance and clothes help them frame the conversation correctly. In SL, the visual features of the interacting avatars also aid in choosing what kind of conversation is exchanged and which thematic course it takes (see Misoch 2008: 54–55). In fact, it has been demonstrated that the process of avatar creation in SL is intimately influenced by the user's (offline) bodily features, even if the user attempts to alter these features completely and create a radically different avatar (Vicdan & Ulusoy 2008), and even though avatar creation offers multiple possibilities for playing with identities and exploring other avatars' reactions.

Hence, even though users may try to create a radically different avatar from themselves, they cannot help bearing their physical bodies in mind when they choose features and qualities that improve the avatar in comparison to their physical bodies (e.g. broader shoulders, flatter belly, more handsome). As D. E. Jones (2005) qualifies, "virtual worlds feed societal fantasies developed within the mind/body discourse of transcending the deficiencies of human flesh. *Second Life*, which allows complete customization of avatar bodies, promises to give users a second skin that can improve on the corporeal and be changed like a suit of clothes."[30] And the body maintains its influence in SL, as intensely as in the physical world. In Stromer-Galley & Martey (2009: 1051), for example, several studies are cited that reveal how users open themselves less to dialogues and keep greater distance with avatars that are not attractive, or how short avatars tend to exhibit less confidence in the dialogues with other avatars in which they participate.

The problem is that, either consciously or unconsciously, users often create the bodies of their avatars according to cultural or sexual stereotype-patterns of the community (in its narrow or wide dimensions). These stereotypes are stored in the form of mental schemas of great strength and easy access, and therefore require little mental effort in the process of the schema reproduction of "embodiment" of the avatar. Given the persistence and depth of the stereotypes of masculinity and femininity rooted in culture and enhanced by the mass media, it is foreseeable that these stereotypes, especially those related to the information *exuded* by the attributes of the body, will also be valid for the information provided by avatars in SL and other virtual worlds. And the same applies to sex roles. Guadagno et al. (2011) concluded that there is a parallelism between offline sex role expectations and the ones performed within SL. Specifically, they tested the *Social role theory*, which indicates that "men and women perform different roles in society with men primarily serving the role as provider and women primarily serving the role as caregiver" (ibid.: 305). Men and women learn different skills and beliefs that fit these roles and also impact their social behaviour. According to these different social roles, men and women are also subject to expectations for behaviour. These factors lead to gender differences in actual behaviour. The analysis of SL avatar behaviour yielded a confirmation of this theory, since

> overall, women reported engaging in more communal activities (e.g. meeting people, shopping) relative to men while using Second Life, and men reported engaging in more agentic activities (e.g., building things, owning and working property) relative to women. Furthermore, when describing their most positive experiences, women reported more communal experiences and men reported more agentic experiences. (ibid.: 307)

Besides, gender stereotypes are part of the encyclopaedic knowledge stored by the user and are updated in daily interactions which, based on relevance, cause certain assumptions to be strengthened or corroborated. The mass media are instrumental in spreading these assumptions, and play a substantial part bombarding the user with patterns of behaviour and models of corporeal identity. Some of these

30. Jones (ibid.) differentiates between *normative avatar bodies*, which behave in similar ways to human beings in the physical world, and *fantastic avatar bodies*, that hardly resemble human beings (e.g. furries).

assumptions can be transferred "vertically" through generations, while others are only valid "horizontally" within a community (although these limits are becoming increasingly blurred by globalization and the ubiquity of access to information from mass media and the Internet). In the second case (horizontal spread), and according to the *epidemiological model* envisaged by Sperber (1996), the transfer of information among humans is essential to make certain archetypical cultural assumptions more or less faithfully stored in the minds of all individuals:

> Most representations are found in only one individual, but some get communicated, transformed by the communicator into public representations and re-transformed by the audience into mental representations. Some even get communicated repeatedly, spread out in a human population and may end up being instantiated in every member of the population for several generations. [...] Each member of the group has, in his or her head, millions of mental representations, some short-lived, others stored in long-term memory and constituting the individual's 'knowledge'. Of these mental representations, some – a very small proportion – get communicated repeatedly, and end up being distributed throughout the group, and thus have a mental version in most of its members. When we speak of *cultural representations*, we have in mind – or should have in mind – such widely distributed, lasting representations. (ibid.: 25, 33)

Gender stereotypes are examples of representations that spread across the population not only through interactions (e.g. parent-child) but also through archetypes that mass media reproduce and propose, and spread easily because the media have reached the status of "source of authority" for many people, especially teenagers. As a consequence, the audience tend to over-emphasize the importance of the body in both sexes, above other factors such as personality. It is not surprising, therefore, that these stereotypes are transferred equally to the three-dimensional world of avatars in SL.[31] This does not mean, of course, that all SL users share the same ideas on what model of masculinity and femininity (and the parallel communication and behaviour patterns by gender) are suitable for the physical and virtual interactions in which they participate, but even if the user does not share gender stereotypes, he/she is likely to be aware of the depth and extent of these

31. In Yus (2001c) a distinction was proposed between *woman-as-signifier* and *woman-as-signified* to refer to how women are portrayed by the mass media. In general, the former has been over-emphasized at the expense of the latter, hardly valued at all. This has generated a kind of *semiotic imbalance*. For example, the comedian Jo Brand (quoted in Wagg 1998: 122) stresses that "there is this attitude towards women which prevails in magazines and on the telly and if a Martian came down to earth and just had to watch telly and read magazines to find out what women were like he'd think that they were all blonde and 25 with big tits, you know. Because that is mainly what you expect on the telly. Also they would think that they were never rude and always looked nice, they always deferred to men, a lot of the time."

stereotypes. To account for this, in Yus (2002c) a distinction was proposed between *private beliefs*, those that the individual has acquired personally through interaction, communication and inference, and *metarepresented cultural beliefs*, those that the individual assigns as prototypical of a community or culture, that he/she assumes as widespread among its members, and that may or may not coincide with his/her their own private beliefs (or overlap in different degrees).

The bodies in SL, therefore, exhibit male and female archetypical qualities[32] and also prototypical clothing, all of them stored as mental schemas and transferred to the virtual environment from physical life (see Brookey & Cannon 2009). The same is true for other "inherited" traits of the body such as the race (see Harris et al. 2009). As Misoch (2008: 60) correctly points out, if the body is the instrument through which we exhibit our presence in physical spaces, the avatar takes on this role in the virtual world, while performing the function of being a visually perceptible and stable receptacle of identity. Furthermore, these stereotypes are often increased in SL (e.g. disproportionate chests in male avatars and big breasts[33] in female avatars), taking the stereotype to extreme lengths. It is not surprising, then, that De Salvador Agra (2009) concluded in her research that in SL

> there are no socially undervalued virtual bodies. In this supposedly simulated world, where almost anything is possible, we can not find any avatar to take the special feature of being disabled, fat, lame, squint or any other typical feature that is object of discrimination in the old offline world. That is, in our incursion into this info-virtual environment, we could see how nobody in this freedom that supposedly defines the environment, opts for a body that is pejoratively valued in our physical society.

---

32. In general, the sexual stereotype of women in SL is "thin, narrow hips, long legs and generous breasts." The male sexual stereotype is "an athletic build, broad shoulders, muscular chest and narrow hips" (Misoch 2008: 59). These correspond to the images that are incessantly communicated to the audience of Western cultures from the mass media.

33. In a survey by the blog *Pixels and Policy* (November 2nd, 2009), 70% of female users surveyed admitted that the design of their avatars' breasts was one of their main concerns during the design of their avatars. Some users showed a clear submission to the woman's sexual stereotype of "exuberant" as a means to be desired and obtain more frequent interactions with other avatars, as can be concluded from the testimony of one of the female users: "At first I played with an avatar that I thought represented me physically, but not many people talked to me. Now [with a big-chested avatar] people go out of their way to IM me and send me friend requests." According to that blog, this is a clear setback for female users, who place the passive and stereotypical role of attractiveness before actively seeking friendship based on personality. Note, however, the study by Ducheneaut et al. (2009), in which informants stressed the importance of the process of hair design for the avatar, over the process of design of sexual attributes.

One consequence of this transfer of body stereotypes from the physical context to the virtual context of SL is what can be called *illusion of corporeal avatar*, that is, the illusion that it is the body of the avatar, and not the user who created it, which holds all the communicative activity in SL interactions. Bente et al. (2008: 134) also comment that "although aware of the avatars' artificial nature, users seem to respond to their appearance much in the same way as they do to humans in real-life encounters." This illusion would explain why many avatars are treated according to their virtual appearance, even though it is obvious that the actual appearance (or even the sex) of the user can vary dramatically compared to that artificial appearance.

De Salvador Agra (ibid.) entered SL with three avatars of radically different appearance: a young and voluptuous woman, a bald old man and an obese black woman (Figure 5.1). Although the user is the same and her physical appearance has nothing to do with these three avatars, she suffered indiscriminate abuse when she used the avatar of an obese black woman, jokes and derision when she played the part of a bald old man (specifically its avatar was heckled by a group of avatar-girls that called him "grandpa" and asked him what an old man like him was doing there) and, finally, sexual approaches when incarnated in the body of a young attractive woman. In other words, participants completely ignore the fact that there may be little or no resemblance between the avatar and the user who created it and, instead, the body shape of the avatar is idealized as an inherent aspect of SL interactions. In similar terms, a professor at Ball State University (Indiana, USA) comments that

> Of course on an intellectual level I realize that the avatars are fictitious, probably quite different from the people who created them, and that there is some real person sitting at some keyboard somewhere in RL [real life] but all of that recedes way into the background while in SL. I identify to a surprising degree with my avatar; it feels as if I'm her, actually visiting places, doing things, and talking to people (that is, talking to avatars, but I don't really relate to them as avatars while in SL; I seem to view them as people). When someone I've been talking to turns out to be a vampire, I feel a little worried, as if the situation were much more real than just images on a screen. (pers. comm., November 26th, 2009)

According to Yee, Ellis & Ducheneaut (2009), this phenomenon is part of a more general tendency: the human desire to reproduce the parameters of the physical world in the virtual setting and, in fact, for these authors SL and other virtual worlds do fit, to a greater or lesser extent, several expectations of duplicity: (1) the *expectation of human embodiment* (to use human-resembling avatars); (2) the *expectation of matched affordances* (the avatars move and do the things that humans typically do in physical contexts, but in SL the avatars can fly and tele-transport themselves to other spaces of this virtual world); (3) the *expectation of congruence* (users, by means of avatars, possess different conceptualizations of what a virtual world is, but these are congruent); and (4) the *expectation of a single avatar control* (each user can control only one avatar at a given moment and each avatar is controlled by a single user).

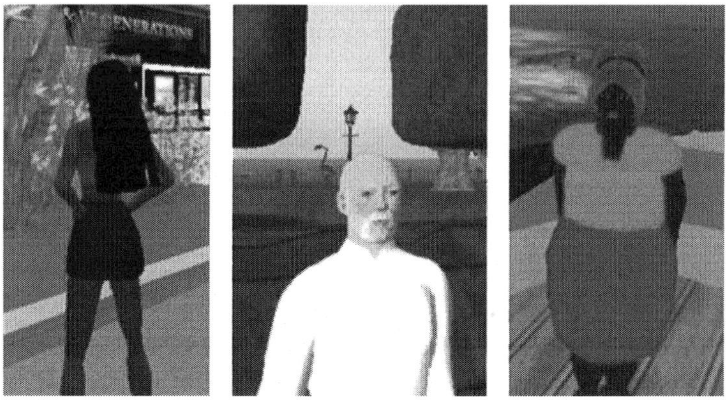

Figure 5.1 Avatars of De Salvador Agra (2009)

4.4 Verbal interaction

There are three main ways to interact verbally in SL: the chat room option, the instant messaging option (IM) and the paid-for voice service.[34] The first two are fairly common in other virtual worlds, but there are also significant differences with pragmatic implications. Thus, when users type their messages in the chat or IM window, the server reproduces the messages on the screen in their entirety after being sent by the "addresser user" (i.e. not word by word as the user is typing it), in the same way as in conventional text-based chat rooms. One cannot, therefore, be generating inferences and interpretive hypotheses as the text is appearing on the screen. These inferences may be corroborated or refuted with the processing of the next part of the message that appears on the screen (thus forcing the reader to backtrack and re-interpret the text according to the "new evidence"

34. Pujolà & Palomeque (2010: 136) summarize most of the forms of interaction in SL: (a) *local chat*: text-based, synchronous and public; (b) *voice chat*: voice-based, synchronous and public; (c) *IM*: text based, synchronous (also asynchronous) and private; (d) *voice call*: voice-based, synchronous and private; (e) *notecard*: text-based, asynchronous and private; and (f) *gestures*: non-verbal communication (the user has a default set of gestures in the inventory but can also create new ones for their avatar).

provided by the text just typed). In SL, while the text appears on the screen, the avatars automatically mimic the action of typing text with their hands on an invisible keyboard. Moreover, if the avatars come within 30 (virtual) metres away from other avatars, then they can "hear" what they are saying, that is, the user can read the conversations that other avatars are participating in if the user draws the avatar close enough.

In the virtual world *There* (recently closed), the protocol of verbal interaction was somewhat different, since it was based on a number of balloons, as in comics, displayed above the avatar's head (when there were multiple utterances, these were inter-connected with each other forming a chain of balloons) and in which the text was being copied as the user was typing it. This quality made it possible for interactions to have overlappings or interruptions that are nonexistent in SL. In addition, the users were able to change the content of their messages while they were reading the other avatars' messages as they appeared in the balloons, and hence the construction of the message obtained a high level of relevance and appropriateness in the context of that conversational interaction.

Of course, in the text typed both in the chat and IM windows, users reproduce the strategies of compensation for the absence of orality and of text deformation that have been discussed earlier in this chapter (e.g. repetition of letters or punctuation marks, capitalization, use of emoticons). However, the use of one or another channel has different connotations, as noted by Boellstorff (2008: 152–154). In general, SL conversations between avatars that have just met are usually held in public using the chat room facility. But it is possible to add an avatar as a "friend"[35] and at this moment the private IM is preferred, especially with friends or contacts who know each other well, just like in conventional IM. Very often, and even if there is no other avatar around, users leave the chat room and continue the conversation through IM when the topic of the conversation becomes more confidential, private or intimate.

Similarly, the use of multiple windows for interactions in parallel is common. The control of these windows, together with the management of the nonverbal behaviour of the avatar, can generate extra effort in SL communication. This effort can be increased by the possibility of engaging in chat room and IM interactions simultaneously. Another extra effort is required to manage the accumulated IM

35. The procedure for doing this is to click on the avatar that one wants to add as a friend with the right button of the mouse and choose "add friend" from the context menu displayed after the click. On the user's screen that controls the behaviour of the avatar the message "X is offering friendship" then appears where X is the name of the avatar, which is always visible above the avatar in SL. If the user enters "yes," from this moment the added user will always be able to locate the avatar in SL and will know when it is online.

messages when the user logs onto SL. In fact, one reason why many users create *alternative avatars* is that it enables them to log on anonymously and not be bombarded by IM messages.

In the case of using the voice option, users definitely get an extra level of contextualization that allows users to "frame" the interaction and identify the underlying communicative intentions efficiently. This is a paid-for option and is not selected by most users, many of whom are satisfied with the options provided by text-based interaction (see Bente et al. 2008: 288–292), as in the case of the researcher at the University of Leeds mentioned above:

> [My avatar] has used voice. I prefer text, especially in my original avatar, as it allows me to be someone else. I also like the ability of text to neutralise social features, such as gender, region of origin, social class and so on. I think it empowers people more than voice and is easier across national boundaries. So I don't think it is a limitation. My avatars can express emotion through actions and expressions and through statements of actions and feelings.
>
> (pers. comm., November 24th, 2009)

In Boellstorff (ibid.: 114), the opinions of several users on the voice option are quoted. For many of them, voice provides a high level of intimacy, while for others the voice "kills the fantasy" of true avatar-mediated communication in SL. For one user, refusing to use the voice was interpreted as proof that she was actually a man. For another user, the voice in SL is a natural extension in SL of interactions that take place in physical contexts. But this same argument is considered negative for others, because it ruins the illusion of interaction among avatars.

4.5 Nonverbal behaviour

In previous pages I mentioned the example of a person yawning and the possible interpretive outcomes depending on the axes of "intentional / unintentional" and "understood correctly / incorrectly." SL avatars are endowed with a great capacity to generate nonverbal behaviours. In theory, it should be clear that these necessarily have to start from an intention by the user that generates them in the avatar. However, as discussed below, the qualities of the computer application that manages SL can lead to interesting situations in terms of avatars' nonverbal behaviour.

Nonverbal behaviours, beyond the basic movement or simple behaviours of the avatar, are usually generated by applications that are available in the environment where avatars interact. For example, in a dance floor context two users can click on the button and select from the menu an animation that gives avatars the ability to dance. If two avatars are getting married (which occurs quite often in SL) and an avatar-priest says "you can kiss the bride" each user clicks on the animations menu and chooses “kiss” so that their avatars can carry out this nonverbal action. In *There*, these animations appeared as blue circles around avatars, also called *action tags*. One clicked on the tag and a menu of possible actions turned up (e.g. “sit next to the nearest avatar”). In *There*, it was also possible, for example, to type a prototypical nonverbal behaviour (e.g. smiling, yawning, blushing) and automatically the corresponding nonverbal behaviour was generated in the avatar. In fact, the nonverbal behaviour available in that virtual world was quite advanced, since the avatar could, for example, sip a drink or offer it to other avatars.

Managing the full range of possibilities regarding avatars’ nonverbal behaviour involves effort and training for which not all users find a reward in communicative terms (cognitive effects). For example, Suler (2007) describes his tiring training in the management of the nonverbal behaviour of his avatar:

> It took me several minutes just to figure out how to move my avatar, and then I was literally walking into walls and trees. I spent most of the first day learning how to move about [in SL] without looking like a complete idiot, how to visually survey and interact with the environment, and, most fun of all, how to fly like superman. The controls for navigating one’s avatar are much more sophisticated than they used to be at the Palace. This posed a rather interesting challenge. Even after several hours, when I thought I was doing reasonably well, a more experienced user who I met in the SL version of Amsterdam commented on me being a newbie. When I asked how she knew, she replied, “By how you walk.”

As noted above, managing the nonverbal behaviour of the avatar entails an underlying intentionality by the user that the avatar should behave in a certain way so that, in principle, we could not find in SL all the unintentionally *exuded* nonverbal behaviour that so often helps people understand each other correctly in physical settings. For example, a prototypical exuded behaviour like *blushing*, uncontrollable in physical contexts, must be generated by the user, and thereby loses much of the “communicative naturalness” that SL aims at, and forces the users to constantly assess the type and intensity of their feelings and emotions. To this we must add the difficulty of making a computer program perform the full range and variability of nonverbal behaviours, which is evident, for example, in the case of facial gestures, as will be briefly discussed below.

However, there are situations in which “unintended nonverbal behaviours” are produced in the avatar. For example, the software can make the avatar behave in a strange way, sometimes due to the user’s inexperience and occasionally due to alterations in the program’s response to the interactive goals of the SL residents. These situations introduce the possibility of an unintended or *exuded* nonverbal behaviour from avatars. Therefore, for any type of nonverbal behaviour in avatars it is important to note the role that the interface can play in the outcome of these behaviours along the intentional / unintentional axis. For example, Antonijevic (2008) proposes the following classification that in my opinion satisfies this requirement:

1. *User-defined cues.* These refer to behaviours that the user deliberately performs, such as separating the avatar from another avatar that is too close by using the keyboard arrows. These cues clearly play a part in interactions, because they reveal communicative intentions and regulate interactions.

The most common behaviour in this first type is the management of proxemics between avatars, that is, the control of the distance between avatars and the interpretations that this distance conveys. In fact, as concluded in Bailenson & Blascovich (2008: 2676), the avatars in virtual worlds like SL often behave in the same way as human beings in terms of interpersonal distance. While there are no "written rules" about what personal distance is the most appropriate in each case, users-avatars consistently identify the most appropriate distance and explicitly show their disagreement when it is not respected by another avatar (Taylor 2002: 42) .

2. *Predefined cues.* They refer to nonverbal behaviours that the user has not generated in the avatar, but it is the computer system that produces them. For example, if the user stops typing text in the chat room or the IM window, the SL system detects user inactivity and forces the avatar to adopt a particular position. Something similar happens to what Boellstorff (2008: 106–107) calls *away-from-keyboard problem.* Users frequently stop controlling the behaviour of their avatars and leave the computer for a while without actually turning off the system, or they stay at the computer but working on another application in parallel. In both cases, the system causes the avatar to adopt a position of complete inactivity and to appear to be dead on the screen, to the astonishment of the other avatars. In theory, after several minutes, the system detects the "sustained inactivity" of the avatar and makes it disappear from the screen, but there are programs available on the Net that invalidate this action.

A sub-group of these predefined cues includes behaviours that seek to regulate and synchronize the interaction among avatars. For example, if a user moves his/her avatar in a certain direction, the heads of nearby avatars automatically turn in the same direction, in an attempt to reproduce the conditions of interaction that take place in physical contexts.

3. *Blended cues.* This category refers to nonverbal behaviours chosen by the user, but carried out by the computer system of the virtual world. All predefined animations displayed in the contextual menu after clicking on the right button of the mouse are included in this category. For example, a user may want his/her avatar to kiss another avatar, but the kiss itself is made possible by an animation

that the system displays, without intervention from the user. In addition, this type of behaviour is heavily influenced by the context and the sex of the avatars. For instance, the action of sitting down is different in animations for male and for female avatars and it is likely that, more or less consciously, archetypical patterns of behaviour according to the person's sex may have been followed (in the form of accessible stereotypical mental schemas) in the design of many of these animations.

4. *Missing cues.* Finally, Antonijevic (ibid.) mentions human nonverbal behaviours that the system is unable to reproduce, although constant evolution in the various versions of the software in these virtual worlds leads to the incorporation of new behaviours in the range of nonverbal behaviours of the avatar. For example, the addition of a voice in SL has been a major change and users are able to communicate, with their avatars' words, the full range of vocal nonverbal behaviours available to humans in face-to-face interactions.

On the other hand, of all the areas of nonverbal expressiveness, it is the human facial expression that probably best illustrates the problems of these virtual worlds to mimic the nonverbal behaviour produced by humans in physical contexts. Currently, at least with the technology available, it is very difficult to reproduce accurately the full range of emotions and feelings that are based on facial gestures, and we can often see that only very general behaviours such as smiling appear in the catalogue of avatar animations. For instance, Donath (2001) stresses that it is not simply a matter of getting more detail in the avatar's facial expression, but there are also many cognitive and cultural determinants that influence the use that humans make of the expressions on their faces, and these are very difficult to transfer to the virtual world. For her, introducing the face in these environments involves a radical reinterpretation of what we consider "personal appearance" and, at the same time, trying to maintain the cognitive and cultural meanings that we associate with the familiarity of the facial gesture. Donath (ibid.) does not consider that avatar expressions should be a duplicate of those found in physical spaces and, in fact, the virtual environment offers many possibilities beyond these natural gestures. However, she considers that we have to be careful when facial gestures are introduced in these environments, because they are full of subtle information, and a poorly designed expression can lead to misunderstandings and cause virtual interactions to fail.

This issue is accentuated by the role of the face in the regulation of conversations. Indeed, facial gestures are *regulators,* in the sense proposed by Ekman & Friesen (1969), of conversations with speakers providing evidence that they are paying attention or sharing an opinion. This role has also been transferred to virtual worlds. In the study by Garau et al. (2001), they concluded that having an avatar whose facial behaviour can be directly related to the development of the conversation increases the quality of virtual communication, compared to those conversations where avatars exhibit random gestures. Put simply, it is not enough for the avatar to be "extrovert" with other avatars, it has to engage facially in the conversation in order to be successful.

However, advances in computer applications that manage interaction in SL and other virtual worlds open up interesting possibilities with clear pragmatic implications. For example, Heike et al. (2009: 207) mention studies that attempt to get the avatar to nod upon detecting the end of the interlocutor avatar's utterance (i.e. to identify the *turn transition place*, in the terminology of conversation analysis). Even more interesting is the advancement cited in Morrás (2009) about the invention of an SL viewer that permits the transference of the user's emotions and gestures to his/her avatar. Therefore, the avatar makes in real time the same gestures and movements that the user performs. This is a very interesting evolution and opens up new areas for pragmatic research in the future, because this interface provides the avatar with greater capacity for facial expressiveness. These advances, together with advances in general avatar design options influence communicative outcomes in SL. For example, Hussain et al. (2011) concluded in their study that there is a high correlation between humanness of avatars and high credibility measures. Therefore, a good design of the avatar may have an important impact on interpretive outcomes and the overall SL user's satisfaction.

Finally, it should be noted that many nonverbal behaviours do not have an inter-cultural applicability. Users come together in a three-dimensional environment that is accessible from anywhere in the world that has Internet access. It is predictable, therefore, that there may be misunderstandings due to ignorance of the nonverbal behaviours that are typical (and taken for granted) in the culture to which the user who is handling the avatar belongs. This is very likely to occur if we consider that on many occasions there are no cultural clues in the final design of an avatar and the real user behind the avatar may be managing its nonverbal behaviour from anywhere (and the corresponding culture) in the world. For example, Koda et al. (2009) compared the expressions of the avatars designed by Western and Japanese users and which depict typical faces of these cultures. The investigation concluded that, in fact, there are cultural differences in interpreting the facial expressions of avatars. It was also concluded that facial expressions of a positive nature have a greater cultural variability in their interpretation than those of a negative kind. Specifically, the subjects' interpretations of negative expressions (sad, disapproving, angry, and confused) were similar to the designers' intentions, regardless of the country, that is, the subjects' answers to those expressions were similar across countries. On the contrary, the subjects' interpretation of positive expressions (happy, approving, proud, grateful, and impressed) varied across countries.

5. Videoconferencing and context accessibility

If we arrange all *cyber-media* on a scale of options for contextualization and capacity to convey vocal and visual nonverbal information typically communicated in face-to-face interactions, videoconferencing would no doubt be ranked highest in contextualization, since it closely resembles face-to-face dialogues even if framed by the computer screen.

It is convenient to start with some terminological clarification. Mouzourakis (1996:22) proposes a terminology to distinguish between (a) *teleconferencing* (any form of communication that involves the use of at least one audio channel, between two participants in a meeting but separated by some distance), (b) *audioconferencing* (teleconference in which the sound is the only channel of communication), and (c) *videoconferencing* (a special kind of conference in which the image is used in addition to the sound). We should also distinguish between *video-mediated communication* and *videoconferencing*. The former involves the use of television cameras that transmit the signal to a satellite which, in turn, distributes the signal to other parts of the world. In *videoconferencing*, by contrast, the signal is transferred through the Internet in digital format and therefore can be analysed from a *cyberpragmatic* point of view. However, the threshold that distinguishes the two forms of signal transmission is becoming diffuse, as broadcasters often use the Internet to distribute video content and rely increasingly on satellites and wireless networks for transmission.

It is noteworthy that achieving close fidelity in terms of contextualization does not necessarily mean that users' estimation of relevance will not be affected by the screen-framed transmission of information. Indeed, the first problem in trying to equate videoconferencing and face-to-face interactions lies in the fact that videoconferencing links spaces that can be very distant from each other, but these spaces are framed by the computer screen. Sometimes, alterations of relevance can be produced due to the mutual influence of the framed area of the room and the one which is out of frame. This is especially clear in the case of dialogues among people who share the same physical space of a room but only one of them is framed inside the screen and, at the same time, all of them in the room have a dialogue with another user through videoconferencing (Raudaskoski 2000). Fayard (2006:154) also comments that the introduction of video in the interactions produces asymmetries that affect everyday communicative practice, thus making the routines and scripts of interactive behaviour (which we use almost unconsciously) less valid. In videoconferencing, the interlocutors do not share the same physical context, which affects the interpretation of the environment of the interlocutor, his/her eyes, body expression, etc. The social context is not shared, either, and that generates a loss in the appropriate contextual clues to interpret the speaker's communicative behaviour. Besides, some technological problems have to be taken into consideration which disrupt communication (Have I heard correctly? Has the interlocutor grasped the meaning of my gesture? Is there synchronization between voice and image?).

Furthermore, the concept of *stage*, in the sense proposed by Goffman (1987), should be emphasized. I referred to this term at the beginning of Chapter 2, when I stated that we must distinguish between the roles we play in society at the *front stage* of interactions and the personal reality that lies at the *back stage* of our identities, the part that hides behind this "social playground." These roles also play a part in videoconferencing. What proportion of the roles that are filmed by the camera belongs to the "social front stage" and how much refers to the real identity of the interlocutor?

In short, videoconferencing offers a fully contextualized environment that conveys users' vocal and visual nonverbal behaviour but suffers from some problems caused by the lack of physical co-presence, screen-framed interaction, and mediation. These problems are more or less obvious depending on how faithfully the filmed scenario is reproduced in the transmission of different utterances exchanged through this type of Internet-mediated communication.

CHAPTER 6

You've got mail

1. Introduction

In Chapter 2 I exemplified the multiplicity of physical-virtual identities with the movie *You've got mail* (1998). At the beginning of this film, the main characters (Tom Hanks, Meg Ryan) are eager to leave their physical relationships to log onto the Internet and check whether they have any electronic messages from their virtual friends. This variety of asynchronous[1] communication was, in fact, a second phase in their relationship after a first encounter at a synchronous chat room in which, allegedly, they met virtually and exchanged their e-mail addresses.

In this chapter I will study electronic mail (e-mail), which has not changed substantially in recent years. The death of e-mail has been announced on several occasions but it is still irreplaceable in certain contexts such as the academic world (e.g. to contact researchers, exchange publications via attachments, see Zimmerman & Bar-Ilan 2009) or specialized discourses (Duke 2001), and the workplace (Skovholt 2009). It is also a key element in the structure of "safe" interactions on Internet portals that specialize in finding partners and romance (*Match*, *Meetic*, and others) and it sustains notifications in social networking sites when someone comments on a photo, asks to be added as a friend, etc.

It is true that certain age groups avoid the use of this form of communication. Young people, for example, consider it too formal[2] and cold, and prefer other

1. Asynchronous because the sender and recipient are not usually online at the same time. However, the development in the quality and speed of message transfer over the Internet often turns e-mail into a medium close to a synchronous virtual conversation, since the parties can exchange messages with the same speed as chat rooms or instant messaging (see Jonsson 1998: Chapter 2, Feenberg 1989: 24, Epperson 1995). It should be noted, in this sense, that the speed of exchanges can influence not only the desire and predisposition to use this *cyber-medium*, but also affects the content of messages. For example, Bertacco & Deponte (2005) concluded that the increased speed of e-mail leads to a reduction in the length of messages and a loss of references to information supposedly shared by the interlocutors.

2. Although my intuition is that e-mails are used for both formal and informal playful interactions (to send jokes, humorous Powerpoint presentations…), Horrigan & Rainie (2002) concluded that the greatest increase in the number of messages exchanged (e.g. from the young to relatives) corresponded to "serious" content such as seeking advice or sharing a problem.

forms of interaction such as instant messaging, SMS, social networking sites, etc.; but e-mail is still very popular and millions of e-mail messages are sent every day through this *cyber-medium*. Furthermore, the apparently "cold" design of the e-mail interface actually makes it a good example of technology that allows users to establish acts of ostensive communication which, by definition, carry a presumption of their eventual relevance (see 4 below). In addition to this pragmatic approach, this chapter includes the analysis of the possible location of e-mail in the oral / written continuum, essential in any Internet discourse and also for virtual conversations (e.g. chat rooms, instant messaging).

E-mail is still one of the most widely used *cyber-media* nowadays. The rise in its popularity from the 90s onwards led initially to a reduction or restructuring in the use of other media, for example the telegram or traditional (*snail*) mail, and even to a reduced use of the telephone (it is well known that at Microsoft's headquarters the phone never rings) and fax.[3] Nevertheless, it should be borne in mind that, although e-mail initially reduced the use of the phone, it is also true that today's use of mobile phones has in turn reduced the use of e-mail. In this regard, as we discussed in the previous chapter, the dramatic increase in SMS *texting* should be highlighted, turning the mobile phone into a sort of hybrid device with multiple options for communication (Benson 2000), a trend that has increased recently with the evolution of 3G technologies and the development of *smart phones* that allow for Internet access via mobile connections.

What is e-mail mainly used for? Basically for the following goals: (1) to communicate with people, regardless of where they are located, rapidly and at low cost (Morrisett 1996); and (2) for advertising products. Wilkinson & Buboltz (1998) and Duran et al. (2005), among others, predicted that these two uses would evolve to connect not only individuals, but also social entities and, above all, predicted that this medium would improve as a tool for teaching which would reduce the teacher's responsibility for the performance of higher-level students and increase communication between both parties. These predictions have come true to some extent, while other platforms that aid in teacher-student interaction, for example the new platform *Moodle*, have emerged in recent years to manage the variety of interactions between teacher and student and are more advanced than e-mail.

Of course, not all scholars have stressed the advantages of e-mail. Some have highlighted its negative effects, such as the computer fatigue and stress that is caused by an excess of messages received on a daily basis,[4] or problems that e-mail

3. See, among others, De la Fuente (1997), Burton & Maitland (1995), Stanford (1999), Dobson (1996), Moran & Hawisher (1998: 84), López Alonso (2003) and Martín (2008).

4. See, among others, Welsh (1997), Gwynne & Dickerson (1997), Burkeman (2001), Freeman (2009) and Soucek (2010).

causes in the lives of its users, as in cases of infidelity caused by virtual relationships, or e-mail harassment (*cyberstalkers*).

2. General characteristics of electronic mail

E-mail is a variety of asynchronous virtual communication that helps people communicate quickly and cheaply. Under this label various types of interaction are grouped with more or less defined attributes, although there is some overlapping among them.[5] Firstly, the most common e-mail type is the *personal message*, a private message sent from one person to another person. But two additional types are worth highlighting: the newsgroup and the Listserv.

2.1 The newsgroup

It is a sort of "bulletin board" where users send messages with news or requests for information, and also where the answers appear on the screen as thematic conversational threads (see Jones 1997b). A good definition of a newsgroup is the following:

> A hierarchically-organized forum open to users interested in a specific topic. Discussions take the form of electronic messages (e-mails), which are sent to the forum by users and filed on internet sites. Users can read the messages filed on the site and also post new messages. Newsgroups are, in fact, a hybrid of interpersonal and mass communication. The newsgroup's usenet address indicates its content[6] (for example, *fr.rec.arts.litterature*) and defines the forum's frame, which is essentially thematic. (Marcoccia 2004: 117)

Among the main features of newsgroups, we can list the following, according to Pérez Sabater (2007: 81 ff): (a) the participants share an interest in the same topic and show opinions only about this topic, at least within that forum; (b) they are inherently asynchronous, although the increasing speed of message delivery might give the impression of a certain synchronicity; (c) they form a closed community

5. As correctly pointed out by Montero-Fleta et al. (2009: 777), the distinction between certain genres of computer-mediated communication is inherently fuzzy, and the qualities of each genre rely more on communicative intentions than on sociolinguistic conventions that have developed from the limitations of the medium.

6. Some examples: *alt.religion.wicca*, *alt.romance.online*, and *alt.support.marriage*. They all indicate their inherent topic in a more or less explicit way.

which often requires a subscription; and (d) frequently, there is a moderator who ensures the proper behaviour and etiquette in the forum (see Smith et al. 1997).

Typically, newsgroups require users to share knowledge on the particular discussion that is taking place (plus knowledge of jargon and vocabulary inherent in the newsgroup), to be aware of its current stage or development, and also of the interplay of overlapping threads generated by the forum.[7] This is why Shank (1993) labelled the newsgroup as *multilogue*, a variety of dialogue in which there is a person who initiates the conversational thread and, from then on, the "addresser user" no longer exerts control over the development of the interaction, since there is no fixed sequence of conversational turns (Harrison 1998a). It is as if in an oral conversation, everybody willing to participate could speak and, despite the simultaneity, all the voices could be heard clearly. Or, as Pano (2008a: 148) describes, in the newsgroup "a double discourse is constructed in which you talk to everyone and have a dialogue with some users, developing a self-image that results from internal representations and from interaction with others, in terms of negotiation of identity."

A typical strategy in certain newsgroups, in order to assure mutual manifestness of topics and information and to keep inter-post congruency, is to fill messages with redundancy. Although, in theory, redundancy provides information which is already part of the users' cognitive environments and hence tends to yield irrelevant interpretive outcomes,

> repetitions may direct the participants to the core of the topic and increase the coherence of the discussions. The repetition of some messages informs participants that the content of these messages is important. In newsgroups, [the practice of] including previous messages in the body of the current one is another form of redundancy. Here, the redundancy preserves the continuity of the context and ensures the consistency of exchanges in which a large number of people participate. (Atifi et al. 2011: 330)

According to Marcoccia (2004: 116), two qualities of newsgroups challenge effective communication: their asynchronous quality and their public orientation. In the former, the problem lies in the fact that users can receive messages and reply to them on different days, thus breaking the normal development of the "conversation" in the forum and substantially increasing the effort to follow the different conversational threads. In the latter, the problem is that many messages sent to the newsgroup are, actually, directed to one single person, but acquire a public dimension due to the arrival at a public area of interaction. General and private messages get mixed up and increase the effort required to follow the

7. See Gruber (1998: 22, 2000a: 37), Matzat (1998) and Gómez (1998), among others.

conversations, a similar phenomenon to the one discussed in the previous chapter regarding public and private messages sent to chat rooms and accumulated in the main area of the web page. An informant's opinion (quoted in Kavada 2010: 365) is illustrative: "you can write what you think is a private e-mail and then someone might send it round to everyone else, or you might press the wrong button and then accidentally send it to loads of people, so there is potential for things to go wrong, communication to go wrong in a negative way with the e-mail."

In any case, the public quality of newsgroups is also useful for spreading general or specific cultural assumptions of the group, in a kind of "electronic word-of-mouth" as it is called in Fong & Burton (2006) or "word-of-mouse," as it is colloquially labelled. Indeed, certain assumptions repeatedly sent to members of the forum become stabilized in the minds of the members of this forum and promote the existence of a more or less similar version of these alleged "group assumptions" among those who are subscribed to it, thereby fostering the stabilization of "the culture" of the newsgroup. Mutual manifestness of these cultural assumptions of the forum requires a familiarity with certain vocabulary and jargon inherent in both the subject under discussion and the overall topic of the newsgroup itself.

There are several proposals for classification of newsgroups. For instance, Fisher et al. (2006) suggest four types: question and answer, conversational, social support and flame. For his part, Himelboim (2008: 164) proposes a dyadic classification into *conversation-centred newsgroups* (opinion-oriented, there are no solutions to problems and no answer is better than the others; it is more a matter of exchanging ideas, especially those with a political, philosophical or ideological connotation), and *information-based newsgroups* (stress the transfer of information, advice or support, for example the ones devoted to health issues). Besides, Fafchamps et al. (1989, quoted in Jonsson 1998: Chapter 2) distinguish three possibilities in a newsgroup: (a) *island* (a message with no reply), (b) *dialogue* (a specific exchange in a wider communicative sequence between two people), and (c) *network* (a complex structure of messages, in which a single message may receive several responses from multiple members of the forum, some of which relate to other areas of discussion).

At present, the newsgroup is in decline, contrary to what occurred in the late nineties. In 1997, for example, as documented in Choi & Danowski (2002), there were 14,347 Usenet newsgroups which generated 6 gigabytes of (text-based) data per day. On an average day, 20,000 users would send an average of 300,000 messages. Today, as noted by Molist (2009), the burden of spam, the lack of moderators' commitment to regulating the newsgroup, or the need for subscription, together with many users' ignorance of their existence, has drastically reduced the number. Nowadays, there are many alternative ways to create graphic forums, more attractive than the ones based on plain text, and whose management is easier. Besides, current forums can be integrated into social networking sites or blogs, thus making them more accessible. As stated in an advertising slogan years ago, in these new forums *access time is not excess time.*

2.2 The e-mail distribution list (Listserv)

A *Listserv* or "mailing list" is basically a collective repository of e-mails. Users need to register on a List, and then they will receive the messages sent by any of its members to its electronic address, under the supervision of a moderator, who is responsible for the "ethics" of all messages flowing through the List. This moderator cannot be identified with the author of the messages. A kind of polyphonic organization is thus created in which there is an underlying semantic framework of the type *A says that X says*, where A is the moderator of the list, and X is the variable of any participant subscribed to the List (Vela Delfa 2005: 369).

Pano (2008b: 29) summarizes some of the attributes of mailing lists: the message that is sent to the address of the list reaches everybody registered to it, but often a subscription is required, which can be cancelled at any time. On some lists, messages distributed among subscribers must be approved in advance by the administrator or moderator of the List. In most of them, there are standards or rules of behaviour (*netiquette*, see Chapter 7), for example to be respectful, to avoid certain words or terms, to be brief, not to send attachments that exceed a certain "weight," and to send messages that really deal with the subject of the List. In others, though, the moderator acts only when discussions diverge too much from the main theme, or when discussions could potentially offend some subscribers.

The mailing list differs in some aspects from the newsgroup. As summarized by Pano (2008a: 145 ff, 2008b: 30–31), the latter (a) establishes limits as to the size of the message, (b) has a limited validity determined by the guidelines (depending on the interest aroused by a topic or the loss of it), (c) favours a "many-to-many" exchange (although often a message is directed to a specific user), while in the list it is usually a "one-to-many" or "one-to-one" (in an inherently public environment), and (d) has an underlying intention different from the one of the list.

My three-fold division of e-mail into personal message, newsgroup and mailing list is rather generic, but sufficient for the purposes of this book. Concerning this triple division, Vela Delfa (2005: 207) comments that it

> is not exhaustive. Its goal does not seem to be a definition of a complete list of the types of messages, but seeks to establish an order and classification in the whole of verbal exchanges. Its interest lies in distinguishing modes of interaction that are inherent in this genre. For this purpose, two classification criteria are considered: the private or group quality of the exchange and the ability to generate dialogue.

Instead, she proposes a more detailed classification of e-mail into personal, professional, institutional, commercial and advertising, mailing-list-centred, forwarded chains of e-mails, spam or unsolicited, automatically generated, and virtual postcards.

López Alonso (2003) also makes a proposal of e-mail classification based on the criterion of relationship between interlocutors and on the functional or objective quality of the message. Thus, two main types are generated: (1) based on mutuality, when the first intention is an exchange between sender and receiver; and (2) without mutuality.

Furthermore, Sanz Álava (2006) makes a proposal based on the relationship between sender and receiver, as well as on the subject of the message: (a) the personal and emotional e-mail, (b) the professional e-mail, and (c) the professional e-mail with features characteristic of the relationship or knowledge sharing among friends or colleagues. The first of these is influenced by the type of relationship between partners, which generates text oralization in the message. By contrast, communication in the second case is justified by a workplace relationship and ceases when the exchange ends. The third type corresponds to a professional e-mail that occurs in the workplace but, due to certain mutuality, ignores certain features of professional e-mail.

Finally, Pano (2008a: 121 ff) proposes to distinguish between (a) *personal e-mails* ("private inter-individual messages between two or more people who exchange them on topics related to business or private life"), (b) *professional or institutional e-mails* ("messages of a private or semi-public quality between two or more identified persons"), and (c) *marketing and advertising e-mail* ("public messages and multiple interlocutors" and used in order to send information about a product or service and encourage purchases).

3. Electronic mail in the oral/written continuum

In the previous chapter I analysed virtual conversations (such as chat rooms, instant messaging, and virtual worlds) as *cyber-media* that often exhibit attributes of *oralized written texts*, hybrids between the stability of written (or typed) texts and the spontaneity and the ephemeral quality of oral utterances. Electronic mail also reproduces aspects of oral and written communication.[8] In this section, I will review some of these qualities following the four dimensions of analysis proposed by Baron (1998a), namely, *social dynamics*, *format*, *grammar* and *style*.

8. See, among others, Baron (1998c, 2000, 2003b), Benito Alcubierre (2003), Gouti (2003), Vela Delfa (2002: 460 ff., 2005), Pérez Sabater (2007) and Pérez Sabater et al. (2008).

3.1 Social dynamics

According to the dynamics of communication that defines the relationship between users, e-mail has attributes typical of written communication, such as absence of physical proximity and temporal distance between the message and its response. But e-mail also exhibits certain qualities of oral interactions.

According to Baron (1998a: 151), the most paradigmatic features of e-mail are: (1) separation in time and space between the transmitter and the receiver, though the increased speed of communication gives users the feeling of a synchronous virtual conversation and even of co-presence; (2) interlocutors often know each other; (3) privacy is taken for granted; (4) the conversational playing field is levelled due to the lack of physical co-presence; (5) there is personal disclosure due to the lack of face-to-face contact; (6) users can write to anyone, but the distribution and arrival of the message by the "addressee user" is not guaranteed; and (7) a certain speed of response is expected, as well as privacy of communication. Other features show a certain evolution, either towards the written end of the continuum, as in the freedom to write to anyone, or to the oral end, and the increasing reduction of waiting time between sending a message and receiving a reply, to the extent that at times e-mail communication almost seems to be more an electronic conversation than an electronic letter. This may be one of the reasons for the urgency with which users often expect a reply from an e-mail recipient.

The inference about the time that is "acceptable" between sending a message and getting a reply is part of the *chronemics* of human nonverbal communication, which includes both universal aspects of time management (for example, human activity in daylight and at night) and specific intra-cultural attributes (for example, different estimations of what being punctual means in different cultures). In the same way as we expect a reply within a specific period of time when we are talking to someone, when we send an e-mail we also expect a response within a certain time and when this intuitive time is exceeded, it generates all sorts of "anticipatory inferences" to understand the possible reasons for the delay and users derive several implicatures in this regard. The recipient's silence generates a restlessness that is generated, similarly, in other forms of Internet-mediated interactions such as chat rooms (see Rintel & Pitam 1997) or instant messaging (although in the latter the delay in the response is generally accepted within certain limits, by the assumption of multi-tasking, as discussed in the previous chapter). Hemp (2009: 12) illustrates this restlessness as follows:

> Our minds go through a series of semi-conscious calculations: How long does this person usually take to answer email? Should I bother her with a followup? Should I escalate my efforts by leaving a voicemail , and at which number? Should I walk over to Building D to see whether she is at her desk? Shout out of the window at the top of my lungs? Meanwhile, you may have to put a project on hold while you await a response that the recipient could provide in no more than a minute or two.

Williams et al. (2000) called *cyberostracism* this feeling of being ignored when the addressee of virtual communication does not reply as expected. Note, however, that "adequate time" does not necessarily mean "as soon as possible." As suggested by Kalman & Rafaeli (2005), an excessively fast reply is often interpreted as "something has gone wrong," for example, that the message is an automatic server message referred to a communication failure or as an "automatic response" that can be programmed on the e-mail account. In general, the literature usually cites one day as the average expected time for response, but there are many factors that influence an expectation of a longer or shorter period of time for a reply.

Besides, some studies address the social dynamics of e-mail. Thomsen et al. (1998) argue that the ability to send messages to multiple recipients simultaneously creates a structure of dialogue between users. Not surprisingly, 74% of messages sent to one of the newsgroups discussed in their research were replies to a previous question, indicating an underlying dialogic interactivity. In line with this, several studies have shown that e-mail does not prevent interactivity, but rather favours hyper-sociability (see Moran & Hawisher 1998: 91).

The pragmatic perspective of *conversation analysis* is particularly suitable in the study of the communicative interaction and social dynamics that take place through e-mail. In this case, the most common strategy when initiating an interaction with another user is *the move,* which is divided into one or more *acts*. The difference is that the former has a function of strengthening the progression in the interaction, while the second indicates the intention underlying the interaction. Strenström (1994) suggests eight major movements or *moves* (1a–h):

(1) a. Summon (calls the interlocutor's attention)
b. Focus (introduces the initiation)
c. Initiate (initiates the exchange)
d. Repair (maintains the exchange)
e. Response (replies in the exchange)
f. Re-open (re-initiates the exchange)
g. Follow-up (resumes or continues the exchange)
h. Backchannel (signals the interlocutor's attention)

As far as acts are concerned, there are three main types: (a) *primary acts* (e.g. accept, agree, reply, apologize, confirm, offer, reject, thank…), (b) *secondary acts* (e.g. emphasize, justify, introduce…), and (c) *complementary acts* (e.g. self-evaluate, frame, resume…).

Harrison (1998a) analysed a corpus of e-mail messages using this terminology. It was hypothesized that since e-mail is a highly interactive medium, most of the moves and acts would be frequent in e-mail communication. The hypothesis was confirmed. Most moves were present in e-mail, even though some moves such as backchannel were absent due to the asynchronous quality of e-mail communication. Furthermore, most acts were also found, except for "reject." Again, this can be explained due to the qualities of e-mail communication, in which a rejection can be performed by simply not sending a reply to the user.

3.2 Format

E-mail has different attributes that place it at both ends of the oral / written continuum. There is some tension between the paradigmatic feature of the lack of grammatical correctness in messages (especially the most dynamic ones, such as the ones sent to chat rooms and instant messaging) and the real tendency of many users to correct their e-mail messages carefully, due in part to the increased number of uses beyond the "informal letter." The same applies to the extension of messages, which are typically quite short, but with a potential for an extension of even several pages.

Directly related to the e-mail format is its (in)formal quality. Many authors stress the informal nature of the e-mail. Frequently, users make an effort to make manifest assumptions regarding the colloquial connotations and group relationships that they wish to communicate through their messages. This is so despite the origin of e-mail, which places it at the most formal end of the continuum. Indeed, the software that allows for the transfer of messages (e.g. *Outlook*, *Microsoft Mail*) generates a strict (and cold) organization of the message, with fixed fields for sender, receiver, subject, etc., and creates a writing environment that does not foster colloquial, informal or humorous communication. This quality is confirmed by the opinion of many adolescents nowadays, who consider e-mail only suitable for communicating with teachers and prefer other forms of interaction (such as chat rooms, SMS, social networking sites, and instant messaging) to interact with their friends.

Nevertheless, some authors have found informal and colloquial features in e-mail communication, among others the following: (a) a frequent omission of pronouns and auxiliary verbs, spelling that reproduces oral pronunciation, and very colloquial greetings and farewells (Lan 2000); (b) messages that are closer to oral communication (ellipsis and colloquialisms) (Tella 1992); and (c) dependency on the immediate context to express semantic relations, simple morpho-syntactic structures as those acquired at the infant stage of speech development, a tendency to repetition and use of similar structures in sequenced sentences (Gimenez 2000, reproducing attributes of oral speech, as proposed by Ochs 1979; for the attributes of informal e-mail attributes see also Gains 1999: 94–95).

Furthermore, Gousseva (1998), analyses the "speech event" of an e-mail following the taxonomy of elements proposed from the ethnography of communication. For analysts such as Saville-Troike (1989), in any speech event we need to highlight the genre of the message, its topic, its purpose or function, the environment (setting) surrounding it, the key of production (how and with what tone or attitude communication is established), the participants in the interaction, the shape of the message, the message content, the structure of the conversation, the rules of interaction and the rules of interpretation. All of these variables, applied to the messages of a mailing list (*Listserv*), would be as follows: (1) *Genre.* Varied, from that of the formal letter to the informal joke. (2) *Topic.* In mailing lists, everyday use defines the topics that users are to follow, at least for a certain period of time. (3) *Purpose / function.* Also varied. The goal of the List is to enrich users' knowledge about a specific topic under discussion. (4) *Setting.* The least identifiable element of e-mail messages. Internet users are always in different physical locations but share the same virtual space of the List. (5) *Key.* In face-to-face communication there are many contextual cues (e.g. nonverbal behaviours) that inform interlocutors of a certain attitude. This is not available in text-based e-mail communication, as was analysed in Chapter 5. Therefore, users resort to typical strategies of oralization. (6) *Participants.* In newsgroups these are typically strangers, only linked by a shared interest in a topic. (7) *Style of the message.* Oralized written texts similar to the ones found in virtual conversations. (8) *Message content.* Defined by the List. (9) *Structure of conversation.* Typically made up of an opening, the main body of the message and a closing. (10) *Rules of interaction.* Subject to politeness rules (*netiquette* or *cyberpoliteness*, see next chapter) that all members of the List have to follow. And (11) *Rules of interpretation.* Certain assumptions belong to "the culture" of the List, which are taken for granted within the limits of this small community of users.

Among the variables mentioned in Gousseva (ibid.), the first one, the delimitation of the status of e-mail communication as a genre on its own, is essential for cyberpragmatics, since the conventionalization of a genre saves effort among users and favours relevant interpretive outcomes. In this sense, several authors claim that e-mail should not be considered an autonomous *cyber-genre.* For example, e-mail users resort to strategies of text oralization similar to those found in virtual conversations, so no distinctiveness is involved. Yates & Orlikowski (1993:320) comment that "although some developments in the structural and linguistic features characterizing electronic mail have been noted, without further empirical study it is not clear whether these features have become sufficiently widespread or stable within smaller or larger communities to be institutionalized as genres." Similarly, Androutsopoulos (2006:420) claims that "it is empirically questionable whether in fact anything like a 'language of e-mail' exists, simply because the vast diversity of settings and purposes of e-mail use outweigh any common linguistic features." And the stabilization of an e-mail genre is not facilitated by today's e-mail handbooks, which "sometimes present starkly divergent views on how to formulate messages" (Baron 2002:410).

However, in my opinion e-mail has reached the status of a conventionalized medium and its fixed areas for sender, receiver, subject, etc. make it a highly identifiable discourse, thus saving production and processing time. Even inside the body of the message, e-mails usually follow strict conventions that aid in their processing. For example, Abbasian & Tahririan (2008) show how academic e-mails follow a strict sequence of *moves* (in Swales' 1990 sense) that help readers in obtaining the right interpretation of e-mail content:

(2) Subject: Abstract Submission

Thank you for your submission to (Conference)	*Thanking the addressee*
The organising committee are pleased to inform you that your abstracts have been accepted for a poster presentation. After all abstracts	*Providing information*
have been received, five will be chosen for oral presentation at each workshop.	*Adding, continuing*
Further details of poster boards will be sent to you.	*Indicating plans*
In the meantime check the meeting web site for regular updates.	*Advising about the message*
We look forward to hearing from you.	*Encouraging further response*
Organising Committee	*Signing off*

Overall, the corpus of e-mails analysed by these authors revealed six major *moves* in the academic e-mail genre, realized in a number of typical strategies (in brackets): (a) establishing the communication information chain (defining participants, opening); (b) establishing the territory (thanking the addressee, introducing a topic, giving a quote); (c) providing information or answers (continuing, adding, updating, indicating wishes or plans, advising about the message); (d) requesting information or action (asking for materials or documents, requesting that something be done); (e) evaluating (giving personal opinion, making comments, (dis)agreeing); and (f) closing (thanking, ending, signing off).

One symptom of increasing conventionalization of an "e-mail genre" of its own lies precisely in the frequency of openings and closings in these messages. As confirmed by Bou-Franch (2011), the ubiquity of opening and closing sequences in the data seems to be interpreted as resulting from institutional and cultural constraints that typically shape genres in these formal settings. In another study dealing with methodological aspects of cross-cultural pragmatic research, Bou-Franch & Lorenzo-Dus (2008) also compared requests. Their analysis revealed that over 77% of the British English and 93% of the Peninsular Spanish e-mails contained opening mechanisms, and all of them had closings in both languages.

Nevertheless, Abbasian & Tahririan (2008) found heterogeneous forms for openings and closings, as well as a wide range of realizations, which reveals that despite the conventional arrangement of *moves* inside the e-mail text, there is still room for creativity and flexibility inherent in this "genre in progress" thus allowing more personal idiosyncrasies to emerge.

On the whole, e-mails and especially academic ones seem to fit the conceptualization of a genre, since, as organizing structures, they "shape beliefs and actions, and in doing so enable and constrain how organizational members engage in communication" (Im et al. 2009: 93). As such, e-mail genre influences the discursive norms of interaction with socially recognized expectations around key aspects of communication (ibid.).

Central to the current discussion on whether e-mail is already a truly conventionalized genre is the place it occupies on the scale between the qualities of a written (i.e. typed) text and the attributes of oral utterances. Some authors stress its oral or written qualities, while others suggest that e-mail is a hybrid of both, thus agreeing with my proposal of *oralized written text* in Chapter 5.[9]

The key to the oral / written status of e-mail lies, like in virtual conversations, in the fact that very often the users type the message but they wish they could be saying it (a possibility available in advanced e-mail programs, but rarely used nowadays, see Outing 2000). Hence there is a possibility of a new discourse halfway between the oral and the written mode, a new register or a special kind of speech event (McElhearn 1996, Baron 1998b). Such discourse, however, raises problems for getting the proper interpretation of the message. Irony, sarcasm, or indirect remarks are difficult to communicate in an environment that prevents us from smiling, using an ostensively ironic tone of voice, or visually showing our smile of complicity (Jonsson 1998: Chapter 2; Yus 2000b, 2000c, 2009a; Sproull & Kiesler 1986). Like in chat rooms and instant messaging, e-mail users have

9. See Yates (1996), Ferrara et al. (1991), Yates & Ornikowski (1993), Wilkins (1991), Gousseva (1998), Danet (2002), López Rúa (2005: 14–15), Pérez Sabater et al. (2008), and Cho (2010), among others.

developed various strategies to compensate for the absence of information from the nonverbal channel (vocal and visual) and those strategies aid to the successful communicative outcome in which the first interpretation of the text chosen by the recipient of the message matches the one intended by the sender.[10] A good summary of the strategies for oralization is found in Danet (1996b), and many of them (perhaps all) are also used in virtual conversations:

(3)	a.	Multiple punctuation	Type back soon!!!!!!
	b.	Eccentric spelling	Type back sooooooooon.
	c.	Capital letters	I'M REALLY ANGRY AT YOU!
	d.	Asterisks for emphasis	I'm really *angry* at you.
	e.	Written-out laughter	hehehe hahahaha
	f.	Descriptions of action	*grins*
	g.	"Smiley" icons	:-) ;-)[11]
	h.	Unusual spacing	J E N N Y
	i.	Verbal/visual puns	A t D h V a A n N k C s E (thanks in advance)

Just like virtual conversations, e-mail can be innovative and connoted with oral features. Danet (1998) summarizes this quality of e-mail as follows:

> Digital writing is strikingly dynamic and playful, and even speech-like. Ordinary e-mail is frequently experienced, paradoxically, as a form of "talking." CMC challenges currently held beliefs among folklorists and students of rhetoric, oral literature, and the history of literacy about the uniqueness of oral culture. They have assumed that its key features are dependent on the personal, face-to-face interaction of individuals, e.g., a storyteller and his or her audience. Oral culture is believed, for example, to be agonistically toned, whereas writing, subject to processes of decontextualization, supposedly neutralizes this component of human interaction. Yet even in ordinary e-mail, both private and person-to-group, we find a striking tendency to sudden flare-ups of anger and insult, known in cyberculture as "flaming."

Besides, users' playful attitude towards their messages is often translated into real iconic images made up of text which users tend to create as part of their

10. Belda Medina (2003a: 47–48) summarizes the strategies of textual oralization of e-mail messages in Spanish: (a) irregular use of punctuation and repetition of characters; (b) irregular line breaks and paragraphs; (c) use of capital letters and asterisks to highlight; (d) simplification of consonants in the same phoneme (e.g. "qu" as "k"); (e) abbreviations and word games; and (f) use of emoticons.

11. According to Harmon (1999), e-mail users prefer the textual description of their feelings and moods, rather than the use of emoticons.

electronic signature. This is a strategy widely used in advertising, in which the text takes on peculiar iconic forms as required by the advertiser. It is also abundant in comics (Yus 1997b, 2008f; Gasca & Gubern 1988), where it is called *lettering*, and used for enriching text with nonverbal connotations. An example from e-mail would be the "textual tree" found in Patterson (1996: Chapter 7) and reproduced in Figure 6.1.

```
              _\|/_
               /|\
                I
               saw
              jolly
             tree of
            christmas
          time in xxx-
          tale and I was
         trying to get it
        decorated with the
       xmas candles but the
      result was not totally
     satisfying so I did make
    my own tree with a star at
   the top so I could follow up
   Ms Widge's "explosive" posting
  and perhaps get you too into the
 Xmas's feeling already in November
```

Figure 6.1 Iconic text in an e-mail

The strategies for increasing expressiveness can turn e-mail into a medium that facilitates the communication of feelings and emotions, because it is less intimidating than face-to-face interaction.[12] For example, Palme (1999) notes that e-mail increases the feeling of togetherness or cohesion and understanding among workers of a company. Not surprisingly, the influence of a person on the group can be measured by the degree and quality of information disclosed to others (Taylor 2000: 97) or, in relevance theory terms, by which assumptions of our personality we manifest to others, some of which may end up as mutually manifest to us and the other members of the group and used as contextual assumptions in discourse comprehension.

12. See, among others, Baron (1998a: 147, 1998b), Mabrito (1991), Livingood (1995) and Moran & Hawisher (1998: 88).

Byron & Baldridge (2007) and Byron (2008) have studied two strategies for the expression of feelings and emotions through e-mail: the connoted use of capitalization and the use of emoticons. Theories such as the *hyperpersonal model* (Walther 1996, 2007) argue that addressee users will pay attention to these aspects as part of the strategy to obtain maximum benefits from the few contextual clues available to them in text-based communication, and it is expected that these recipients will establish certain assumptions about the sender's personality, mood, feelings, attitudes or emotions of the sender depending on the use of capitalization or emoticons.

The use of capital letters can be neutral (for example, to differentiate copied text from personal text). Capital letters can also be used to communicate positive and negative emotions, although the latter are often prohibited in the various e-mail "etiquette manuals" that can be found in bookshops and on the Internet. In theory, it is assumed that a sender who makes proper use of capitalization will be regarded as more favourable, because improper use can generate additional mental effort to assess the origin and intensity of emotions that justify the appearance of this capitalization. Emoticons (specifically the conventional one of happiness) in a message can promote a more favourable impression in the addressee, for example by smoothing with a veil of irony the initial connotation of "seriousness" of the text that the emoticon accompanies.

However, although undoubtedly "strategies of oralization" connote e-mail messages with certain expressiveness, the lack of physical co-presence of the partners and the time that elapses between the coding of the e-mailed emotion and its subsequent reception produces many misunderstandings of feelings and emotions. One consequence is the so-called *neutrality effect*, whereby positive emotions tend to be interpreted as lighter or more neutral than were intended by the sender. A second consequence is called the *negativity effect*, according to which e-mail messages with negative emotions are often interpreted as more intense than were intended by the sender, a conclusion also reached by Kato et al. (2007: 1903). The third and final consequence was also concluded in Yus (2005a), and also mentioned in the previous chapter for emotions in chat rooms, namely that it is difficult to communicate degrees of intensity in emotions communicated by e-mail.

As happens in chat rooms and instant messaging, many of the strategies mentioned above are intended to make messages more expressive, so that they can communicate the whole range of the user's feelings and emotions, impossible to reproduce by using only words on the keyboard. According to Goleman (2007), such an undertaking inevitably promotes misunderstandings:

> Sitting alone in a cubicle or basement writing e-mail, the sender internally "hears" emotional overtones, though none of these cues will be sensed by the recipient. When we talk, my brain's social radar picks up that hint of stridency in your voice and automatically lowers my own tone of exasperation, all in the service of working things out. But when we send e-mail, there's little to nothing by way of emotional valence to pick up. E-mail lacks those channels for the implicit meta-messages that, in a conversation, provide its positive or negative spin.

3.3 Grammar

When compared to word processors, which have advanced systems of spelling and style, e-mail appeared in the past to be a rather rudimentary medium (Harrison 1998b), closer to the informal letter than to a well revised printed document. However, today's e-mail programs (e.g. *Outlook Express* or the recent *Windows Mail*) offer the user various options, both for spelling checks and for variation of font size and text appearance. Therefore, it should not be surprising to find in e-mails typical grammatical constructions and lexis of written texts and also features of informal letters.

As far as grammatical features of formal texts are concerned, Baron (1998a: 153) finds in e-mails a somewhat high density of constructions (subordination over coordination), and sustained use of connectives of a disjunctive kind (e.g. *however, by contrast*).

Concerning the grammatical features typically associated with oral communication, Baron (ibid.) found in e-mails a frequent use of first and second person pronouns, present tense and verbal contractions. Abbasian & Tahririan (2008) also found a considerable number of discourse markers, abbreviations and contracted forms in their corpus, which provided evidence for the stylistic similarity of electronic messages and informal spoken discourse. In Kirkgöz (2010: 342), the corpus of e-mails (from a textile company) exhibited a surprisingly high percentage of abbreviations, some of a standard kind (4a–b), some more creatively constructed (4c–d), and some specific of that industry (4e–f), with different amounts of processing effort involved and variations in the pieces of supposedly shared information, and which is needed to interpret them correctly:

(4) a. Pls (please)
b. Info (information)
c. rgds (regards)
d. thnks (thanks)
e. gmt (garment)
f. mts (meters)

It is also interesting to mention the study by Biesenbach-Lucas & Weasenforth (1998), in which they compared the grammatical and orthographic features of e-mails to those of a conventional message. Four hypotheses were raised, as commented in Table 6.1.

Table 6.1 Hypotheses and results of Biesenbach-Lucas & Weasenforth (1998)

Hypothesis	Results
Sentence connectors, such as *moreover* and *therefore*, conjoin sentences / clauses and explicitly mark logical relations in discourse. Several researchers have associated sentence connectors with planned, formal written discourse. Therefore, they expected to find more sentence connectors in off-line than in on-line writing.	No dramatic differences in the use of these connectors were found in offline texts and e-mails.
Clause coordinators are typically associated with spoken discourse where ideas are chained together. They expected clause coordinators to be more frequent in the on-line writing than in the off-line writing.	Confirmed. There are more coordinators in e-mails (stressing their oral connotation) than in offline texts.
Many researchers agree that clausal subordinators have different functions in pre- and postposed positions, but the reasons for this distribution are not clear. It has been suggested that postposed *because* is more frequent in spoken discourse. They hoped that their investigation would shed light on the potentially differential use of clause subordinators.	There is similar use of subordinators in e-mails and offline texts, but there is certain higher frequency for e-mails in the use of adverbial subordination and conditionals.
Since nominalization is more characteristic of written language and because phrase subordinators are, by definition, followed by a noun phrase, they expected phrase subordinators to be more frequent in off-line writing.	No valid conclusions. Too few examples of nominalizations were found in the corpus, which prevented drawing clear conclusions.

3.4 Style

As already mentioned, e-mails are usually more informal than traditional texts, despite the fact that the current increase in the range of e-mail applications places them closer to more formal discourses (Baron 1998b, Yates & Ornikowski 1993,

Wallace 2000).[13] Among the attributes that indicate the presence of an informal style in the electronic message there is the use of informal forms of greeting and leaving (Watson 1996, Lan 2000, Yongyan 2000), and the frequency of interactions, largely phatic in nature, in which humour is the main reason for communication. Again, we can see how an interactive typed text can adopt many forms and strategies typical of oral discourse. The comments made in the previous chapter regarding oralized written texts in chat rooms and instant messaging are, to some extent, of similar applicability to e-mail, even though, in principle, it seems to be a much more traditional variety of online communication in the use of written text.

In her analysis, Baron (1998a) also found the prototypical features of informal oral communication in e-mails. But the growing range of e-mail uses increases its formality in certain contexts (e.g. academic, scientific, specialized). She also found extensive use of the first and second person pronouns as forms of addressing that enhance the feeling of informality. Besides, regarding the use of greetings and farewells, e-mail users frequently type generic greetings such as "hello!," even with strangers. The (formal) signature is often omitted, especially if the message is sent from an upper to a lower level in a hierarchy of users (e.g. manager to employee in a company). Furthermore, the unrestricted and uncontrolled expression of feelings and emotions, so typical in oral interactions, is also valid for e-mail communication, with *flaming* as a common instance favoured by the lack of physical co-presence (see Chapter 7). Finally, e-mails are often humorous, phatic, aiming at strengthening a social bond.

Clearly, and predictably, the style of e-mail and its (in)formality are influenced by a number of exogenous aspects such as the power relationship exiting between the interlocutors. To illustrate this point, Bou-Franch (2011: 1779) concluded in her analysis of e-mail closings that "email users in the corpus were especially careful in elaborating their comments in order to begin conversations, particularly when addressing users of higher institutional standing. In contrast, a more casual, less elaborated style seemed appropriate in closing conversations with non-dominant users."

13. For example, in a survey made in South Korea (Fitzpatrick 2007), it was concluded that the perception that e-mail requires more communicative effort than SMS or instant messaging has led young people to shy away from using e-mail. Among other reasons, respondents noted that it is difficult to know whether the message has arrived and to determine the reasons for the lack of immediate response (which they do get with *messenger* and SMS). For one respondent, SMS is like playing table tennis, while e-mail is like "doing one's homework." The immediacy, privacy and customization of SMS, combined with the trend of mobile phones as personal non-transferrable items has resulted in a loss of interest by young people in this *cyber-medium*.

4. Elements of an electronic message

4.1 The (ostensive) call for attention

One of the most interesting pragmatic features of e-mail is that it is an ostensive technological medium. As outlined below, the electronic message reproduces the characteristics of acts of ostensive communication, and therefore, e-mails carry the presumption of their eventual relevance.

In ostensive communication, the sender makes manifest the intention to provide some information for the addressee (a number of assumptions). For example, if someone sniffs the air ostensively, that person may make manifest his/her intention to communicate that there is a smell of gas in the house. In Sperber & Wilson's (1986, 1995) terminology, sniffing the air would be the *communicative intention*, which is useful to underline the *informative intention* to convey the warning of a smell of gas.

Ostension is useful in human communication because it helps to reach successful interpretative outcomes. Often, if the addressee does not identify the communicative intention he/she may not access the information intended by the speaker. This is what would happen if, in the example above, the addressee believes that the other individual is sniffing the air because he suffers from some kind of allergy, i.e. without any communicative intention. The identification of intentions underlying the production of stimuli is essential for effective communication, and the definition of ostensive-inferential communication (S&W 1986: 63) reflects this emphasis on intentionality: "The communicator produces a stimulus which makes it mutually manifest to communicator and audience that the communicator intends, by means of this stimulus, to make manifest or more manifest to the audience a set of assumptions."

E-mail is also ostensive and includes a relevant "call for attention," because the computer produces an audible signal (sometimes also visual, depending on the software) that alerts the user that someone intends to communicate some information (the same applies to instant messaging, but in this case, with an additional window arising from the computer taskbar warning of a user's communicative intention). The sender produces, in this case, an auditory-visual stimulus that makes manifest to the recipient that the sender intends, through an e-mail, to make manifest or more manifest a set of assumptions.

Computers offer many possibilities for ostensive communication, all of them aimed at bringing to the attention of the addressee (*communicative intention*) the information that the sender wants to communicate (*informative intention*). For example, in the movie *You've Got Mail* (mentioned above), every time the main characters logged on, they were greeted with an icon of a mailbox with an envelope poking out of it, and in addition, the program generated the audible message "You've got mail." Most e-mail programs used nowadays include some form of call for attention. One of the most widely used, *Outlook Express* (and its successor, *Windows Mail*), generates the sound of musical notes while, at the bottom of the screen, it displays a small icon of an envelope. In both cases the function of the signal is the same: to draw the recipient's attention to a *communicative intention* behind which lies an *informative intention*.

As in any ostensive act, the arrival of an e-mail carries the presumption of its eventual relevance. Since interpretation involves effort, it makes no sense – in theory – to call the addressee's attention with sounds and images unless what is going to be communicated is considered relevant.[14] However, to the effort involved in any interpretation of stimuli we have to add the effort inherent in this *cyber-medium*, that is, the effort required to know how to use the technology to access the information that the sender wants to communicate and subsequently devote cognitive resources to distinguishing relevant messages from "spam." Therefore, we refer to a presumption of relevance that the electronic message carries and which has to compensate, in its ostensive quality, for the effort required to use the technology that allows us to access the message and, simultaneously, the effort of processing the assumptions that the message makes manifest and to obtain a relevant interpretation.[15]

Several studies deal with the effort involved in the management and mastery of the different media and their potential impact on the quantity and quality of interpretations. Clark & Brennan (1991) analysed this "communicative cost" in the use of media. Some variables were taken into account, including the cost of establishing communication (writing requires more effort than the spoken word), the time that this demands from the sender (for example, to find a phone number and dial it, in the case of the phone) and the time required for asynchronous

14. Interestingly, these acoustic-visual "calls for attention" can generate problems related to the priorities of the different communication sources in everyday communication. Baron (2000: 236), for example, wonders where e-mail fits, whether it has preference over face-to-face interaction. Computers that are primed to sound when new mail arrives are more like phone calls than letters, especially if one is anxiously waiting for the arrival of a message. Messages that are received with no visual or auditory cues are more like letters that lie in the mailbox, waiting to be collected.

15. Dertouzos (2000) comments that "prolific e-mail authors should think of each message they send as an instrument that reduces the recipient's life by two to three minutes. They should send it only if they judge that the effect justifies the cost. This may sound unreasonably harsh, especially since all human work involves invasions into other people's time. But e-mail differs from face-to-face encounters where everyone's time is equally taxed. That's because with only a flick of a finger you can send copies to a huge number of people."

communication (a letter takes longer than a phone call). But these variables do not influence everybody in the same way. In the case of e-mail, for example, the effort depends on the ease or difficulty with which the user accesses the e-mail account and the degree of skill in dealing with the software that manages it. Concerning the time required by different media, e-mail may even be more economical in terms of communication cost, than the phone call if the caller is at a great distance and it is difficult to establish the connection.

Another theory that studies the processing effort in media discourses is *Information Richness Theory* or IRT (see Ngwenyama & Lee 1997 for an application to e-mail). The variable of information richness is measured according to how information is processed within an interval of time. The richest communicative exchanges are those that allow for more effective interpretations with much contextual support, the possibility of immediate feedback, the use of different channels of communication, the variety in the language used, etc. From this point view, face-to-face communication would be the richest medium, because it provides immediate feedback, multiple channels (verbal / non verbal), use of oral language, etc. (Donath 1997).

Regarding the use of e-mail, several studies have refuted this theory. Users tend to e-mail other people more frequently than IRT predicted. Despite having poor informational features such as the absence of immediate feedback or the existence of a single communication channel, e-mail users use it very frequently (see Huang 2002). This is because this theory, and similar *information richness theories* (as they were generically labelled in Yus 2007b: Chapter 3) establish a rigid pairing between the contextual richness of the medium, in this case e-mail, and user satisfaction, the desire to continue communication with others through this medium, and maintenance of interest. In fact, although it is true that certain qualities of the medium can generate extra effort to access the desired information, the relevance of electronic communication does not depend solely on these fixed parameters, and users constantly surprise analysts with sources of interest and communicative satisfaction that were not predicted by these theories, such as the feeling of belonging to a group, the benefit of generating content for other users, and the *ambient awareness* of being present in the conversation and sharing a network of interactions.

This is coupled with the users' ability to extract from few contextual clues a number of inferences about the sender (e.g. nonverbal communication from *oralized written text* in e-mails) and to pay attention to features that are barely detected in face-to-face communication, as predicted by *Social Information Processing Theory* (Walther 1992) and the *Hyperpersonal Model* (Walther 1996, 2007). According to these models, as the relationship develops partners open themselves up to each other and establish more channels of communication and reveal more intimate information about their personalities. As text-based virtual communication is devoid of nonverbal information and contextual support from the environment, which typically foster privacy and therefore strengthen relationships, self-opening of the user to the other person in virtual environments is more essential for developing relationships than in physical environments. Such opening-up, however, may require more time due to the nature of this electronic medium (Walther & D'Addario 2001: 325). This is why Walther argues that different social and psychological feelings associated with communicating online give users a benefit that compensates for the loss of contextual information. Therefore, e-mail communication can compete in satisfaction with face-to-face interactions. Internet users quickly learn how to exploit the few contextual sources available to generate a more adequate virtual self-presentation. Thus, many of the contextual sources that have little or no importance in face-to-face communication situations acquire vital importance in virtual environments such as e-mail.

4.2 The sender

In relevance terms, it is not the same to receive a message from a person that presumably shares a mutual cognitive environment with the user than from an (impersonal) mailing list. The expectations of relevance are altered from the moment we identify the origin of the message and, in the case of known senders, the human cognitive system has the ability to select, from the vast array of assumptions stored in long-term memory, those which are relevant to a communicative exchange with that person. Every day, almost without realizing it, we exchange utterances with others and in the course of conversations we select topics which refer to predicted mutual cognitive environments (e.g. mutual friends, shared tastes and interests, awareness of the same physical environment). Often mutual manifestness of certain assumptions is essential for a correct interpretation and may even be the only reason for interaction.

The same applies to the expectations generated by identifying the sender of the message. Irrespective of whether the sender is known or unknown to us, the message carries a presumption of relevance, but the inferential operations to make sense of the message involve an access to different contextual information in each case.[16]

---

16. Of course, the access to certain assumptions from contextual sources (paralanguage, facial expressions, etc..) is reduced in e-mail, as in text-based virtual conversations. This fact can lead to misunderstandings (see Philips & Barnes 1995, Jiménez Gómez & Vela Delfa 2004).

It is pertinent to note that the expectations of relevance generated at the moment of identifying the sender make an impact on further processing of the message, that is, accessible contextual assumptions at the time of making inferential hypotheses regarding the content of the message vary according to the expectations initially generated by the name and e-mail address of the sender. For example, in Williams (1999), some participants in an experiment acknowledged the influence of the sender's name in the subsequent attention given to each message:

> I look at who it's from first, and that really determines whether or not I read it straight away.
>
> Because I can see who it was from, which everyone obviously can, I knew that it was really going to be quite important because she's the manager
>
> If it's from someone in my immediate area then I'm not too worried about it, because if it was urgent they would have come up and talked to me anyway.

The changing expectations of relevance in the addressee, provoked by the sender's name (expectations of the addressee about the sender) are complemented with other alterations, also important, in the expectations of relevance generated by the addressee from hypotheses about what purpose underlies the production of the message by the sender (Suler 1998).

Finally, the change that has occurred in the "sender-location" duality in e-mail communication is noteworthy. When e-mail became popular in the early nineteen-nineties of last century, there was a strong bond between the e-mail account and its user's location. Thus, people used to talk about their "home" and "workplace" e-mail accounts. And although this is still the case, in the past when someone received a message sent from either of these accounts, it was inferred that the sender was at the location associated with the account (home or work) and the addressee would be surprised to get a message from the "workplace account" on a Sunday (it would mean that the user was physically at the workplace). At present, however, mail accounts are accessible from websites for e-mail management (*web mail*) and, therefore, from anywhere in the world. Besides, 2009 has proved to be a key year in the dissemination and popularization of "third generation" mobile phones (*iPhone, Blackberry*...), colloquially called *smart phones*, with which the user can manage e-mail without a traditional computer logged onto the Internet.

4.3 The addressee

As already noted, one of the actions that influence estimations of relevance is the confirmation by the addressee of whether the message is personal, comes from a mailing list (Beke 1998) or is plain *junk mail*. One of the reasons for the loss of interest in messages from mailing lists is that messages, many of them totally useless, accumulate in the user's inbox, generating the annoying spam and the user's inability to cope with the massive list of e-mails accumulated there (see Szóstek 2011). Hence, it is understandable that the mere identification of such useless messages will lead to an estimation of irrelevance.[17]

Once the message is opened, the addressee's task is, of course, to look for an adequate context in which to process the message in a relevant way, that is, to find an interpretation that produces a high number of cognitive effects in exchange for minimal processing effort. For this task, it is essential to identify the sender's propositional attitude underlying the production of the message and which can be made manifest either explicitly or implicitly. For example, for a correct interpretation of the utterance in (5a), it is not enough to identify the basic-level explicature (5b), but the listener must also obtain higher-level explicatures such as (5c–f) in which the proposition expressed in (5a) is incorporated into assumption schemas that incorporate the speaker's attitude when pronouncing (5a) (S&W 1986: 11):

(5) a. Mary [to Peter]: "You're going."
b. Mary has said "you're going."
c. Mary is informing Peter that he has to go.
d. Mary guesses that Peter is going.
e. Mary is asking Peter to confirm whether he is going or not.
f. Mary is angry at the fact that Peter is going.

One of the easiest ways to identify the sender's propositional attitude of the message is to check what speech act is performed with the message and how such an act is expressed, either explicitly with textual attitudinal markers (modal verbs, questions, imperative mood, etc.) or implicitly. (6a) is an example of a request expressed explicitly, while (6b) expresses the same request implicitly. In addition, it is possible (and frequent) to use indirect speech acts, as in (6c), in which the speech act of a question is in fact an indirect act of request for the interlocutor to hurry:

(6) a. Open the window, please.
b. It is hot in here.
c. Do you know what time it is?

In e-mail communication, senders often make clear at the beginning of the message what they intend to achieve with the message, and often provide textual markers that identify the speech act performed in the message. Some analysts have studied which speech act is typically performed in e-mail communication (see Hassell 1998).

17. About this topic see, among others, Arthur (1994), Edlington (1995), Tanaka (2000), Harmon (2001), Mika (1998) and Barron (2006).

4.4 The e-mail address

The address is of little use when processing the e-mail message, with the exception of the initial filtering to check whether the message is addressed personally to the user or generated from a mailing list. However, some elements of the address generate inferential assumptions about the identity of the sender. For example, Portillo & Hartza (1995: 23) assert that

> with e-mail addresses the myth of the difference between the individual and the family is repeated. Indeed, in <eloy@upm.univ.es> the individual name is placed before the "at" sign (@) and the family name (Polytechnic University of Madrid) is placed after it. Lastly, the family is placed inside a bigger community, the country (in this case, Spain).
>
> Mere mortals send their addresses with a national ending: es, fr, it, etc. Only the gods can violate this rule: there are exceptional roots such as .com (which brings together U.S. companies and multinationals), .int (for international organizations). The only country that has no root is the USA.
>
> The "at" sign (@) plays the role of separation between individual identity and group identity, between the individual and society.

In general, then, the e-mail address influences the perception that the addressee has about the sender. The address allows us to guess, for example, if the sender works at a university or is an employee in a company. A typical example is academic journals. Sometimes in academic journals editors infer, from the e-mail address, and not always correctly, not only whether the author of the manuscript belongs to a university, but also the kind of university post the author currently occupies. Thus, the author of a manuscript who uses a university e-mail address (@ucl.ac.uk, @ua, @ual, etc.) will be considered, almost unconsciously, a member of the research staff at that university. By contrast, authors who provide popular addresses (@hotmail, @gmail, @yahoo, etc.) will be labelled, inevitably, only *potential* university staff, or authors who are still in a phase of training (e.g. writing a doctoral thesis). Of course, this almost unconscious inference often turns out to be incorrect.

4.5 The subject line

The subject line, i.e. those words that appear on the list of messages when accessing the e-mail program, has a vital importance for the inferential strategies of attribution of meaning by the addressee. Two functions of the subject line discussed in Vela Delfa (2005: 594–595) are (a) to allow for the inclusion of the message in a specific category or type of e-mail, that is, to facilitate its classification as a member of a subclass of e-mails, and (b) to reconstruct the chain of turn sequences plus replies by using the nomenclature "Re:" which, among other elements, relates a message to the rest of the e-mails belonging to the same thread of the "e-mail conversation."

The relationship between the subject line and the message is very close, and e-mail users are aware of it. For example, Gains (1999: 90–91) describes how the apparent formality of this element frequently becomes a barrier when senders wish to communicate an informal message. To alleviate this asymmetry between the subject line and the body of the message, users usually seek some informal text also for the subject line.

In addition, senders are aware that, when the recipient is faced with a long list of messages to read, on many occasions an original or effective subject line can save the message from being deleted even before it is read (McElhearn 1996). As noted by Suler (1998),

> The subject line is a tiny microcosm unto itself. Often people use it to just summarize or introduce the major idea/s contained in the body of the message. But experienced e-mail users understand the more subtle techniques for communicating meaning and emotion in the titles they bestow to their e-mail. The subject line can lead into, highlight, or elaborate a particular idea in the message. It can ask a definitive question, shoot back a definitive answer, joke, tease, prod, berate, shout, whisper, or emote. Sometimes its meaning may blatantly or discreetly contradict the sentiment expressed in the body of the message. A creative application of caps, commas, slashes, parentheses, and other keyboard characters adds emphasis and complexity to the thoughts and emotions expressed in the subject line.

An interesting study on the reasons for paying attention to a subject line and not others is found in Wainer et al. (2011). Initially, and predictably, they concluded that one of the reasons people may attend to an email is because they are curious about the content. However, and surprisingly, people attended to messages that had the largest information gap in terms of the least amount of information about the content in the subject line, which is counter-intuitive. In relevance theory terms, users' mental effort should be reduced (and expectations of relevance more accurately built) if the subject line provides detailed information about the subsequent content of the e-mail itself.

In any case, users clearly play with the expectations of relevance which, at a subsequent stage, the recipient expects to be confirmed by reading the message itself. Given a list of messages on the screen, the recipient will not devote much interpretive effort to messages which do not guarantee that this "cognitive expenditure" will be compensated for. In this regard, in recent months, an evolution in unsolicited messages (or *spam*) can be detected. These unsolicited messages accumulate in the inbox of the program and share the space of the screen with other e-mail messages and, in the case of *phishing*, their authors try to get information about bank accounts or make recipients perform bank transactions. All these useless messages require additional mental effort because they end up mixed with truly interesting messages in the inbox, an effort that is added to the effort involved in reading the messages themselves. This is an tedious task that increases the so-called *e-mail stress*, which alters workers' performance. As noted by Campbell (2007), the problem is that, after managing the huge amount of messages, one has to return to the task at hand, but the user has probably lost track of what he/she was doing and hence the employee becomes less productive. The human mind gets tired of having to stop work on a task to check and read e-mail every so often.[18] Therefore, e-mail programs in general and the e-mail inbox in particular should be designed to aim at relevance, that is, at reducing the mental effort involved in sorting out and grading e-mails according to their potential relevance, and also at dismissing useless spam and offer users potentially interesting messages. Software companies are making a number of efforts in this direction. For example, Szóstek (2011: 724) reviews the following:

> Outlook offers elaborate means to mark email priority and to view messages according to their due date. Gmail™ automatically threads messages with the same subject line. Gmail™ also enables message labelling and starring to further support grouping and sorting of related emails. Recently Yahoo made an attempt to automatically prioritize incoming messages based on the importance of the sender and also on the frequency of email exchange.

As Richtel (2000) points out, the senders of unsolicited messages tend to use subject lines that carry a high presumption of hypothetical eventual relevance to the recipient. Instead of predictable lines, senders (often advertising companies trying to sell a product) are now trying to make connections to the hypothetical cognitive environments of the target users, connections which carry, in themselves, expectations of relevance. In (7) there are some examples:

(7) a. About the information you asked for (in the body of the message: an advertisement for Viagra).

---

18. Professor Thomas Jackson, from Loughborough University, determined that an average of 64 seconds are needed to resume the task being carried out after a break to read e-mail, so a worker who checks the inbox every five minutes will have lost eight hours a week just thinking about what he/she was doing before checking the mail (quoted in Charman-Anderson 2008). See also Soucek & Moser (2010) for an analysis of "e-mail overload" specifically related to large amounts of incoming information, inefficient workflow and deficient communication quality.

b. Sack that worm you call boss (in the body of the message: an advertisement for investments).

c. Do you want to know how? Here you'll discover how (in the body of the message: an advertisement for software).

Another example of this kind of e-mail arrived in my inbox on December 28th, 2009. For users who were expecting a parcel to de delivered by this company (DHL), the subject line would no doubt generate expectations of relevance that might lead the user to click on the suggested link and hence be cheated by the authors of the fraudulent message:

(8)	Subject:	DHL Services. You have a parcel pending to be delivered to your address.
	Text:	Dear client, We could not deliver a parcel to your address.
	Reason:	Error in details of recipient's address. You can get the parcel yourself at the nearest post office to your address.
	Warning!	There is a postal tag attached to this message. You have to print it to get the parcel. Please, click here
	Thanks	DHL Services

This kind of fraudulent message includes a hypothetically relevant subject line that might obtain an action by the recipient. The same applies to *spam* messages, whose subject lines also create expectations of relevance for certain users. Some of these are listed in Barron (2006: 889), and these intend to create expectations of relevance by (a) requesting attention (*Just watch out for this!*), (b) addressing credentials of supplier (*Hate driving all the way for a doctor's visit* (addresses credentials)), (c) indicating awareness of prospect's needs (*Guaranteed lowest prices* (addresses credentials by assuring prospects of the price competitiveness of the organisation)), (d) addressing the offer itself / benefits of offer (*Viagra and Diet Pills Prescribed on line! US doctors and pharmacies! Overnight Shipping*), (e) soliciting a response (*Get prescribed Viagra, Diet Pills and much more online! Click on…*), (f) addressing unrelated, general content to capture attention (*You left your umbrella*), and (g) including no subject line except for an abbreviation common to e-mail communication (re: FWD).[19]

19. After drawing the user's attention, the body of the message usually follows a hierarchy of successive moves that aim at corroborating the expectations of relevance raised initially: (1) establish credentials; (2) introduce the offer, (3) use pressure tactics, (4) solicit response, and (5) give a polite way-out (Barron ibid.).

Another example took place in late 2009, when hackers obtained relevant data about many users, including personal details in social networking sites and e-mail messages sent or received by these users. All of this information was valuable to construct relevant subject lines for friends and contacts of the users, and the information contained in the body of the message included personal information that convinced recipients of the non-fraudulent quality of the messages. Messages such as the one quoted in (9) were supposedly sent by a friend and included the subject line in (9a), and further credibility was added with the (mutually manifest information) quoted in (9b), especially the last sentence:

(9) a. [subject line] I need your help.

b. [body of message] I'm on a trip in London. I left my wallet in a taxi, I have no money to pay for the hotel bill and cannot return to USA. Please, transfer 950 dollars to my account below. As soon as I arrive in New York I'll return your money immediately. Thanks, my friend and my apologies for the inconvenience caused. I look forward to meeting you again in Alicante, after my last visit.

One of the techniques used to increase expectations of relevance raised by the subject line is to make an explicit reference to the sender's message when replying. E-mail programs offer users the possibility of replying to the sender by simply clicking on a "reply icon." Another window is then created (either empty or with a copy of the sender's message, depending on the default options) with the same subject line but with the nomenclature "re:" at the beginning. In this way, the program makes it easier for the sender and the addressee to be mutually aware of the existence of the message and, in theory, to obtain mutual manifestness of its content. This intersecting area between the sender's and the replier's cognitive environments makes it possible to assume and take for granted the assumptions initially communicated by the sender.

4.6 The body of the message

In the past, e-mail texts used to be short due to restrictions imposed by companies for Internet access. At present, by contrast, with the trend of flat rates and increased bandwidth, users can type long texts and attach "heavy" files (a term typically used in computing jargon) such as photos and videos (see Weasenforth & Lucas 1997, Baron 2000: 241).

Messages are usually structured coherently and their different elements are recognizable by readers. For example, Herring (1996a) proposed to distinguish between (a) *informational expository schema* (divided into the following steps:

identification of the problem, proposal of a solution, evidence in support of the solution, and evaluation of the solution); (b) the *persuasive schema* (same as before, but the last step differs: appeal to give credence or to adopt certain values); and (c) the *interactive schema* (divided into these steps: link to previous discourse, contentful message, and link to following discourse). As a general rule, the author of a message is expected to organize the message according to a structure, and also organize its elements so that it is clear which information is supposedly mutually manifest and which is presented as "new" (Gruber 1998: 30, Williams 1999).

However, the most original way (and inherent in e-mail genre, if already stabilized) to provide coherence to an e-mail message and obtain maximal mutuality of assumptions is to insert one's words into the sender's message that one is replying to, that is, to type inside the quoted initial message. This is part of a more general strategy of including a message or part of one in the reply to a message and which Severinson Eklundh (2010) calls "contextual quotation." In my opinion, this "type-inside-the-quote" strategy is better explained by commenting on the relevance-theoretic discussion on the resemblance of two propositions, as discussed below.

A useful term in relevance theory is *interpretive use*. In general, stimuli only *resemble* the assumptions that the speaker intends to communicate. I have already commented upon the different inferential operations that hearers have to perform to turn schematic logical forms into fully contextualized propositions that, supposedly, *resemble* the information that the speaker wanted to communicate (see the analysis of the interpretation of *tweets* in Chapter 4). And there are informative gaps (between what is coded and what is finally interpreted) that the hearer has to fill inferentially with the aid of context. There is also resemblance between two stimuli that share some attribute. Consider the Example (10):

(10) a. Mary: What did the policeman tell you at the police station yesterday?
b. Peter: Still nobody has reported the loss of a Rolex.
c. Peter: That nobody had reported the loss of a Rolex.

In (10b) Peter quotes literally what the policeman told him. Peter's utterance and the policeman's are identical in linguistic structure (semantic structure and logical form). It is a direct quote, an extreme case of similarity between stimuli that is so high that it almost becomes identity of stimuli, although the propositions expressed by the same statement in different circumstances might differ slightly (for example, the temporal scope covered by the adverb *still* may be more restricted when the policeman uttered it than when, the next day, Peter has uttered it). Furthermore, in (10c) Peter's utterance is a representation of the policeman's utterance, not a direct quote: both utterances contain different linguistic structures, different semantic structures and different propositional forms. However, they communicate similar information. Peter's utterance *interpretively* resembles the policeman's but does not reach the level of similarity of (10b) (see Sperber and Wilson 1986: 227–231).

E-mail programs have an option that generates, automatically, an exact copy of the sender's message when it is replied to. To differentiate the sender's reproduced text from the user's reply, this copy has some special graphical attributes, basically marked by the sign ">" at the beginning of each line and, in some programs, by the words in italics (Suler 1998). The interesting aspect of this specific function of e-mail is that users often insert their own words within the initial message that the program has copied automatically, as in (11):

(11) Dear Mary,
> Thanks for taking the trouble to make our stay in London so pleasurable.
> You needn't have made such an effort to show us the whole city.
You're welcome, but it was no trouble, but pleasure. Come back whenever you want
> Now we have to return to work. Just the next day after returning from
> London and we have to be up by eight!!!!! We'll have no time to recover
> from the trip!!!!
You are telling me!!! Can't you just pretend you're ill or something?
Lotsa kisses, Peter and Ann

When the text is cited again, users typically add a second ">" to make clear that this is a re-transcription. From the second level of quoting, however, there is no convention for quoting , and users rely on a variety of strategies for clarification as to which part of the text is a quote, how much is "re-quoted" (a quote of a quote), and what part is new, as in (12a–c) (see Suler 1998, Bavelas et al. 1997), a strategy that, in extreme cases, annoys users due to the mental effort involved in decoding and inferring this multiplicity of "markers of quotation":

(12) a. *>> I know what you mean. He told me the same.*
> How did you react?
I didn't know exactly how to react.

b. *>> I know what you mean. He told me the same.*
> How did you react?
-----------> I didn't know exactly how to react.

c. *>> I know what you mean. He told me the same.*
> How did you react?
.... I didn't know exactly how to react.

Interestingly, e-mail quotations are not usually part of the propositional content that the sender wishes to communicate when replying to a message within the quoted text or, in other words, the user who replies does not intend to communicate a series of assumptions that are totally irrelevant to the addressee (the initial sender), because he/she was the one who originally typed the words of the quote (it does not even happen in the messages sent to newsgroups and read by multiple recipients, because again the sender does not have the intention to communicate the information contained in the quoted text). The quote is, rather, meant to emphasize mutuality between sender and recipient of the assumptions made manifest by the quote, the function to check that there is an area of the cognitive environments of both users that is mutual (Baym 1998: 45, Giese 1998), and the function of generating a preliminary context from which the user can provide further information such as an argument for or against (Mabry 1997).[20]

However, although this is a relevance-enhancing aspect of e-mail communication that saves mental effort while stressing mutuality, Severinson Eklundh (2010) considers that there are potential problems with quoting:

> Including a previous message makes the reply longer and may cause less than ideal reading conditions on the computer screen. Moreover, verbatim repetition of an interlocutor's utterance may be perceived as a violation of everyday norms of conversation. Many respondents in the Usenet survey characterized quoting as impersonal or impolite, and they reported that they avoided it in personal email communication.

E-mail users can choose between letting the program generate an exact copy of the message that the users wants to reply to, and using discursive strategies of quoting. One of the most common ones, indirect speech, has already been discussed with regard to Example (10c). These indirect quotations make manifest propositions with higher or lesser interpretive resemblance to the paraphrased message (Baron 1984). González-Bueno (1998: 62–63) notes that sometimes propositional resemblance is achieved by quoting part of a previous message, and then proceeding to provide new information, as shown in (13a–b):

(13) a. For the future? I want to get married.
  b. Your husband is sad because The Cowboys won the Superbowl?
  I'm very happy! I love Cowboys!

The reference to a previous discourse can be made in a subtle way, simply assuming that the recipient will be able to provide a context in which the statement will

20. Gruber (1998: 36) analysed the quote from another point of view, *systemic functional linguistics* (Halliday 1995). After identifying the *theme* and *rheme* of each message, he concluded that the quote is useful for (1) checking whether the reference to a previous message is correct, and (2) tracing a discussion back to its starting point. See also Gruber (2000b).

reach optimal relevance, as in (14a), or the "Yes" in (14b), which links the text to an alleged previous question. However, the use of "No" with the same function in (14c) is surprising: besides being a response to a question, sometimes "No" seems to give the recipient an opportunity to elaborate, clarify or correct erroneous assumptions which had been made manifest in a previous message (González-Bueno ibid.):

(14) a. My grandmother is much better, thanks.
 b. Hello. Yes, I got good marks in English and Maths.
 c. No, I do not have a pet in my apartment. I do not have room for animals. I want a dog. I like pit bulls. You like cats a lot, right?

Finally, it is relevant to mention the *epidemiological role* that this "writing inside the quote" generates. Indeed, the mutuality of assumptions that this strategy promotes is combined with the greater or lesser alteration of the text by the recipient. Thus, the message maintains its initial qualities, but at the same time it is varied in its re-distribution, just like virus epidemics, in which the virus mutates constantly while maintaining its "infecting idiosyncrasy." Authors such as Kibby (2005) see similarities between this strategy of quoting in e-mails and other social strategies of viral dissemination of information such as urban legends or folklore.

4.7 The signature

The signature is a brief text that the sender attaches to the message automatically and shows his/her affiliation, address, phone number, personal web page, and other. All of them are markers of real identity transferred to the virtual realm (Greenhill & Fletcher 1996). The quantity and quality of information contained in the signature varies depending on the number of assumptions about their lives that users wish to communicate (i.e. make manifest).[21] Undoubtedly, this information goes beyond the mere reference to the identity of the sender, invalidating statements such as Sherblom's (1988, quoted in Baron 1998a: 148) concerning the signature as redundant to the field "From:" in the e-mail message. This author, along with others like Ziv (1996), studied the possible hierarchical structure that underlies the use of the signature. In his study, he concluded that the messages

21. Donath (1999) applied Goffman's (1987) distinction between information (intentionally) communicated (*given*) and information (accidentally) exuded (*given off*) to e-mail signatures. Possibly the author of the message makes manifest more assumptions about his/her identity and social status than he/she initially wished to communicate. For example, the reader may interpret a certain narcissism in showing one's professional status when, in fact, the author of the message just wanted to provide neutral information about it.

sent from a higher to a lower level in a hierarchy (within a company, for example) tended to enclose no signature, while it was often used when the message was sent in the opposite direction, from a lower to a higher level.[22] The same applies to openings of e-mails within a hierarchy, as mentioned above (see Bou-Franch 2011: 1778).

Other authors such as Herring (1996a), Harrison (1998b) and Wallace (2000), although not specifically focussed on the signature, reach radically different conclusions on the use of e-mail. Rather than promoting the hierarchy, e-mail has, for these analysts, the ability to leverage social differences:

> A worker can communicate directly with a member of the senior management; a student can e-mail a professor [...] e-mail does not indicate the status of the writer. This can give a lower status correspondent a greater opportunity of being heard. The absence of non-verbal information about people's age, gender, race, and social status can benefit minority groups, while the lack of synchronicity can improve access for participants who might find difficulty in a face-to-face discussion. (Harrison ibid.)

Frequently, signatures exhibit rules of use that are accepted within a virtual group or community. In newsgroups, it is easy to find signature patterns inherent in the subject under discussion and to general attributes of participants. For example, in the discussion forum *soc-couples.wedding*, whose central topic is the preparation of marriages and the pros and cons of being married, participants typically sign by making it clear that they are married and for how long (Donath 1999), as in (15):

(15) a. Joan (& Mike, May 22nd, 1995).
b. Amy (& Chris, September 7th, 1996).

Some users add to the signature drawings made with typographical ASCII code (so-called "ASCII art") that visually connote the message (see examples in Di Segni 1997a). The signature shown in Figure 6.2 shows the options of the ASCII code for generating images in signatures.

22. Of course, not only the signature is an indicator of the level of the sender within the hierarchy of a company. Other features allow us to label the sender according to his/her status (see Stephens et al. 2011). As discussed in Headlam (2001), those of an upper level often send short messages to minimize contact with subordinates and reinforce their authority, and rarely use the option "copy" (CC:) to demonstrate that communication is established only with that employee, while messages from a lower to a higher level are often long and full of explanations and comments. In addition, messages from a top level tend to have more spelling mistakes, as if to make clear that the authors have more important things to do than checking their messages.

```
|➧
 ~~~~_____    |~~~~~~~~~~~~~~~~~~~~~~~~~~|
 Y_,__| [] |  | email ---> jbbixler@mail.uccs.edu |
{|_|_|_|PU|_,_|__________________________|
//oo---oo=oo   ooo   ooo          ooo        ooo
```

Figure 6.2 E-mail signature made with ASCII art

Concerning the elements of the signature, some users habitually add additional text to the signature like proverbs, famous quotes, parts of poems, etc. Rains et al. (2009) call this strategy *electronic bumper stickers*. These texts resemble stickers because they are ideas that are added to a device (a signature to the message and the sticker to a car bumper, respectively) and under the control of a particular person. Moreover, both are placed at a specific location (end of message and end of the car) and are offered to a broad audience with no control over who reads them. The authors sense that the stickers have some communicative value since they reveal certain aspects of identity and create certain impressions in the readers, just as happens with the signature that they accompany (see Rains & Young 2006).

CHAPTER 7

Politeness on the Net

1. Introduction

Politeness is a typical human strategy that aims to foster human relations and mitigate the imposition of certain actions on other people. Despite being a universal attribute, politeness is expressed differently in every culture. Not surprisingly, every language, apart from differences of lexicon, exhibits different ways of conceptualizing the world and the relationships among individuals within speech communities (and also within virtual communities and networks). A social behaviour that clearly indicates the existence of such a community is politeness. In fact, it is often a strategy that reveals which speakers do not belong to a community and ignore the inherent rules of (im)politeness which are, nevertheless, taken for granted by its members.[1]

But what is politeness? According to the literature on this topic, we can deduce that politeness refers to a speaker's manifestation of attempted appropriate social behaviour, but this is a term without clear boundaries. Neither Leech (1983) nor Brown & Levinson (1987), despite devoting many pages to the study of politeness, define it explicitly. It seems to be more a necessary attitude for the smooth flowing of life in society, a heterogeneous set of features that we usually associate with good manners, elegance and good taste (Sell 1991: 208).[2]

Escandell Vidal (1996a: 136) defines politeness as "a set of social norms, established by each culture, that regulate adequate behaviour of its members, prohibiting some and favouring others," but she also stresses its importance in the specific context of conversational interaction. Besides, politeness is usually related to the type of utterance chosen by the speaker (Fraser 1990: 221), a circumstance

1. This entails a picture of politeness as *learned* within a specific speech community and with an *intra-cultural* connotation of the conventional uses of politeness, which is observed in strategies such as the choice of certain forms of address, which differ from one culture to another. There are, in this direction, many studies on cultural differences in the use of politeness, but reviewing them is beyond the goals of this chapter.

2. According to Escandell Vidal (1998: 46), there are three main meanings of the term politeness: (a) what is socially correct, a view supported by Fraser (1990); (b) being nice or friendly, as Leech (1983) defends; and (c) being diplomatic, that is, not being aggressive, a stance that Brown & Levinson (1987) support.

that entails a greater or lesser degree of politeness inherent in certain phrases (Lakoff 1973, Leech 1983), as happens with the typical strategies used to make polite requests (*could you please*...?). Ultimately, politeness seems to be added to utterances so that the friction in personal interactions is reduced (Lakoff 1977) as well as the threat of certain acts (Brown & Levinson 1987).

The role of politeness in verbal interaction is so essential that other communicative strategies are placed on a secondary, less relevant level if politeness is threatened. For example, Grice's (1975) *conversational maxims* are generally not followed if the expression of a polite attitude is required (e.g. lying about the taste of a meal simply because we cannot be impolite to our host).

Leech (1983: 82), in the same lines, even formulated what he called the *Principle of Politeness*, motivated by the desire "to maintain the social equilibrium and the friendly relations which enable us to assume that our interlocutors are being cooperative in the first place." Cooperation and politeness appear, therefore, at the same level of importance during interactions, since both "are largely regulative factors which ensure that, once conversation is under way, it will not follow a fruitless or disruptive path" (ibid.: 17).

On the Internet, the expression of politeness is also common and often compulsory, which indicates its importance beyond face-to-face interactions. Typically, politeness on the Net is called *netiquette* (from *net* and *etiquette*), and thus the theoretical models analysed in this chapter are also applicable to Internet-mediated communication.[3] Below there are some examples (adapted from Slembrouck 2000) of messages where the sender uses the *maxims of politeness* proposed by Leech (1983) in his *Principle of Politeness* (the text in italics reflects the polite strategy used):

(1) a. *The Tact maxim* (minimize the interlocutor's effort; maximize the interlocutor's benefit).

b. Hello [name],
This message is from the organizer of a Seminar that will take place in Italy between September 26th and September 27th. I would like to invite you to participate in a round table about research on multimedia translation (*at last translation studies have realized that they need discourse analysis!!!*).
Would you like to participate in this event? *The tickets and hotel are on us and, if you wish, you can also give a plenary lecture.*

(2) a. *The Generosity maxim* (minimize personal benefit; maximize personal effort).

3. In some books, the alternative term *chatiquette* (from *chat* and *etiquette*) is proposed for the specific rules of politeness in chat rooms (see Jonsson 1998: Chapter 3). *Cyberpoliteness* is a term also proposed in the bibliography (e.g. Maricic 2000, Mariottini 2008).

b. Hello [name],
Sorry for not replying before. In fact, I have already read the article and made some comments. *The hard bit is to insert them in the file… It would be ideal to have a chat session through Telnet, if possible. But since you are at home, I will insert them and send you the file before going home. It's a promise.*

(3) a. *The Approbation Maxim* (minimize criticism of others; maximize praise of others).

b. Hello [name],
I am glad to hear from you. I've just had a look at your personal web page. *It looks fantastic!!*

(4) a. *The Modesty Maxim* (minimize self-praise; maximize self-criticism).

b. Hello [name],
It seems like the letter from Belgium has taken an eternity – although it is downhill from where you are towards us. Many thanks for the dedication. *I am very envious that you've got a hardback book already!*

(5) a. *The Agreement Maxim* (minimize disagreement with others; maximize agreement with others).

b. Dear [name],
Thanks for your comments. *You are, no doubt, right in your opinion about ideologies or work and accumulation in capitalism.* I wish I could have read the draft!

(6) a. *The Sympathy Maxim* (maximize sympathy to others).

b. Thanks [name], I'd be interested in the second Conference on your list, I hope you're fine, *incredibly busy like all of us of course!!!* I wish you the best, and thanks for the report.

On the Internet, polite strategies also operate in the production of electronic messages and utterances in environments such as chat rooms. These strategies can be chosen by the user, or they can be imposed by a moderator of the system used for online communication, as in the newsgroups discussed in the previous chapter (see Collins & Berge 1997, Smith et al. 1997). For example, participants who first enter a newsgroup are usually greeted with a message from its moderator (or an old member) which indicates what type of behaviour is considered (un)acceptable for that particular newsgroup.[4] In Patterson (1996: Chapter 6) there is an example taken from the newsgroup *AGM.net*, in which WrenZ is welcomed:

4. Interestingly, it has been demonstrated that the presence of a moderator in newsgroups increases participation by its users: "The presence of the moderator encourages to participate, perhaps by the simple fact that the participants know that someone, other than their peers, will read and check their messages; the moderator also provides guidance on how to participate, how often and asks questions involving more postings to answer or clarify" (Espinosa Villareal 1999).

(7) Greetings to you WrenZ!
Welcome to *alt.good.morning*, normally refered to as AGM here, making you one of our newest agmer's! As I said, welcome to the friendliest and most caring newsgroup around.
We accept anyone who wishes to post here who follows our simple guidelines:
a.) Post often
b.) Post morning posts
c.) No flame posts
We are certain you can meet these guides, but to make sure we remind you that here often means as often as you can; and morning is defined as anytime it is morning anywhere on the globe we call earth. As to the last one, we mean it, and ignore flame messages, or invite the author to join our friendly group.
You will be a wonderful addition to our growing community of posters!
Take care and have a good morning, and may your tomorrows be glorious!
P.s. here we use '{' '}' to denote hugs so ==> {{{WrenZ}}}

In this sense, the extensive use of the imperative in the listing of norms of politeness to be followed in Internet-mediated interactions, as collected in books and online, is surprising. Many manuals of *netiquette*, for example Shea (1994), assume that anyone who participates in any of the interactive possibilities of Internet (e.g. chat room, instant messaging, newsgroup, mailing list and e-mail) must comply with the official standards of social etiquette.[5] Some tips of netiquette are quoted below for e-mail in (8) (Escribano Otero 1998), for newsgroups in (9) (Laborda Gil 2005: 104) and for the mailing list in (10) (Escribano Otero ibid.):[6]

(8) a. Avoid excessively formal address forms, especially when the message is directed to users from other cultures. On the Net a more informal language is permitted.
b. Sign your messages.

5. These official rules have, in principle, a trans-cultural attribute that universalizes politeness on the Net against intra-cultural specificity of offline politeness that has been extensively analysed in the bibliography. The same does not apply to sex differences in the use of politeness, even though some analysts have viewed the disembodiment of virtual communication as prone to sexual equality in the use of language. As shown by Herring (1994, 1996b), among others, strategies of power and aggression (in men) and solidarity and support (in women) are also reproduced in computer-mediated communication, and the same applies to the use of politeness strategies. See also Witmer & Katzman (1997).

6. See also Shapiro & Anderson (1985), Shea (1994), Escribano Otero & Peña (1999), and Pérez Sabater (2007: 100 ff).

c. Be as brief as possible.
d. When attaching files, try to keep them small. Otherwise, compress them.
e. When you reply to a message, don't quote the whole original text, unless strictly necessary.

(9) a. Don't participate anonymously, but with your name and surname.
b. Write about topics of interest to the newsgroup.
c. Don't send the same message to different newsgroups.
d. Respect other users' opinions.
e. Don't quote other users' messages without permission.

(10) a. Before sending messages to the list, read the FAQ (frequently asked questions).
b. Devote one or two weeks to reading subscribers' messages before sending yours. That will give you a picture of "the culture" of the list.
c. Be brief and avoid messages that might be offensive to any subscriber to the list.
d. Send a summary of the messages when other users have finished replying to your query.
e. Choose carefully the subject line of your messages.

As we have seen, on the Internet the estimation of the suitability of a certain politeness strategy no longer applies only to interactions between a sender and a recipient, but also to the participation of a third person, such as an external moderator of the virtual interaction who may, in turn, assess positively or negatively the sender's strategy of (im)politeness positively or negatively (extending to the extreme case in which the user might even be expelled from the list or newsgroup if he/she is too impolite, rude, abusive or insulting). In this way, eight possibilities as regards the estimation of politeness are generated, as summarized in Table 7.1 (O'Sullivan & Flanagin 2000).

Table 7.1 Possibilities in the estimation of politeness according to O'Sullivan & Flanagin (2000)

Sender's perspective	Recipient's perspective	Third party perspective	Comments
appropriate	appropriate	appropriate	Messages clearly devised following norms of politeness that all users involved accept as appropriate.
appropriate	appropriate	transgression	Messages clearly devised following norms of politeness that sender and recipient accept as appropriate. However, a third party (e.g. the moderator) finds them inappropriate.

Table 7.1 (*continued*)

Sender's perspective	Recipient's perspective	Third party perspective	Comments
appropriate	transgression	appropriate	The recipient finds the sender's message inappropriate, for instance because of lack of mutuality on cultural or interactive norms. However, the third party does not think that any norm of *netiquette* has been disobeyed.
appropriate	transgression	transgression	Both the recipient and the third party (moderator) consider that the sender's message is inappropriate. The sender seems to lack knowledge of the rules of behaviour relevant to that specific interactive environment.
transgression	appropriate	appropriate	The sender's attempt to produce an impolite message (i.e. *flaming*) is unsuccessful, since both the recipient and the third party find the message appropriate.
transgression	appropriate	transgression	The sender's attempt to produce an impolite message (i.e. *flaming*) is unsuccessful, since the recipient does not find it inappropriate. However, the third party does consider it inappropriate. This may be produced due to an asymmetry in the users' conceptualization of the general norms of behaviour in their own cultural contexts or due to a simple case of misunderstanding.
transgression	transgression	appropriate	The sender tries to violate the norms of behaviour in the virtual interaction and the recipient identifies this intention correctly. The third party, however, does not share this opinion. Again, not sharing the same cultural background may favour this asymmetry.
transgression	transgression	transgression	An inappropriate strategy is identified as such by both the recipient and the third party: a successful flaming message.

2. Some approaches to the study of politeness

Besides Brown & Levinson's (1987) study, which will be reviewed in Section 3 below, other authors have addressed politeness from different points of view. These are summarized below.

2.1 (In)direct speech acts and politeness

Traditionally, it has been accepted that the more indirect a speech act is, the more highly its potential politeness is expressed textually in the utterance. At the same time, it is assumed that the more highly we climb on the "social ladder," the more indirect utterances tend to be. In this claim lies the belief that there is an inherent politeness component in certain grammatical structures, and that an increasing complexity in sentence structure parallels increasing politeness (see Ogiermann 2009).

It is true that many utterances suggest, from their verbal content, the speaker's willingness to be polite, but in the analysis of politeness we must include other elements that connote utterances with their eventual (im)polite interpretation (e.g. tone of voice). We can speak, then, only of potentially (im)polite utterances, which acquire their final effect in a particular speech situation. No sentence is inherently polite or impolite. We often believe that certain expressions are impolite, but it is not the expressions themselves, but the conditions in which they are used that determine the judgement on politeness (Fraser & Nolen 1981:96). As Haugh (2010:26) correctly qualifies, there are several points of view involved in an interpretation of impoliteness:

> first, evaluations may focus on the speaker's behaviour as impolite, rude and so on, or alternatively may involve the recipient's response to the speaker's behaviour, namely (feelings of) offence; second, the degree of perceived impoliteness/offence itself can vary. Indeed, such evaluations are clearly open to discursive (re)negotiation as well as dispute.

Haverkate (1988:63) stresses the role of the interlocutors when they make sense of indirect utterances in a dialogue, because it requires more effort to interpret phrases such as "*it's cold in here*" than more direct ones such as "*close the window.*" However, it is always assumed that the speaker wishes to be polite when choosing an indirect utterance. For example, concerning directives, if we move up along the scale of indirectness, the listener will gradually abandon a negative attitude towards performing the action asked for by the speaker (Blum-Kulka 1990:269). In the case of directive speech acts, speakers must use politeness to ensure the realization of their intentions. More direct speech acts such as "*can you pass the salt?,*" indirect ones such as "*it is cold here*," or intermediate ones such as "*those comments should not be made in public*" (Haverkate 1988: 64) are acts that threaten the interlocutor's freedom (or *face*, see Section 3 below). Therefore, the speaker must mark the utterance with politeness. Obviously, the more direct an utterance is, the fewer options the interlocutor is given to decide whether to carry out the required action or not.

Similar assessments regarding the use of a particular (in)direct speech act occur in virtual synchronous conversations (e.g. chat rooms) and asynchronous electronic messages (e.g. e-mail). Whenever the utterance is aimed at achieving a personal goal that forces the listener to perform some action, a "veil" of politeness must be included in the structure of the utterance.

However, the lack of physical co-presence on the Internet can generate more or less prominent alterations in the use of those (in)direct speech acts that we perform in physical settings (Baron 1984: 130). An example is e-mail communication between university students and their teachers, especially in inter-cultural contexts (Bjorge 2007). Several informal comments by teachers suggested that most of them had received from their pupils a request devoid of politeness-markers (as reported in Knupsky & Nagy-Bell 2011).[7] One possible explanation is that students are unaware of e-mail *etiquette*: "Communication styles and conventions are typically shared in speech communities and learned by new apprentices over a course of years. However, email, as a relatively recent development, is not yet governed by clear conventions and expectations" (Biesenbach-Lucas 2007: 61).

Besides, in this case of requests (and more specifically requests between students and teachers), we can also include the role of contextual support (e.g. the presence of contextual support of nonverbal behaviour) in the choice of a specific (im)polite strategy. Indeed, on the one hand it seems that a reduction in the amount of contextual support and the lack of physical co-presence should favour a lack of self-control and parallel lack of linguistic markers of politeness. However, several studies seem to refute this claim. For example, Duthler (2006) concluded that, if e-mail was complemented with richer contextual support (specifically *VoiceMail*), users tended to include fewer politeness markers, against the aforementioned hypothesis. An explanation may be that when typing the message

7. Bou-Franch (2006) comments that requests are acts that can be perceived and produced as *face-threatening acts* or *face-enhancing acts* (in Brown & Levinson's 1987 terminology) depending on the circumstances in which they are produced and on the kind of requests. They are threatening in the sense that they demand a reaction from the addressee and impose on his/her freedom of action. But they can also be positive by showing the addressee that someone is interested in something that he/she has (opinion, object, advice….). In any case, the speaker has to choose carefully among the range of politeness strategies used in formulating the request so that it fits the specific context of its performance.

with no nonverbal support, the user has more time to plan, design and choose the markers that are more convenient in each communicative situation.

Something similar happens in e-commerce, studied in Garcés-Conejos Blitvich & Bou-Franch (2008). In this case, there are requests from both parties: from the provider of services or goods asking for an action of purchasing (in a more or less covert way), and from the user asking for services, explanations or advice about the product. Again, the lack of physical co-presence and reduced nonverbal contextual support influence the choice of a specific politeness strategy. Besides, e-commerce is peculiar in the sense that the level of politeness "is not negotiable, as in traditional face-to-face business meetings or in other electronic discursive genres. Therefore, the provider should meta-represent the client's needs, establish the parameters of the relationship and achieve the levels of empathy and appropriate deference towards the customers" (ibid.: 473).

Requests are also an important element of web pages that promote services via combinations of visual and verbal elements. Concerning the latter, the request of some action constitutes the nucleus of the communicative act and, therefore, of the choice of certain politeness strategies. Montañés Brunet et al. (2005), in an analysis of web pages promoting tourism services, call this phase *exhortative centre* (*núcleo exhortativo*), which would be preceded by a *pre-sequence* of offers and incentives, and followed by an optional *post-sequence* to add motivations and eventual benefits.

2.2 Politeness in transactional and interactive discourse

Some studies have suggested two varieties of discourse depending on the linguistic behaviour of interlocutors (Kasper 1990: 205 ff): *transactional discourse* and *interactive discourse*. In the former, the optimal transmission of information is salient. Therefore, speakers will follow, where possible, the requirements of Grice's (1975) Cooperative Principle. In the latter, speakers want to preserve social bonding with others, and therefore they will adopt politeness strategies, even if Grice's maxims cannot be followed.

Thus, we could establish a kind of discursive scale (see Lakoff 1989), ranging from the most transactional utterances, characterized by a minimum level of politeness (e.g. lectures, legal discourse, therapeutic discourse) to the most interactive ones, in which politeness, i.e. the proper maintenance of social relations and phatic language, predominates over informativeness.

On the Internet, the transactional / interactive scale varies depending on the *cyber-medium* used. Undoubtedly, e-mail tends to fit formal communication, although there are many messages focussed on the playful aspect of human communication. By contrast, chat rooms and instant messaging are closer to the interactive end of the scale and in these *cyber-media* sociability is more important than an effective exchange of information.

2.3 Politeness and rudeness

Lakoff (1973, 1977) established two main rules of pragmatic competence: (1) Be clear, and (2) Be polite. From these two rules, which may conflict in certain situations, Lakoff proposed three sub-maxims, which we could paraphrase as follows:

a. Don't always impose your opinion on others.
b. Let your interlocutors express their opinions.
c. Make your interlocutor feel comfortable in the interaction.

Rudeness, on the other hand, is a deviation from what is understood as polite in a social context, and it is inherently confrontational and disruptive to social equilibrium. Kasper (1990: 208 ff) proposed a dual classification of rudeness:

1. *Unmotivated rudeness*. This refers to a failure to follow the rules of politeness due to ignorance. This is the case of speakers outside the community who are unaware of specific patterns of behaviour, and who do not make their utterances fit the intra-cultural peculiarities of that community. It is also frequent in children, whose language development involves a progressive internalization of politeness as demanded by their social context. It is frequent, for example, in child discourse, to find directives with an imperative mood and devoid of grammatical elements that denote a desire to be polite, such as "thank you" or "please."

2. *Motivated rudeness*. In this case the speaker wants the utterance to be interpreted as rude. Kasper (ibid.) differentiates three cases:

a. Rudeness for the lack of control of one's feelings. Excessive public expression of emotions (e.g. joy and anger) can be interpreted as rude. But in this case, the excess of emotion is not considered rude if it is expressed in response to previous rudeness or aggression from the interlocutor. It is considered rude if the action arises from the speaker with no prior aggressive act.

b. Rudeness aimed at achieving a purpose, which Kasper (1990: 210) calls *strategic rudeness*. An example of this type of rudeness is found in legal discourse (Lakoff 1989), an environment where the prosecutor is allowed to verbally "attack" the defendant with an attitude that would be incompatible with the rules of politeness in normal conversations.

c. Finally, Kasper proposes what she calls *ironic rudeness*. It is similar to strategic rudeness in that it also aims at a purpose. Contrary to Leech's (1983: 143) view of irony as an aggressive act without the form of a confrontation, Kasper believes that irony may (or not) be as rude as a direct remark.

Internet also reflects the duality between politeness and rudeness. On the one hand, the moderators of the various discussion forums (newsgroups) or mailing lists offer a series of exhaustive rules of politeness that are mandatory for all users, and complemented by multiple netiquette manuals available on web pages. These rules favour an inter-cultural levelling of politeness beyond the peculiarities of each community. On the other hand, the lack of physical co-presence and face-to-face communication has also led to the existence of uncontrolled conversational strategies, the so-called *flaming*.[8] The testimony of a user, quoted by Hauben & Hauben (1998), is illustrative:

> There is something else I've discovered that is really rather fascinating. People can be incredibly rude when communicating through this medium. For example, some time ago, I posted a question to lots of different news groups, and many people felt my question was inappropriate to their particular group. They wrote to me and told me so, using amazingly nasty words. I guess it's easier to be rude if you don't have to face a person, but can say whatever you want over a computer.

There is no absolute agreement on the definition of this phenomenon, although all authors consider that it is an unacceptable communicative attitude that prevents normal polite communication on the Internet. In O'Sullivan & Flanagin (2000) some definitions are quoted:

> direct, sometimes gratuitous, criticism (Deuel); hostile verbal behaviour (Thompsen & Foulger); hostile, provocative posting (Kollock & Smith); hostile expression of strong emotions and feelings (Lea et al.); verbal aggression, blunt disclosure, and nonconforming behavior (Parks & Floyd); emotional outbursts (Korenman & Wyatt); antisocial interaction (Thompsen); a form of social aggression (Colomb & Simutis).

Interestingly, flaming online is sometimes comparable to insults in their conventional vs innovative quality and their use as a strategy for social bonding. Indeed, as analysed in Mateo & Yus (2000, 2010), in some contexts speakers intend to express strong feelings of camaraderie, friendship or admiration for the interlocutor, but they cannot resort to the usual flattering or adulating repertoire. As

8. See, for example, Collins (1992), Belson (1994), Aycock (1995), Mabry (1997), Millard (1995) and Bellamy & Hanewicz (1999).

this repertoire might sound "soft" or "feeble," speakers will choose instead more "manly" insults which convey the intended emotional load. This is achieved by transforming the negative and insulting original overtones into positive and praising ones. Similarly, coded flaming on the Net exhibits bonding attributes. As concluded by Moor et al. (2010: 1538), the users under analysis

> developed communication styles in which flaming was quite common. Although outsiders might think that group members were being offensive to each other, a closer view showed that flames were in fact meant to be funny. Whereas students in one group seemed to enjoy insulting one another, other groups only rarely flamed, indicating that flaming can indeed be normative behaviour within a group.

Nevertheless, the traditional view of *flaming* as an uncontrolled display of rude behaviour is also typical of text-based Internet-mediated interactions, where users typically use imperatives, capitalization and direct insults while taking advantage of the lack of physical co-presence, as in the excerpt from a chat room conversation (Hastrdlová 2009: 103), quoted in (11), and the e-mail messages that scored highest in flaming connotation in Turnage's (2007) corpus, quoted in (12):

(11)	james6:	hi ladies	
	SpecialED:	great the room is broken	(inappropriate)
	SpecialED	kicks the edge of the room	(aggressive, impolite)
	james6:	broken?	
	SpecialED:	YES JAMES ITS BROKEN	(shouting, impolite)
	james6:	what happened	
	SpecialED:	dont question me	(imperative, inappropriate)
	Wiky:	room?	
	SpecialED:	dont as questions	(imperative, inappropriate)
	james6:	i always question	
	SpecialED:	ask	(imperative, inappropriate)
	james6:	i just did	
	Wiky:	question	

(12) a. I am on top of this. I am simply waiting for the goddamn quote. I will call the company again.

b. Now how in the hell would I know that if nobody tells me??? GEEZ! :0

c. CAN YOU PLEASE GIVE ME THE FINAL ON THIS AND DO YOU KNOW WHO WAS TO SEND THIS OUT????????????????????

d. Do these guys actually know what the curriculum is or are they making it up as we go along?

Moreover, all the strategies for text oralization that were discussed in previous chapters as useful for *microblogging* (Chapter 4), virtual conversations (Chapter 5) and e-mails (Chapter 6) are now useful to make manifest the neutral, polite or explicitly rude intention that underlies the posting of messages. Anger, joy, or the explicit mitigation of the threatening quality of a request can be manifested in a clearer way by using the strategies for oralization. In Darics (2010: 136), Example (13) is quoted:

(13)	1 Sarah:	its so great you are back
	2 Liz:	it is good to be back
	3 Sarah:	I am so sorry i have been crap at keeping in touch!
	4 Liz:	and thanks for the text messages you sent me
	5 Liz:	nooooooooooo
	6 Liz:	no worries at all

In this dialogue, Sarah expresses her apology (line 3), and by doing so she threatens her own image (*face*) by implying that not being in touch is her own fault. Liz in her response (5–6) types an exaggerated sound effect by oralizing the word "no" (line 5). This text oralization is useful to minimize the effect of the act committed by Sarah, and to reduce a possible tension from the *face-threatening act* (in Brown & Levinson's 1987 terminology, see 3 below).

Di Segni (1997b) analysed the verbal and typographic resources to which Internet users turn to connote their messages with an unacceptable or explicitly rude meaning, and that are generically labelled *verbal aggression*. As expected, flaming is usually communicated with an explicitly connoted use of punctuation, particularly capitalization and exclamation marks. For example, in an inter-racial newsgroup studied in McKee (2002) it was easy to find exalted utterances about the role of race (or superiority of one over another) typed as capital letters:

(14) YOUR VIEWS SHOWED TRUE IGNORANCE
HOW CAN YOU COMPARE A CAR TO A HUMAN BEING?
HOW!!!!!!!!!!!
STOP POINTING YOUR DIRTY FINGERS AT MINORITIES!
How can you say such a ridiculous thing?

Another form of unacceptable message is verbal sexual assault, ranging from the tedious repetition of a question about sex and age of the interlocutor in chat rooms, to the use of obscene words in the messages, and even virtual rape. In (15) an example of the first of these possible sexual assaults is quoted, i.e. the tedious question (Di Segni 1997b):

(15) [brrrrr] sexmaker-where frm?
* greenguy <----- is 18+/m/israel !!! msg me !
[PRINCE__] looking for girls
[BEST] PRINCE : go and kiss frogs
[IchyBung] ilo m/f?
[IchyBung] how old r you?

Chat room users tend to use the full range of resources for oralization discussed in Chapter 5 to communicate their intense feelings and emotions. Conventions are often taken from other media such as comics. As discussed in Yus (1997b: III.1), there is a graphical resource that I called *iconic euphemism*. Given the ethical constraints imposed on comics in the past, cartoonists sought alternatives to insults and curses, so that the reader could grasp the intensity of the feeling or emotion without having to resort to prohibited words or phrases. Likewise, chat room users rely on various signs (e.g. $#@%!#) which, when typed together, provide that same intensity. Moreover, users exhibit a great capacity for reproducing feelings and emotions by using text deformation (oralized text), and even compose visual images with their keyboards (Di Segni 1997b):

(16) [IgalAmir] * (__) BBBBBB--EEEEEEE-EEEEEEE-FFFFFFF
[IgalAmir] * /oo\\BB---BB-EE------EE------FF-----\\
[IgalAmir] * \\ / BBBBBB--EEEEEE--EEEEEE--FFFFF-- |
[IgalAmir] * oo BB---BB-EE------EE------FF----- |
[IgalAmir] * BBBBBB--EEEEEEE-EEEEEEE-FF----- ^
[IgalAmir] * | | | |
[IgalAmir] * ^ ^ ^ ^

2.4 Metalinguistic and linguistic politeness

Haverkate (1988: 53 ff) proposes two types of politeness because a polite attitude can be conveyed both verbally and nonverbally. The first type of politeness, *metalinguistic*, serves two main purposes: to create or preserve sociability (by using phatic language, for example when faced with the interlocutor's silence) and to preserve the rules of social etiquette (with constraints such as "do not shout," "do not show lack of attention" or "do not interrupt").

The second type of politeness, *linguistic*, refers primarily to directive speech acts (with which the speaker intends the interlocutor to perform some action). Haverkate makes a preliminary division between *macro speech acts* and *micro speech acts*. The former tend to be preceded and/or followed by a pre-sequence or post-sequence that can also serve to establish a phatic language that mitigates the illocutionary force of a request, as in this utterance: "*Yes, she is also well, by the way, could you go and pick her up at the airport?*"

Haverkate (ibid.: 61 ff) focusses on *micro speech acts*, specifically on the relationship between directives and the expression of politeness. In this case, there are four sub-acts to be taken into account: articulation, predication, illocution and reference. Haverkate studies especially the last two, regarding the choice of direct or indirect utterances by the speaker. The context in which the conversation takes place will help to determine the correlation between (in)direct acts and politeness. On the Internet, the reduced quantity and quality of such contextual assumptions entail a hyper-reliance on purely textual attributes for the expression of politeness.

2.5 The context of speech situation

Any study that seeks to explain politeness in communicative exchanges must take into account a number of factors that affect its use, but are not strictly speaking linguistic (i.e. cannot be inferred directly from the semantic content of the utterance). We cannot claim that politeness is permanent, unchanging, as there must be variations according to specific parameters of social behaviour. Among others, the following should be highlighted:

1. *Trans-cultural differences*. People tend to think that the interactive features of the language that speakers use in their community are more or less the same as in other languages and cultures. However, several anthropological studies have concluded that speech in the world is divided into two well defined areas: the Western, on the one hand, including Europe, USA, etc., and the rest of the world, on the other. In addition to this initial division, within these two areas there are also abundant differences. These local features include, of course, the exclusive, distinctive and specific use of politeness.

There are interesting studies of other cultures that reveal the existence of behaviour understood as polite that would not be valid for a Western culture. From these and other studies we can deduce that the idiosyncrasy of a community includes the specificity of its politeness. Kasper (1990: 198) states that the linguistic encoding of politeness strategies is derived from the linguistic system and conventionalized rules of use, such as everyday formulas and idioms that tend to be specific to the language. Therefore, we cannot expect formal or even functional equivalence with other languages. To this statement, we could add differences which are not exclusively linguistic, but related to the way people conceptualize the world and the relationships between individuals within their community.

Internet is particularly appropriate for an analysis of trans-cultural differences in the use of politeness. This network of nodes allows multiple users from all over the world to interact asynchronously (e.g. e-mail) and synchronously (e.g. chat rooms). This entails making an effort to understand the idiosyncratic use of polite strategies for each speech community. Generally, the pressure of English, the *lingua franca* of Internet-mediated communication, makes many users adopt not only the Western pattern of politeness but, specifically, the *hyper-polite* Anglo-Saxon pattern.

2. *The individual and society.* In addition to trans-cultural traits, each person, as an individual, has specific and unique qualities that influence the linguistic treatment he/she receives from others. We can make a first distinction between *intrinsic characteristics* (e.g. age, sex, place in the family) and *extrinsic* or *acquired social properties* (e.g. title, rank, social position).

In the first case, we are faced with characteristics that prove to be of great importance when choosing a politeness strategy. In the second case, we find a similar situation, since the social status that the individual has achieved in society is one of his/her distinctive properties. The difference lies in the origin of such attributes. In the first case, these were intrinsic or "inherited" qualities of the person, while in this second case we are dealing with acquired attributes through one's personal development in society (e.g. one's job). In these social circumstances, it is normal to impose a certain politeness strategy on others simply because of holding certain social attributes. This imposition can be either explicitly or implicitly assumed by the interlocutor. Once more, this applies to Internet-mediated communication, in which users take into account the features of the individual when devising their electronic messages as happens, for instance, in e-mails from workers to bosses or from pupils to teachers.

3. *The specific context of speech.* The context where the communicative exchange takes place is crucial, so that an act performed by one person might be considered rude, while the same act, in similar circumstances and by an almost identical person, might be considered polite. The contextual elements, including politeness, are suddenly reorganized, as in human interaction in general, in a way that we cannot clearly foresee (Sell 1991: 217). As already mentioned, the "multi-geographical location" of Internet users entails supplementary effort when sending and interpreting messages on the Net, effort mainly directed at the search for an adequate (im)polite strategy in the context where the interaction is taking place.

Leech (1983) also stressed the importance of the conversational context when he distinguished between *relative* and *absolute politeness*, the former being related to a particular speech situation, and the second being a consequence of a particular speaker's act. This distinction, however, seems to presuppose that there are inherently polite utterances, with a more or less accepted meaning within a particular community. In fact, one cannot speak categorically about polite utterances in any situation, since the strategies and means of polite actions are not endowed with absolute polite value. Their polite potential of a strategy seems instead to be determined mainly by contextual constraints that are at work in different kinds of discourse and speech acts. Even though the main focus of analysis may be a specific speech act, the discursive context must be taken into account to properly describe and evaluate the distribution and function of mitigating and aggravating politeness strategies (Kasper 1990: 201). So instead of talking strictly about polite or rude strategies, it is more advisable to establish a continuum of possibilities, as Watts (2003) does when proposing a range of strategies between "polite" and "rude" but with "appropriate" and "inappropriate" in between, and all of them linked to the peculiarities of the context, since it greatly influences the different perceptions that interlocutors have of what being (im)polite means. As Graham (2007: 744) correctly points out,

> If we accept the fact that notions of Speaker intent (and therefore notions of politeness) are open to differing interpretations by the Speaker and Hearer, then it is a worthy area for exploration to attempt to understand how the different interpretations of (im)politeness and (in)appropriateness affect interaction and rapport within groups. If interactants have differing interpretations of whether a given utterance is (im)polite, conflict is likely to result, which has a direct impact on rapport.

But besides the variables of the environment of the exchange, there are factors related to the linguistic act itself that usually indicate the degree of politeness. We are not talking, in this case, about the content of the utterances, but about their illocutionary features. It is generally accepted that the type of discourse that is to be delivered influences the speaker's linguistic choices (see Mariottini 2004: Chapter 5). Of course, inter-cultural differences are also relevant.

4. *Social power, social distance* and *rank of imposition*. These three variables have been proposed by Brown & Levinson, and will be reviewed in the next section.

3. Brown & Levinson's model

Brown & Levinson (1987, henceforth B&L) start from a *Principle of Politeness* of a different conceptualization from Grice's *Cooperative Principle*. In their model, they give politeness a new character by suggesting that it can be communicated (or *implicated*, in Grice's terminology) like any linguistic message. Thus, certain utterances that are seen as polite not only contain a specific linguistic message, but also reflect the speaker's explicit desire to be polite in a given situation.

The key concept in their analysis is *face* (ibid.:61), taken from Goffman's (1987) terminology on the public image of a person. Morand & Ocker (2002) offer the following description of the term: "Face, the positive social value each person effectively claims for him or herself in the public arena, is proffered and thus exposed throughout interaction. Face is the very reflection of self worth; upon this presentational aspect hangs individuals' self-esteem, self-identity, and their credibility as a member of the social group."

During verbal interactions, one's face can be maintained or lost, it is vulnerable and unstable, and each partner in the conversation will seek to preserve his/her image (social status, power relationship, position on a scale, and other) in relation to the other partners. As everyone is aware that certain acts threaten the interlocutor's (and one's) face, they will choose in their speech acts (and expect from others) strategies that soften the threat, and this is where politeness comes in. Thus, the emphasis is on the politeness of the conversational act, rather than on the actual content of the utterances, against the opinion of Leech (1983), among others.

B&L distinguish two types of politeness:

1. *Negative politeness*: the desire that the speaker's intentions are not prevented by interlocutors (e.g. that an order gets performed, that a request gets done).
2. *Positive politeness*: the desire that the speaker's intentions are accepted by others (e.g. that one's ideas are agreed upon, that one's opinions are accepted).

The first case is reminiscent of the traditional concept of politeness, namely not to impose a particular behaviour on others. The dilemma of threat to the interlocutor's face that an utterance constitutes, on the one hand, and the desire that the interlocutor performs a particular action, on the other, leads the speaker to use negative politeness strategies.

The speaker who wants to alleviate, in some way, the imposition that a language act makes on the listener, will look for positive politeness strategies. By contrast, those who want to impose their wishes on the interlocutor, will prefer a sharp utterance, in imperative, that leaves no doubt of the compulsory action required. As noted by Sell (1991:212), there is a certain scale depending on the polite option chosen by the speaker. The most polite act would be one that is carried out through a veil of metaphor or irony, to mitigate its illocutionary force. The least polite act would be a straightforward directive. In an intermediate position, we would place the actions in which the speaker and the listener recognize the threatening nature of the act and use some politeness strategy.

B&L (1987) link types of behaviour to the variety of face threatened and to whether the threat affects the speaker or the listener: (a) speaker's face, negative, (b) speaker's face, positive, (c) listener's face, negative, and (d) listener's face, positive. In Park (2008a: 2055) there is a clear example of utterances that communicate similar information (an invitation to a party) but entail radically different politeness strategies:

(17) a. I know you're not that crazy about parties. But come to the party tonight! You'll love it.

b. It would be nice if you could come to the party tonight if you have got the time. Well, we'll understand if you can't make it.

In (17a) we can perceive a casual, intimate interpersonal relationship beyond the literal meaning of the invitation. This is achieved through a number of positive politeness strategies such as the establishment of shared knowledge (*I know that…*), the use of informal vocabulary (*not crazy about…*), the use of a directive speech act in imperative (*come…*), and the expression of optimism about the good time that the listener will have (*you'll love it*), all of them aimed at highlighting closeness and solidarity between the two partners. By contrast, (17b) satisfies the listener's negative face by stressing his/her independence, freedom from imposition and respect for his/her space, time and personal decision. This is achieved through negative politeness strategies such as giving the option to reject the invitation (*we'll understand…*), the use of the conditional sentence denoting insecurity (*it would… if you could…*) and underlining the lack of expectations (*…if you can't make it*) marked by a word indicating doubt (*well*). In this sense, it should be stressed that different politeness strategies and their direct or indirect expression are sources for strengthening feelings of community and also for solidarity building, as concluded by Garcés-Conejos & Lorenzo-Dus (2010) in their study of political campaigns through e-mail.

3.1 Parameters for weighing politeness

B&L (1987) suggest three variables that influence the choice of a politeness strategy:

Relative power	(P)	of the addressee towards the speaker.
Social distance	(D)	that includes intimacy and familiarity between the interlocutors.
Ranking of imposition	(R)	of an act on the interlocutor's face.

In general, one can speak of two main axes: one vertical, which articulates power relationships, and one horizontal, establishing exchanges between equals, but this distinction is not clear-cut because in many cases there may be a mixture of these parameters or variables. This division is, however, useful for understanding the nature of the choice of an (im)polite strategy by the speaker.

The three parameters influence the speaker's choice of the type of discourse that is more appropriate to achieve his/her goals. For example, power relationships between interlocutors are often related to a more or less direct utterance (Blum-Kulka et al. 1985). In addition, some studies suggest that the power relationship over the interlocutor should not necessarily entail more politeness (Holmes 1984, Preisler 1986).

There also seems to be some relationship between the degree of social distance and the choice of more indirect utterances. In any case, the power relationship, the social distance, and the ranking of imposition are all subject to changes in the ongoing interaction itself (Kochman 1984, B&L 1987:74 ff). The importance of the speech situation, and the qualities of the cultural environment favour specific variations in each conversation.

B&L (ibid.) propose that the three variables (P, D, R) should be combined, and the result would be the potential threat posed by a *face-threatening act* (FTA).[9] According to B&L, speakers can choose among a range of strategies according to the potential threat of their utterances (and expect, in the same way, that certain politeness strategies will be used by their interlocutors when they threaten their own face):

1. To make the FTA on record, without redress.
   [e.g. "go to the shop to buy some meat for the meal"].

According to Haverkate (1994:25), this strategy usually takes place in three situations: (a) when the speaker is in a power relationship with the interlocutor, (b) when the speaker is emotional or angry at the behaviour of the interlocutor, and (c) when external circumstances require the listener to react immediately to the order.

2. To make the FTA on record, with redress, with positive politeness.
   [e.g. "are you going to the shop to buy some meat?"].

The positive politeness strategies are based on the expression of understanding and appreciation towards the ideas and wishes of the interlocutor and on the fact that both share them. They exhibit an attitude of familiarity, intimacy and friendship

9. The weightiness of an FTA is calculated as follows: Wx = D(S,H) + P(H, S) + Rx, where Wx is the numerical value that measures the weightiness of the FTA x, D(S, H) is the value that measures the social distance between speaker (S) and hearer (H); P(H, S) is a measure of the power that H has over S and Rx is a value that measures the degree to which the FTA x is rated as an imposition in that culture.

with the speaker. Normally, positive politeness is directed at complimenting the interlocutor's face and for that purpose it is normal to use different linguistic markers of politeness, of a grammatical, lexical or even prosodic kind (Meier 1992: 17, Condon & Cech 1995).

3. To make the FTA on record, with redress, with negative politeness.
   [e.g. "would you mind going to the shop to buy some meat?"].

Negative politeness is intended to make clear that the FTA is not seeking a limitation on the interlocutor's freedom or opinions, and every language has a repertoire of expressions that serve this purpose, for example the use of indirect requests (*could you…? Would you mind…?*).

4. To make the FTA off record.
   [e.g. "there's no meat for today's meal"].

With the off-record strategy, the speaker tries to conceal the true intention underlying the utterance and, in turn, leaves the whole responsibility for the act to the interlocutor, who is free to interpret it in any way. In the above example, the listener may think that the speaker is simply informing that there is no meat, without concluding that the intended interpretation is, in fact, an indirect request.

3.2 Face on the Internet

Undoubtedly, the politeness model that is more often applied to Internet-mediated interactions is B&L's (1987) which, as already discussed, is based on mitigating the inherent threat that the different communicative strategies exert on the positive or negative face of the interlocutor.[10] On the Internet, there is an absence of contextual cues that normally facilitate, in physical environments, the choice of a particular (im)polite strategy. This may lead to an overabundance of overt expressions of politeness.

One of the most interesting applications of B&L's model to Internet-mediated communication is the study by Maricic (2000) on the use of requests on *The Linguist List*. On this list, we can find examples of the different strategies studied by B&L, with the exception of the off-record FTA, almost inexistent (Table 7.2).

10. See, among others, Simmons (1999), Harrison (2000), Stone (1995), Bays (1998) and Park (2008a, 2008b).

Table 7.2 Politeness strategies on *The Linguist List* (Maricic 2000)

Bald on record	**Linguistic realizations**	**Examples**
Making the FTA on record	– imperative – direct questions – elliptical questions – straightforward, assertive requests	– *Help!* – *Is anybody aware of…?* – *Any hints?* – *I need help*
On record, positive politeness	**Linguistic realizations**	**Examples**
1. Claiming common ground		
– Noticing/attending the addressee	– informal salutation	– *Hi everyone!*
	– informal thanking	– *Thanks for the input!*
– In-group identity markers	– markers as address forms	– *Dear fellow linguists!*
– Seeking agreement	– common opinions, views	– *Am I right in thinking…?*
– Avoiding disagreement	– hedges	– *seem to, tend to, perhaps*
– Exaggeration	– overstatement	– *We'd be delighted to…*
– Jokes, acronyms	– emoticons, graphic signs	– *;-)BTW (by the way)*
2. Focussing on cooperation		
– Assuming cooperation (be optimistic)	– use of optimistic expressions of response anticipation	– *Thanks in advance*
On record, negative politeness	**Linguistic realizations**	**Examples**
1. Giving freedom of action (being generous)		
– Not assuming that H is able/willing to help (being pessimistic)	– questions, hedges, if-clauses – tentative modals *could/would*	– *maybe/perhaps/I guess…* – *Could/would anyone help?*
2. Minimizing threat		
– Showing deference (being respectful)	– use of formal politeness formula and politeness markers	– *Dear listers/subscribers* – *Please help me…*
3. Minimizing imposition		
– Going on record as incurring debt	– expressions of appreciation	– *I'd be very grateful if*
– Apologizing	– use of direct or implied apology	– *I'm sorry…*
– Dissociating the requester from impingement	– avoidance of personal responsibility	– *My colleague needs help*
	– use of 1st person plural pronouns	– *We need some help*
	– passive constructions	– *Any help will be appreciated*

In (18a) and (19a) below there are examples of politeness strategies followed by explanation in (18b) and (19b) (positive = "+"; negative = "–"), respectively (adapted from Maricic ibid.):

(18) a. Hi everybody [1],
About a month back I wrote that I was putting a book together on later ancient Hebrew culture. I asked if someone could write me a quote which showed that English is not related to Hebrew, and I would quote it in the book. *I was astounded* [2] by the number and in depth responses I got back from many people. Some even saying that Hebrew was related to English. *Thank you for all the people that wrote back* [3]. Since then, my drive crashed, and I've lost the answers. *I feel very bad about asking people again* [4], but *if* you did write me a reply you wanted in the book, *could you please send it again*?? [5]
Maybe you could check your mail program for a mail sent to [e-mail address] and send it again? Or, *if* you have any other ideas, *could you write me? I'm kicking myself* [6] here for losing so many *wonderful* [7] comments on the subject. *Pleeeze* send again! [8] :) [9] (*The Linguist List* 9: 687).

b. [1+] Informal greeting. [2+] Exaggeration related to everybody's need to be accepted. [3+] Appreciation of other's help. [4–] Indirect apology. [5–] Use of conditional that introduces a tentative appeal; modal verbs for requesting, marker of politeness (*please*); hedge (*perhaps*). All of them give the addressees freedom of action. [6+] Self-blame to generate sympathy. [7+] Exaggeration. [8–] Direct request with marker (*pleeeze*). [9+] Use of emoticon to introduce a humorous or playful atmosphere.

(19) a. Dear LINGUIST *friends*, [1]
My wife [2] is working on *her* [2] thesis dealing with nasal sounds. *She* [2] recorded several people and analyzed the recordings in order to discover how many bilabial, alveolar, velar and so on nasal sounds those people said. It has been a very tough job. *We* [3] were *wondering* [4] *if* there *happens to be any* [4] sort of computer program that, once fed with the recordings, could do a transcription into phonetic symbols. *We* thought there *probably* [5] is something. *Could any fellow* linguist give *us* [6] a hint or some reference regarding this point? *Thanks* for the *time* and *effort* [7]. (*The Linguist List* 9: 388).

b. [1+] Respectful greeting to minimize the threat of the subsequent request. The use of *friends* generates a positive atmosphere of proximity. [2–] Avoidance of responsibility in the request (keeps his own face). [3+/–] Pluralization as a negative strategy of dissociation from the FTA and as a positive solidarity strategy. [4–] Tentative grammatical constructions that minimize imposition over addressees. [5–] Hedge. [6–] Politely constructed request that gives addressees freedom of action. [7–] Negative politeness strategy that minimizes the imposition generated by the request.

There are also several studies that focus on textual markers of positive and negative politeness. An example is a study by Mariottini (2006), where she analyses the use of diminutives in Spanish and Italian chat rooms. According to B&L (1987: 103), diminutives "may at large be seen as in-group identity markers that fall into the sphere of our notions of familiarity, intimacy, and decreased psychological distance. Besides being indices of social closeness, diminutives may also function as 'accelerators' of intimacy among strangers."

Mariottini shows how diminutive suffixes in Spanish (-ito / -ita) can either refer only to the modified object, thus diminishing its importance or quantity, or they can modify the whole utterance. Diminutives, in short, are used in chat rooms both to reduce the force of requests when they are added to the accompanying noun, and to reduce the threat of rejection, when their scope covers the whole utterance. Examples are the diminutives quoted in italics in (20) (Mariottini ibid.: 121):

(20)	rnr-cu-cu-taz:	hay *turnitos*? [*are there speaking turns?*].
	mr-cu-cu-taz:	umm.
	rnr-cu-cu-taz:	y mi *turnin*? [*and my turn?*].
	tomb_raider_oOkaro10o:	pido turno al audio *lueguito* de mine [*I ask audio for a turn after mine*].
	el-fantasma-de-la-opera:	tomb_raider_oOkaro10.o … NO ES CHISTE.. NECESITO TU AYUDA .. PER0 BUENO YO L0 HARE *SOLITO* COMO VINE AL MUNDO [*This is no joke, I need your help… but well I'll do it myself, as naked as the day I was born*].
	escritora-9:	*turenito* pal uliiiiiiiiiiiii
	marceloespaa:	españa
	escritora-9:	*turnito*

Furthermore, Lorenzo-Dus et al. (2011) analyse impoliteness in new forums that, beyond traditional newsgroups, have appeared recently in web-based environments, specifically the forums around *YouTube* videos (see also Moor et al. 2010). They claim that the study of (im)politeness in mediated settings, into which web-based Internet forums would fit, is significantly more complex than in dyadic, face-to-face situations. Impoliteness strategies are affected by the public nature of communication in *YouTube*, and they stress that "a more fruitful modelling avenue might be found within recent debates in the field of broadcast discourse about the traditional role of the overhearing audience" (ibid.). Garcés-Conejos Blitvich (2010) also stresses the differences of impoliteness in modern Internet forums compared with traditional dyadic interactions. Her account draws interesting conclusions from forums in *YouTube*, where "impoliteness is multifunctional: it is used against the out-group to create a sense of 'us versus them' by making their attributes undesirable, and to heighten a sense of membership in the in-group" (ibid.: 541). She stresses that current theories need to be extended to account for intergroup communication such as the one taking place in Internet forums, in which the pairing of social identity and individual identity is salient. In fact, the former seems to be more salient in these open portals, as claimed also by Garcés-Conejos Blitvich (forthcoming):

> assessments of the in/appropriateness of a given contribution may vary depending on factors such as the identity of the person involved, special circumstances, ideological (dis)alignment and emotions. The influence of these factors on assessments of im/politeness is heightened by the deindividuated nature of the *YouTube* environment which fosters the saliency of social rather than individual identity.

4. Politeness and relevance

As noted earlier in this book, Sperber & Wilson's (1986, 1995) emphasis on inferential operations for extracting interpretations from utterances has led to an individualistic view of relevance theory, opposed to an explicit interest from other disciplines such as sociolinguistics or the ethnography of communication in more social issues (see Jary 1998a). This would entail difficulties in using this theory to approach inherently social strategies such as politeness. However, some studies have suggested possible applications of relevance theory to politeness.

One of the studies is Jucker's (1988), who starts with a rejection of classical theories of politeness such as that of Leech (1983), especially of his emphasis on a parallelism "+ indirect = + politeness," and of the excessive dependence on Grice's (1975) maxims. Instead, he explains politeness as the speaker's choice of the most appropriate way to communicate assumptions in a given context. For example, Mary's utterance (21) provides less information than required, but this is clearly explained by relevance theory, as described by Jucker (ibid.: 381–382) in (22):

(21) Peter: That was a marvellous concert.
Mary: Well, the first piece was quite nice.

(22) Mary's utterance appears to be less than sufficiently informative, but may be more relevant than an answer covering the entire concert. If Mary found all the pieces of the concert with the exception of the first one less than marvellous, she could obviously have said so, but such an utterance would have given rise to additional assumptions, for example, assumptions about her evaluation of those pieces but also assumptions to the effect that Mary does not share Peter's judgement or that she does not approve of his opinion, a clear threat to Peter's positive face in B&L's terminology.

Therefore, all that matters to the speaker, according to relevance theory, is to make predictions of mutual manifestness and choose an appropriate utterance that obtains the desired interpretation (from multiple possible interpretations of that utterance) in a specific context. The assumptions that this utterance makes manifest include those relating to the content of the utterance and referring to the relationship between the interlocutors. By choosing the utterance, the speaker must decide what he/she wants to achieve in a particular conversational environment, and to what extent the need for respect for social and personal requirements influences the choice of a particular (im)polite utterance.

Escandell Vidal (1996b) suggests that the alleged impossibility of creating a universal theory of politeness (given the inter-cultural variability of the strategies used) can be resolved if we appeal to a cognitive approach to politeness. To this end, she stresses three assumptions commonly accepted in cognitive psychology: (a) that the mind is a symbolic system that transforms people's perceptions, objects and events in the world into mental representations, (b) that human knowledge is highly structured, so that our internal representations are not simply a list, but are grouped into highly organized entities, and (c) that perception, behaviour and understanding depend on the individual's background knowledge. In (c) we can distinguish a *general* variety, which allows us to understand the actions of others simply because of their membership of the human race, and a *specific* one, which allows us to generalize on other individuals' actions as archetypical patterns of behaviour (e.g. frames, schemas).

Among the assumptions that we use daily in our interpretation of utterances, cultural assumptions about one's community and society are very important. These assumptions, many of which are commonsense and highly accessible due to their stereotypical quality, are not easily modified or altered by other information that contradicts them, since the individual tends to be certain of their veracity.[11] Therefore, despite the cultural differences that affect understanding,

11. According to Sperber (1996: 33), some cultural mental representations are communicated repeatedly, and end up being distributed by the community until all its members share a more or less faithful version of these representations. Sperber calls *cultural representations* these

the universality of inference that Sperber & Wilson (1986, 1995) proclaim in their theory is easily applicable to all communities if we postulate that in each of them different stereotypical patterns guide addressees in the construction of an appropriate context in which to process utterances optimally. Within these patterns, we would find the frequent use of politeness strategies, specific to each community, and used by its members as part of their interest in communicating not only messages, but also the attitude with which such messages are produced. For example, the message in (23):

(23) Mary [to Peter]: Oh! You had your hair cut! (Escandell Vidal 1996b: 642)

This message seems to be irrelevant, given that Peter already knows that he had his hair cut. However, the utterance itself is combined with Peter's background assumptions, specifically those *cultural assumptions* that refer to stereotypical behaviour in situations that are similar to that of a phatic quality in which (23) is uttered. A corroboration of this is the fact that Peter would probably have been surprised if his new haircut had raised no comment. Escandell Vidal (1996b) even proposes a kind of *politeness by default value* concerning our stored assumptions about what principles govern interpersonal relationships.

Together with general expectations of polite usage, speakers constantly judge what textual markers of (im)polite attitudes should accompany utterances issued in specific conversational contexts and which assumptions of (im)politeness these textual markers will make manifest (Kuiper 1997, Haugh 2003). To do so, they evaluate the threat or imposition that the utterance could cause in the interlocutor (the threat to the speaker's positive and negative face, in B&L's 1987 terms) at a particular stage of a conversational exchange. From the listener's point of view, it is also essential to assess the intention (or lack of it) to communicate a certain (im)polite attitude towards the utterance that the speaker has uttered. In this regard, several studies have addressed, within relevance theory, the intersection between the "polite / impolite" and "intentional / unintentional" axes:

M. Jary (1998b) proposes a taxonomy of cases depending on the degree of compliance with the expectations that are generated in the conversational exchange. If the speaker's behaviour conforms to these expectations, no assumptions

mental representations of a social quality. Verbal interactions are ideal vehicle for transmitting these representations which tend to reinforce existing social stereotypes, often without the individual being aware. As was mentioned in the first chapter, Žegarac (2007) states that this kind of information is consistent with what he calls *central cultural representations,* in the sense that they are valid in different contexts of everyday life without any danger of misunderstanding. Intra-cultural communication is often characterized by "cultural proximity" and by an almost unconscious transference of central representations (see Mateo & Yus 2009).

are generated worth paying attention to (interpretive route 1). If it does not fit expectations, such behaviour is relevant to the listener because it is a reflection of the fact that the speaker has a greater or lesser appreciation of the listener than he/she believed. In this case, four possibilities are generated by the attribution or non-attribution of intentionality by the speaker: (a) greater appreciation, without attribution of intentionality (route 2); (b) greater appreciation, with attribution of intentionality (route 3); (c) lower appreciation, without attribution of intentionality (route 4); and (d) lower appreciation, with attribution of intentionality (route 5).

Escandell Vidal (1998) takes a somewhat different direction, but she also stresses the importance of intentionality in the expression of politeness. The point of departure is the typology of ostensive possibilities pointed out by Wilson & Sperber (1993). In general, every utterance makes manifest a number of assumptions, some of which deserve processing (and interpreting) by the hearer. However, not all of the assumptions made manifest by the utterance – not even those which are eventually considered relevant – have to have been communicated ostensively (ibid.: 50). For effective communication to take place, it is necessary that the transmission of assumptions be made intentionally and ostensively, and that these assumptions become mutually manifest. In general, therefore, several types of transmission of information are possible: (a) unintentional, what Wilson & Sperber (1993) called *accidental transmission of information*; (b) intentional and covert; and (c) intentional and overt. Of these, only the third can really be labelled effective communication.

This three-fold division of information-transmission is applied, subsequently, to the use of politeness. In general, the use of politeness strategies that fit the requirements of the conversational context often goes unnoticed and does not lead to the (expensive) processing of assumptions about their use. If these strategies only correspond to their archetypical conventional use, it is likely that the speaker did not want to communicate any additional assumptions with this strategy. Therefore, we cannot speak of authentic communication of mutually manifest assumptions.

By contrast, the use of politeness strategies that do not fit the requirements of context and are performed ostensively result in the effective communication of assumptions on the possible reasons for their use. The cases in which one can say that an utterance communicates politeness (or impoliteness) are limited, since only when (im)politeness is deliberate and manifest (overt) can the utterance convey an assumption, in cases where it is accompanied by a presumption of relevance, which makes the listener pay attention because it might lead to an optimal balance of processing effort and cognitive effects. Later in this study, she explains that the inadequacy of a particular (im)polite strategy in a particular conversational context (and sometimes even the appropriateness of this strategy) can generate inferential operations of relevance attribution regardless of whether the (in)adequacy is intentional or unintentional (Escandell Vidal 1998: 53). In any case, it seems to be assumed that intentional and ostensive communication is the most paradigmatic type of effective communication of assumptions.

More recently, Escandell-Vidal (2004) has proposed that human cognition is capable of processing, almost simultaneously, both the specific information from utterances, and the social information obtained from the processing of verbal stimuli. Basically, we are dealing with two "cognitive skills" (or faculties) of the human mind, carried out by specialized cognitive modules, in such a way that one module (which will be called the *inferential cognitive system*) is responsible for processing the utterance (the filling of informational gaps that has already been mentioned recurrently in this book) while the other module (called the *social cognitive system*) contrasts the information obtained in the interpretation of utterances to already stored social information, inside which information about politeness strategies forms a sub-group.

Of course, these cognitive tasks are carried out independently of the source or origin of the utterances, which may be of various kinds: oral face-to-face utterances, oral via mobile phone, typed e-mail messages, etc. In each case, a double cognitive activity is at work, one devoted to the interpretation of verbal stimuli and the other aimed at updating socially-connoted information. Repeated interactions produce, according to this model, an increasingly accurate picture of the rules, interests, common goals, etc., which underlie behaviour within the community, and also interactive behaviour in physical and virtual environments.

These modules or cognitive systems, labelled, as we have seen above, the *inferential cognitive system* (for processing utterances) and the *social cognitive system* (for cognitive storage of social information, including that which concerns politeness strategies) undoubtedly are, as shown in Figure 7.1, different but interdependent. However, both systems (or cognitive faculties) have, at the same time, the same universal status (i.e., both systems exist in all humans and we systematically tend to optimize the information that these systems process). The inferential system is oriented towards the maximization of relevance, that is, to obtaining the most relevant information from the utterances that, ordinarily, we infer from the schematic words that our interlocutors utter. The social system, on the other hand, is devoted to obtaining and stabilizing social information from these everyday communicative interactions, features which include, of course, the use of politeness strategies.

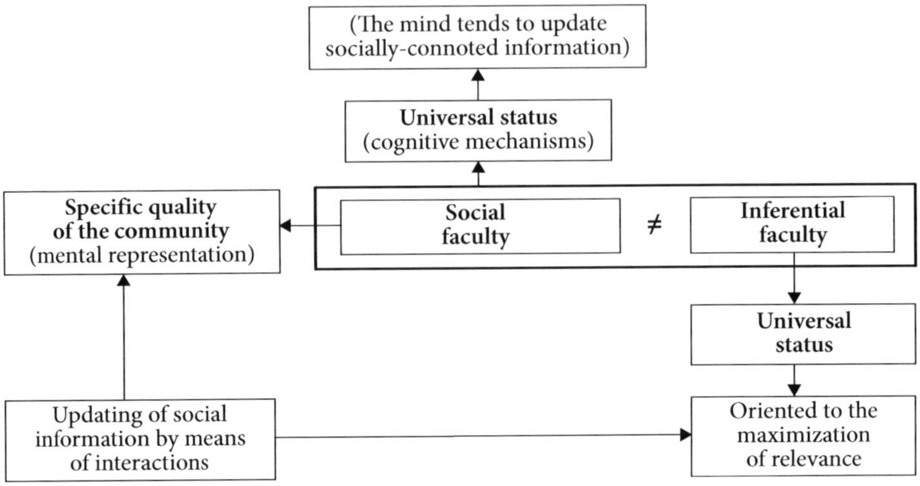

Figure 7.1 Inferential and social cognitive systems (adapted from Escandell-Vidal 2004)

Both systems are activated, simultaneously, which is very interesting to explain how speakers fit their requirements of politeness. Take the utterance (24b) in situation (24a):

(24) a. [*Two men sitting at a bar. They do not know each other, but one of them talks to the other*].
b. The salt! Pass the salt!

In this example, the hearer of the utterance in (24b) would use his inferential cognitive system to conclude that the speaker is asking for the salt. At the same time, the social cognitive system would detect an anomaly in the use of politeness rules accepted by the community, concluding that the speaker is not following these rules. Normally, the activity of the social cognitive system generates a store of information on politeness strategies which, with greater or lesser differences, is shared by the community. As argued in Yus (2007b), it is not a case of "duplicated storage" in all the members of the group, but of information that overlaps with other people's stored information, forming a kind of *communal intersection* of people's background knowledge on the use of politeness. The stored information will be, in turn, the starting point for further contrasts and adjustments between the information provided by the various utterances that we interpret and the information of a social kind that is filtered and stored from these interactions.

Internet users also type various politeness strategies in their electronic messages and evaluate the adequacy of their messages to the communicative and social activities in which they participate on the Net. As already mentioned, these strategies may reflect a desire to manifest (overtly) the intention to communicate assumptions and a particular propositional attitude towards the utterance. On other occasions, the hearer will draw relevant conclusions on the adequacy (or inadequacy) of a certain (im)polite strategy in a particular conversational context, sometimes beyond the addresser user's intention. In any case, politeness is essential to keep relationships on the Internet within acceptable margins, something which is particularly delicate in a multi-geographic interactive environment. Indeed, users are often placed in different physical locations and within very different speech communities which, as we have noted, often possess specific politeness strategies that are non-exportable to other communities. In this sense, the multicultural quality of the Internet has led users to an abandonment of specific intra-cultural strategies in order to ensure effective communication with addressee users from other communities. This effective communication is carried out assuming what we could call, following Escandell Vidal (1998), a *default level of politeness*. However, we must admit that this alleged default politeness is, in fact, the Western use of politeness or, more specifically, the hyper-polite Anglo-Saxon conceptualization, a fact that is a consequence of the increasing use of English as a *lingua franca* on the Internet.

CHAPTER 8

Conclusion: Prospects for cyberpragmatic research

In the last few years, some *cyber-media* designed for Internet-mediated interactions have changed very little. This is the case of e-mail or text-based chat rooms. However, other *cyber-media* have changed considerably, and new forms of interaction and socialization on the Net have emerged and consolidated. Several of these novel and traditional forms of Internet-mediated communication (among them blogs, web pages, social networking sites, *Twitter*, avatar-mediated interactions in 3D virtual worlds, and instant messaging) have been analysed in this book from pragmatic perspectives, especially those of cognitive pragmatics and relevance theory. This approach is principally interested in tracking down the users' intentions when communicating on the Net, in analysing *cyber-genres* as public evidence of users' underlying communicative intentions, and in predicting certain inferential strategies intended to interpret information and messages on the Internet. These three possible research areas are influenced by the qualities of the different discourses on the Net and by the availability of contextualization that users are offered by the medium.

Internet has changed a lot and has changed us a lot in recent years. The impact of this network of nodes in the daily lives of citizens in advanced societies has been enormous in the first decade of the 21st century and this impact will continue in ensuing years when, again, some current forms of Internet-mediated interactions will become obsolete (e.g. the traditional newsgroup, almost extinct nowadays) while new options of interactions will emerge and become popular. In any case, future *cyberpragmatic* research envisages a series of analytical goals or challenges, and pays special attention to a number of factors that will affect virtual communication and language use on the Internet in the near future. Some of them are summarized below.

1. The oral/written and visual/verbal dichotomy

At the beginning of the popularization of online interactions, the kind of discourse exchanged among users was predominantly textual, but in recent years the options for the contextualization of messages with the support of nonverbal information (visual and vocal) have increased enormously. Currently, interactions in chat rooms offer the option of using a camera (web cam) and voice. Videos are easily uploaded on *YouTube*, and users can engage in phone-enabled conversations on the Internet (e.g. *Skype*), etc.

From *cyberpragmatics*, following the theoretical assumptions of relevance theory (Sperber & Wilson 1986, 1995), it is assumed that there are no differentiated ways of processing information or obtaining, from the addresser user's message, the intended interpretation in a particular context. Humans have a biologically rooted cognitive ability to pay attention to what hypothetically can be more relevant (the *cognitive principle of relevance*) and a cognitive tendency to enable the processing of information as soon as the communicative intention of the speaker is detected, since according to the *communicative principle of relevance* all ostensively communicated utterances carry the presumption of their eventual relevance. This processing always follows the same pattern of enrichment of coded information (i.e. the schematic logical form) to obtain or assess, in parallel, an explicit interpretation and/or an implicit or implied interpretation (i.e. *implicature*), together with a parallel selection of as much contextual information as necessary to obtain such interpretations. By default, it is assumed that one interpretation of all the possible interpretations that an utterance has in a particular context, offers the best balance between the interest it arouses (in the form of *cognitive effects*) and the mental effort required to process it, and the addressee will invariably tend to choose that interpretation as the most relevant one. This pattern of inferential strategies leading to an interpretive choice is applied to the interpretation of all kinds of utterances or written (and typed) messages, whether in situations of physical co-presence (face-to-face conversations) or in Internet-mediated interactions. There is therefore no difference between the inferential steps taken to turn spoken utterances into meaningful interpretations in physical scenarios and the way typed messages are processed in online interactions.

However, this choice of interpretations based on ranked balances of cognitive effects and mental effort can be altered by the discursive qualities and availability of contextual information that underlies the production of the utterance or message on the Net. The same information can be processed more or less easily depending on the context and the qualities of the medium, which can influence the estimation of relevance and the very choice of an interpretation. Wilson (2002) offers the following illustrative example:

> Imagine exactly the same information being presented, first in a clearly printed form; second as a faint photocopy; third as an illegible handwritten scrawl; fourth translated into a language you read only with difficulty. Each of these versions will demand different amounts of effort from you. Though they carry exactly the same information, you will have to work harder to retrieve it from one version than from another, and this may affect your willingness to attend to it at all (and your intuitions of how relevant it is).

Therefore, although the inferential procedure we use to interpret utterances is always the same, the availability of contextual information and the attributes of the utterance or message can influence the evaluation of interpretations and how the balance between cognitive effects and processing effort is assessed while aiming at a relevant interpretation. A central goal of *cyberpragmatics* is to analyse the role of this contextualization in the interpretation of utterances transferred through the Net and will remain central in the future.

Nowadays a striking habit on the Internet challenges our conceptualization of the availability of contextual information (as tested empirically in Yus 2001b): that users do not always use the resources for vocal and visual contextualization at their disposal, which has prevented a complete popularization of the different applications for an enrichment of Internet-mediated communication with nonverbally produced information. Indeed, although instant messaging and chat rooms, for example, offer the possibility of using voice and image, many users still prefer the text typed on the keyboard and choose to enrich it with the different strategies for oralization that have been discussed in this book. Furthermore, although in avatar-mediated interactions within *Second Life* there is an option to use the user's voice, most users prefer text-based instant messaging or a chat application (both embedded in *Second Life*). One possible explanation is that the text offers users safety, control over how much information is disclosed and how much will be interpreted by the interlocutor, whereas vocal and visual nonverbal information might provide valuable information about the users that, perhaps, they are not willing to communicate explicitly. In coming years we will see if the different options for the enrichment of plain text become popular, or if typed text remains an option massively chosen by users despite this availability of richer *cyber-media*.

2. The ubiquity of the Internet

In the early years of the 21st century the Internet was still for many people something one "had to log onto," an addition outside the daily life of citizens. But with the advent of the second decade of the century, the Internet is now ubiquitous and accessible from multiple, geographically dispersed devices. For example, once it was assumed that there was an "e-mail account / e-mail place of access" duality. A few years ago, student users would be surprised if a teacher replied to an e-mail message on a Sunday from his/her university account, immediately inferring that he had gone to his/her office on a Sunday to work. Today, with the increasing use of *web-mail* applications and the popularity of free e-mail accounts such as *Yahoo*, *Hotmail* or *Gmail*, that duality is no longer assumed. What is indeed assumed is, at least in advanced societies, that the citizen *always* has access to the Net by one means or another (e.g. from home, workplace, Internet café, wi-fi connection, mobile phone, tablets) and this is taken for granted in several kinds of discourse, such as advertising, which often suggests a website where more complete information about the product can be found ("for further information enter www…").

In Chapter 2, I mentioned the accurate prediction by William Gibson, who coined the term *cyberspace* and was the author of the cult novel *Neuromancer*, that in the future people would no longer pay for Internet access but, rather, would pay to get disconnected from the Net. That prediction will be confirmed in the near future. The connection will be taken for granted, in the same way as nowadays users turn on the computer with the certainty that they will be automatically connected, and the range of sources for accessibility (from mobile and fixed locations) will be multiplied to the point where, as is currently the case with television, no one will be allowed to resort to the excuse of not being aware of some information that is distributed exclusively online. Humans, more than ever, live and process information in front of the screen and through the Net. As pointed out by Lipovetsky & Serroy (2009, quoted in Grau 2010: 31), "the network of screens has transformed our way of life […] It has become an instrument of communication and information, almost inevitable intermediary in our relationships with the world and with others. Living is, increasingly, to be attached to the screen and connected to the Net."

In this respect, 2009 was the year when there was a turning point (or *tipping point*, as Gladwell 2002 would call it), from a technology used by a minority of users, to a mass and ubiquitous technology. Indeed, 2009 was the year of popularization of mobile connection to the Internet through USB-enabled devices, in parallel to the widespread use of laptops and netbooks (in the summer of 2009, over 40% of users who were on holiday took their laptops to the hotel and logged onto the Internet there, either through the wi-fi connection at the hotel or through USB devices). Moreover, 2010 was the year in which *smart phones* became popular for Internet access. And 2011 has been the year of tablets such as the famous *iPad* to log onto the Net. This phenomenon of mass mobile access definitely removes users from the anchorage of their physical location when surfing the Net. At the same time, many of these applications to access Internet via USB include the option to send text messages (SMS) to mobile phones using software that comes with these devices. The incorporation of these text messages, typical of mobile phones, to Internet communication will force *cyberpragmatics* to include them into its scope of research in the near future.

Within a pragmatic framework of analysis of interactions on the Internet, this ubiquity and the presumption of constant accessibility to the Net can result in alterations or changes in the effectiveness of virtual communication, for example with regard to the management of mutually manifest information on how to handle certain Internet protocols for interactions and use of language therein (together with inherent jargon), which are presumably conventionalized (but perhaps not for all users), or to the creation of a preliminary context upon which virtual interactions are constructed and whose source is found on the Internet. For some users, ease of access to this preliminary context may be taken for granted, but this may not be the case for all interlocutors. Moreover, the presumption of "always on, always available" will generate many problems and communicative disruptions due to the different attitudes towards the Internet and its importance in everyday life. *Cyberpragmatics* must provide a pragmatic account of all those aspects related to universal access to the Net.

3. The consolidation of hybrid networks of interactions

In several chapters of this book and, in more detail, in *Virtualidades Reales* (*Real Virtualities*, Yus 2007b, a development of the ideas outlined in Yus 2003c and 2005b), a comparison of interactions in physical and virtual settings was carried out. In the evolution of the importance of physical vs. virtual interactions, several stages can be isolated:

1. At the beginning of the 1990s of the last century, when Internet started to become popular, Internet connections, the interactions therein and the communities that developed virtually were clearly a supplement, something fictitious and incomplete that was added to physical interactions and communities that were truly satisfactory to people and considered by everybody to be *the real scenario* of their lives. At the same time, interactions and social gatherings on the Internet were clearly limited, deficient in the quantity and quality of contextual information that was available to the few inhabitants of the Net. At that time, therefore, the only interest of Internet was the capacity to interact with users located in faraway places.

2. In this 21st century, by contrast, virtual relationships through the Internet (and mobile phones) are subject to an increasing process of *materialization*, a process of increasing importance in people's everyday lives, as well as a growing loss of the physical anchorage in daily interactions. Nowadays, Internet-mediated and mobile phone interactions possess a *local* and ordinary connotation, and have become essential for many users, especially for the so-called *digital natives*, the current generation of adolescents and young people who were born with an assumed permanent connectivity to the Net. And the ubiquity of *cyber-media* underlies, as a fundamental foundation, the acceleration of this unstoppable process of materialization of the virtual into everyday life.

In this sense, as has been commented upon throughout this book, interactions and communities in physical settings have undergone an evolution in which they have "virtualized" by losing their physical anchorage (as happens with mobile phones) and their presence in everyday communicative goals (a decrease in the amount of interactions and community fostering in public physical places such as bars, squares, etc.). On the contrary, virtual interactions and virtual communities are undergoing a parallel evolution into *materialization* or *physicalization*, they have become even more *real* than traditional physical interactions or communities. These two evolutions are currently in a process of hybridization or imbrication, a collision between increasingly virtual physical settings and increasingly real virtual settings.

3. In the next few years, the outcome of this collision will be a multiplicity of possibilities of interactions and social groupings and of several roles of technology in the formation, development and stabilization of different forms of interpersonal relationships and communal gatherings. We are undergoing a process of *physical-virtual hybridization*, according to which in each person several types of interactivity co-exist forming *personal networks of interaction* in each person. Some of them are sustained exclusively in physical scenarios with the aid of technology. On other occasions, the Internet completely satisfies users' communicative needs, and there are multiple possibilities of physical-virtual interactions.

These *personal networks of interaction* and communities of geographically dispersed users may form a dense matrix of intersections in the individual as his/her unique *personal anchorage* within this growing range of possibilities for interaction and community in physical and virtual settings, especially at a time when both settings tend to mesh. An example is the social networking application *Foursquare*, which combines the attributes of social networking sites such as *Facebook* (e.g. profile, comments, friends and contacts) with geo-localization through mobile phone positioning (GPS): "a mobile social networking game that encourages people to 'check in' online to places they visit in the real world – bars, restaurants, Starbucks – in order to accrue points. More adventurous types, with lots of check-ins, win more points. The more places you visit, the higher you rank on a league table made up of you and your friends" (Halliday 2011).

Faced with the impossibility of resorting to easily identifiable *exogenous referents* that provide individuals with social and communal stability (limits of the neighbourhood, the street, bars, community centres…), the individual has to take full responsibility for the different physical-virtual interactions and communities in which he/she participates in or belongs to. Moreover, the individual is the only stable entity in an increasingly hybrid form of communication, to the extent that the term *local* is now applied to the user, wherever he/she is located (workplace, home, a bench in a park with wi-fi connection, a cybercafé, a chat room…). *Cyberpragmatics* should analyse the impact of this hybridization of physical-virtual interactions and communities on the language and expectations of mutuality exhibited by users in their everyday socialization on the Net.

4. The transference of information from the Internet to the mobile phone

As already mentioned, 2009 was a turning point in the popularization of mobile access to Internet through USB devices attached to laptops and netbooks. Moreover, 2010 was the turning point in access to Internet through mobile phones. There are currently many mobile phones with Internet access, but the ones which really provide consistent access to the Net are *smart phones*, which allow for "natural" Internet access and allow the user to take full advantage of the different forms of interaction on the Internet discussed in this book (e.g. e-mail, instant messaging, social networking sites, and Twitter). These phones are still expensive at present and the real turning point will only come with the popularization of their use. Nevertheless, the prediction is that by 2015 all phones sold will be "smart."

The mobile phone has evolved radically since it became popular a few years ago. As summarized in Muñoz (2009), increased bandwidth of mobile Internet access has enabled the transfer of content equivalent to that obtained through the computer. Moreover, almost one third of users of the Spanish social networking site *Tuenti* access it from a mobile phone (*Tuenti-Mobile*). Iñaki Cabrera, who works for one of the companies that provide Internet access, has recently said that "the mobile phone is replacing the PC as a 'multipurpose device' for using the Net. Social networking sites and mobile phones tend to a natural symbiosis: the phone, by its nature, encourages the network effect and the possibility of sharing information in multiple formats with immediacy" (quoted in Muñoz ibid.: 24). Oscar Rodriguez, a member of another mobile company, adds that

> social networks started with the PC, but their natural use, which is immediate communication, knowing what is happening in your environment at that moment and through the computer, can only be done when you're at home or at the office. Mobile phones are their natural vehicle, they give meaning to the immediacy of social networking sites. Users' virtual lives are becoming very complex – calendar, e-mail, social networks, etc. – and the mobile phone can bring together, more than any other device, all of this complexity." (ibid.)

Currently, many Internet documents to which users have access are filtered and adapted to the peculiarities of the small mobile phone screen, with many small texts linked to one another, rather than lengthy documents that the user has to scroll down manually to read on the screen. This attribute of specificity of content for mobile phones is an interesting object of study for *cyberpragmatics*, because the different ways of presenting information and the cutting and re-shaping of texts to adapt web content to mobile formats certainly affect the eventual balance obtained between the interest of the information (cognitive effects) and the mental effort required to process it, an effort that may increase due to the size of the mobile phone screen and may even affect the user's willingness to process the information in the first place. Again, smart phones have much larger screens and even full keyboards, minimizing the aforementioned effort.

5. The transference of content to the web

In Chapter 3, I commented on two examples of discourses that were created offline and which have been transferred to the Net: the printed newspaper and the printed advertisement. As happens with mobile phones, the adaptation of the printed format to the screen entails a new segmentation and linking of chunks of information, a new accessibility to contextual information and, therefore, a chance of alterations in the way the relevant interpretation of these discourses is obtained.

In the next few years, the phenomenon of transference of content will increase parallel to increased bandwidth and improvements in the capacity of the infrastructure of Internet nodes and cables to transfer the dense information that audio and video files contain. A good example is television. It is currently common practice that, once a TV episode has been broadcast using the traditional television medium, this episode will almost immediately be available for downloading on the official television channel website. This is what happened, for instance, with the TV series *Lost*, whose episodes were available on the TV company's website the next day).

From a pragmatic point of view, if the episode is uploaded onto the Net without any alterations, and taking into account the quality of current screens, the processing of that episode will not differ substantially in either case, and only certain aspects will be interesting from the sociological point of view, for example analysing why young people have abandoned traditional television and prefer a television "on demand" in their most friendly environment, the Internet (Grau 2010). However, if the transference involves an adaptation of content and, especially, if the episode is complemented with an option for interactivity between the company and its users, or between users and additional contextual information, *cyberpragmatics* will, again, have an interesting object of study and the cognitive perspective may shed light on how the processing of televised information varies in each case.

6. The consolidation of Web 2.0, participatory culture and user-generated content

Considerable space in this book has been devoted to studying the possible consequences of the flood of information that users generate every day and the pragmatic effects of the existence of such information, as well as the problems associated with processing it. This user-generated content, sometimes involving a significant expenditure of energy which is not immediately compensated for (except for the *ambient awareness* of being part of a group and contributing positively to it), are a real challenge for a cognitive pragmatics analyst, since from this perspective we ask ourselves why an activity entailing unrewarded effort may nonetheless be relevant to the "addresser user."

This trend towards mass production of information by users, as happens with the universally-used *Wikipedia* phenomenon, will continue in the future, and *cyberpragmatics* should provide an answer to the puzzle of cognitive satisfaction that often defies the equation of "cognitive effects against processing effort" predicted by relevance theory. While we wait to see what the immediate future will bring and what new *cyber-media* will be created, in this book I have attempted to offer the reader a preparatory overview of the current state of *cyberpragmatic* research.

References

Abbasian, Fateme and Tahririan, Mohammad H. 2008. "The discoursal and formal analysis of e-mails: A cross disciplinary genre analysis." *Iranian Journal of Applied Linguistics* 11 (2): 1–22.

Abels, Eileen G., White, Marilyn D. and Hahn, Karla. 1998. "A user-based design process for web sites." *Internet Research* 8 (1): 39–48.

Abril, Guillermo. 2006. "España encabeza la pasion por el 'Messenger'," *El País*, July 16, 2006, p. 37.

Adams, T. 2006. "Goodbye, cruel world…" *The Observer*, 29–10–2006, Review, pp. 8–9.

Adrian, Angel. 2008. "No one knows you are a dog: Identity and reputation in virtual worlds." *Computer Law & Security Report* 24: 366–374.

Agren, Per-Olof. 1997. "Virtual community life: A disappearance to third places for social capital." *Virtual Communities* (project). Department of Computing, University of Umea, Sweden.

Akman, Varol. 2000. "Rethinking context as a social construct." *Journal of Pragmatics* 32 (6): 743–759.

Alamán, Teresa. 2003. "The banner ad as a netvertising subgenre: An analysis of form and content." In *Internet in Linguistics, Translation and Literary Studies*, Santiago Posteguillo, Elena Ortells, José R. Prado, Alicia Bolaños and Amparo Alcina (eds), 375–390. Castelló (Spain): Jaume I University, Servicio de Publicaciones.

Alandete, David. 2009a. "Internet, al asalto de su última frontera." *El País*, April 9, 2009, pp. 28–29.

Alandete, David. 2009b. "Los arrepentidos de Facebook," *El País*, November 11, 2009, pp. 30–31.

Alby, Francesca and Zucchermaglio, Cristina. 2008. "Collaboration in web design: Sharing knowledge, pursuing usability." *Journal of Pragmatics* 40: 494–506.

Alcaraz Varó, Enrique. 1990. *Tres Paradigmas de la Investigación Lingüística*. Alcoy: Marfil.

Alcaraz Varó, Enrique, Mateo, José and Yus, Francisco (eds). 2007. *Las Lenguas Profesionales y Académicas*. Barcelona: Ariel.

Aldama, Zigor. 2011. "QQ le tose a Facebook," *El País*, March 18, 2011, EP3, p. 6.

Ali-Hasan, Noor F. and Adamic, Lada A. 2007. "Expressing social relationships on the blog through links and comments." Paper presented at the *International Conference on Weblogs and Social Media*, Boulder, Colorado, USA, March.

Al-Maskari, Azzah and Sanderson, Mark. 2010. "A review of factors influencing user satisfaction in information retrieval." *Journal of the American Society for Information Science and Technology* 61 (5): 859–868.

Allott, Robin. 1998. "Group identity and nation identity," Paper presented at *ESS Conference*, Nottingham, England.

Amichai-Hamburger, Yair, Kaynar, Or and Fine, Adi. 2007. "The effects of need for cognition on Internet use." *Computers in Human Behavior* 23: 880–891.

Anderson, Jeffrey F., Beard, Fred K. and Walther, Joseph B. 2010. "Turn-taking and the local management of conversation in a highly simultaneous computer-mediated communication system." *Language@Internet* 7.

Andreevskaia, Alina, Bergler, Sabine and Urseanu, Monica. 2007. "All blogs are not made equal: Exploring genre differences in sentiment tagging of blogs." Paper presented at *International Conference on Weblogs and Social Media*, Boulder, Colorado, USA.

Androutsopoulos, Jannis K. 2000. "Non-standard spellings in media texts: The case of German fanzines." *Journal of Sociolinguistics* 4 (4): 514–533.

Androutsopoulos, Jannis K. 2006. "Multilingualism, diaspora, and the Internet: Codes and identities on German-based diaspora websites." *Journal of Sociolinguistics* 10 (4): 520–547.

Androutsopoulos, Jannis K. 2008. "Potentials and limitations of discourse-centered online ethnography." *Language@Internet* 5.

Androutsopoulos, Jannis K. 2010. "Localizing the global on the participatory web." In *The Handbook of Language and Globalization*, Nikolas Coupland (ed.), 203–231. Oxford: Blackwell.

Antheunis, Marjolijn L., Valkenburg, Patti and Peter, Jochen. 2008. "The quality of online, offline, and mixed-mode friendships among users of a social network site." Paper presented at *2a Conferencia Europea de Comunicación ECREA 2008*, Barcelona.

Antonijevic, Smiljana. 2008. "From text to gesture online: A microethnographic analysis of nonverbal communication in the Second Life virtual environment." *Information, Communication & Society* 11 (2): 221–238.

Araujo, Maria H. and Melo, Silvia. 2003. "Del caos a la creatividad: Los chats entre lingüistas y didactas." In *Nuevos Géneros Discursivos: Los Textos Electrónicos*, Covadonga López Alonso and Arlette Séré (eds), 45–62. Madrid: Biblioteca Nueva.

Ardèvol, Elisenda and Vayreda, Agnès. 2002. "Identidades en línea, prácticas reflexivas." Paper presented at *Seminario sobre la Identidad en la Era Digital*. Las Palmas de Gran Canaria (Spain).

Armentia, José I., Caminos, José M., Elexgaray, Jon and Merchán, Iker. 2000a. "La información en la prensa digital: Redacción, diseño y hábitos de lectura." *Zer* 8.

Armentia, José I., Caminos, José M., Elexgaray, Jon and Merchán, Iker. 2000b. *El Diario Digital. Análisis de los Contenidos Textuales, Aspectos Formales y Publicitarios*. Barcelona: Bosch.

Arthur, Charles. 1994. "A spammer in the networks," *New Scientist* 144, November 19, 1994, p. 16.

Arthur, Charles. 2009. "Blogging's long tail is dying off," *The Guardian*, June 25, 2009, Technology, p. 4.

Askehave, Inger and Nielsen, Anne E. 2005. "Digital genres: A challenge to traditional genre theory." *Information Technology & People* 18 (2): 120–141.

Atifi, Hassa, Mandelcwajg, Sacha and Marcoccia, Michel. 2011. "The co-operative principle and computer-mediated communication: The maxim of quantity in newsgroup discussions." *Language Sciences* 33: 330–340.

Aycock, Alan. 1995. "'Technologies of the Self': Foucault and Internet discourse." *Journal of Computer-Mediated Communication* 1 (2).

Baehr, Craig and Schaller, Bob. 2010. *Writing for the Internet*. Santa Barbara: Greenwood Press.

Bahr, G. Susanne and Ford, Richard A. 2011. "How and why pop-ups don't work: Pop-up prompted eye movements, user affect and decision making." *Computers in Human Behavior* 27: 776–783.

Bailenson, Jeremy N. and Blascovich, James J. 2004. "Avatars." In *Encyclopedia of Human-Computer Interaction*, 64–68. Great Barrington: Berkshire Publishing Group.

Bailenson, Jeremy N. and Blascovich, James J. 2008. "Self-representations in immersive virtual environments." *Journal of Applied Social Psychology* 38 (11): 2673–2690.

Bainbridge, William S. 2007. "The scientific research potential of virtual worlds." *Science* 317: 472–476.

Baldwin, Beth W. 1996. *Conversations, Computer-Mediated Dialogue, Multilogue and Learning.* Unpublished PhD thesis. University of North Carolina, USA.

Baltes, Boris B., Dickson, Marcus W., Sherman, Michael P., Bauer, Cara C. and LaGanke, Jacqueline S. 2002. "Computer-mediated communication and group decision making: A meta-analysis." *Organizational Behavior and Human Decision Processes* 87: 156–179.

Barnes, Susan B. 2003. *Computer-Mediated Communication. Human-to-Human Communication across the Internet.* Boston: Pearson Education.

Baron, Naomi. 1984. "Computer-mediated communication as a force in language change." *Visible Language* 18 (2): 118–141.

Baron, Naomi. 1998a. "Letters by phone or speech by other means: The linguistics of e-mail." *Language & Communication* 18: 133–170.

Baron, Naomi. 1998b. "Writing in the age of e-mail: The impact of ideology versus technology." *Visible Language* 32 (2): 35–53.

Baron, Naomi. 1998c. "Email as a contact language: The evolution of a modality." Paper presented at *International Conference on Speech, Writing and Context*, University of Nottingham.

Baron, Naomi. 2000. *Alphabet to Email. How Written English Evolved and Where It's Heading.* London: Routledge.

Baron, Naomi. 2002. "Who sets e-mail style? Prescriptivism, coping strategies, and democratizing communication access." *The Information Society* 18: 403–413.

Baron, Naomi. 2003a. "Language of the Internet." In *The Stanford Handbook for Language Engineers*, Ali Farghali (ed.), 59–127. Stanford: CSLI Publications.

Baron, Naomi. 2003b. "Why email looks like speech: Proofreading, pedagogy and public face." In *New Media Language*, Jean Aitchison and Diana M. Lewis (eds), 85–94. New York: Routledge.

Baron, Naomi. 2008a. *Always On. Language in An Online and Mobile World.* Oxford: Oxford University Press.

Baron, Naomi. 2008b. "Adjusting the volume: Technology and multitasking in discourse control." In *Handbook of Mobile Communication Studies*, James E. Katz (ed.), 177–193. Cambridge, Mass.: The MIT Press.

Baron, Naomi. 2008c. "Instant messaging by American college students. A case study in computer-mediated communication." In *Netlingo: The Metamorphosis of Language*, Salonee Priya (ed.), 141–161. Bangalore (India): Icfai University Press.

Baron, Naomi. 2009. "The myth of impoverished signal: Dispelling the spoken-language fallacy for emoticons in online communication." In *Electronic Emotion*, Jane Vincent and Leopoldina Fortunati (eds), 107–136. Berlin: Peter Lang.

Baron, Naomi. 2010a. "Discourse structures in instant messaging: The case of utterance breaks." *Language@Internet* 7.

Baron, Naomi. 2010b. "Are instant messages speech?" In *International Handbook of Internet Research*, Jeremy Hunsinger, Lisbeth Klastrup and Matthew Allen (eds), 1–21. Berlin: Springer.

Baron, Naomi. forthcoming. "Instant messaging: Rethinking theory and practice." In *Handbook of the Pragmatics of CMC*, Susan Herring, Dieter Stein and Tuija Virtanen (eds). Berlin: Mouton de Gruyter.

Barron, Anne. 2006. "Understanding spam: A macro-textual analysis." *Journal of Pragmatics* 38: 880–904.

Barthes, Roland. 1977. *Image-Music-Text*. London: Fontana.

Bartle, Richard A. 2010. "From MUDs to MMORPGs: The history of virtual worlds." In *International Handbook of Internet Research*, Jeremy Hunsinger, Lisbeth Klastrup and Matthew Allen (eds), 23–39. Berlin: Springer.

Bauman, Zygmunt. 2005. *Amor Líquido. Acerca de la Fragilidad de los Vínculos Humanos*. Mexico: *Fondo de Cultura Económica*.

Baumer, Eric, Sueyoshi, Mark and Tomlison, Bill. 2008. "Exploring the role of the reader in the activity of blogging." Paper presented at the *ACM Conference on Human Factors in Computing Systems (CHI) 2008*, April 2008, Florence, Italy.

Bavelas, Janet B., Hutchinson, Sarah, Kenwood, Christine and Hunt, Deborah. 1997. "Using face-to-face dialogue as a standard for other communication systems." *Canadian Journal of Communication* 22 (1).

Bayles, Michelle. 2000. "Just how 'blind' are we to advertising banners on the web?" *Usability News* 2 (2).

Baym, Nancy K. 1995. "The emergence of community in computer-mediated communication." In *Cybersociety*, Steven G. Jones (ed.), 138–163. London: Sage.

Baym, Nancy K. 1998. "The emergence of on-line community." In *Cybersociety 2.0*, Steven G. Jones (ed.), 35–68. London: Sage.

Baym, Nancy K. 2010. *Personal Connections in the Digital Age*. Cambridge: Polity Press.

Bays, Hillary. 1998. "Framing and face in Internet exchanges: A socio-cognitive approach." *Linguistik Online* 1.

Bays, Hillary. 2010. "Visual iconic patterns of instant messaging: Steps towards understanding visual conversations." In *International Handbook of Internet Research*, Jeremy Hunsinger, Lisbeth Klastrup and Matthew Allen (eds), 41–64. Berlin: Springer.

Beamish, Anne. 1995. *Communities On-line: Community-Based Computer Networks*: Cambridge, Mass.: The MIT, Master in Urban Planning.

Becker, Jennifer A. H. and Stamp, Glen H. 2005. "Impression management in chat rooms: A grounded theory model." *Communication Studies* 56 (3): 243–250.

Beer, David. 2008. "Social network(ing) sites… Revisiting the story so far: A response to danah boyd & Nicole Ellison." *Journal of Computer-Mediated Communication* 13: 516–529.

Beke, Tibor. 1998. "Fending off automated mass electronic mail: Or, how to distinguish yourself from a computer." *First Monday* 3 (2).

Belda Medina, José Ramón. 2000. *Estudio Morfológico Contrastivo Inglés-Español sobre Terminología de la Informática y de Internet*. PhD thesis. University of Alicante (Spain), Department of English Studies.

Belda Medina, José Ramón. 2003a. "La comunicación en español a través del correo electrónico." *Revista de Investigación Lingüística* V (2): 39–59.

Belda Medina, José Ramón. 2003b. *El Lenguaje de la Informática e Internet y su Traducción*. Alicante (Spain): University of Alicante, Servicio de Publicaciones.

Belinchón, Mercedes, Riviére, Angel and Igoa, José M. 1992. *Psicología del Lenguaje. Investigación y Teoría*. Madrid: Trotta.

Bell, Emily. 2010. "The long and the short of print and internet content," *The Guardian*, January 11, 2010, Media, p. 4.

Bellamy, Al and Hanewicz, Cheryl. 1999. "Social psychological dimensions of electronic communication." *Electronic Journal of Sociology* 4 (4).

Belson, David. 1994. *The Network Nation Revisited*. Hoboken: Stevens Institute of Technology.

Benito Alcubierre, Ricardo. 2003. "Análisis pragmático del discurso e-mail para la delimitación de género." In *Internet in Linguistics, Translation and Literary Studies*, Santiago Posteguillo, Elena Ortells, José R. Prado, Alicia Bolaños and Amparo Alcina (eds), 355–374. Castelló (Spain): Jaume I University, Servicio de Publicaciones.

Benson, Richard. 2000. "The joy of text," *The Guardian*, June 3, 2000, Weekend.

Bente, Gary, Rüggenberg, Sabine, Krämer, Nicole C. and Eschenburg, Felix. 2008. "Avatar-mediated networking: Increasing social presence and interpersonal trust in net-based collaborations." *Human Communication Research* 34: 287–318.

Bente, Gary, Krämer, Nicole C. and Eschenburg, Felix. 2008. "Is there anybody out there? Analyzing the effects of embodiment and nonverbal behaviour in avatar-mediated communication." In *Mediated Interpersonal Communication*, Elly A. Konijn, Sonja Utz, Martin Tanis and Susan B. Barnes (eds), 131–157. London: Routledge.

Benway, Jan P. 1998. "Banner blindness: The irony of attention grabbing on the World Wide Web." In *Proceedings of the Human Factors and Ergonomics Society 42nd Annual Meeting*, Vol. 1, 463–467.

Benway, Jan P. and Lane, David M. 1998. "Banner blindness: Web searchers often miss 'obvious' links." *Internet Working* 1 (3).

Berger, Charles R. and Calabrese, Richard J. 1975. "Some explorations in initial interaction and beyond: Toward a developmental theory of interpersonal communication." *Human Communication Research* 1: 99–112.

Bertacco, Massimo and Deponte, Antonella. 2005. "Email as a speed-facilitating device: A contribution to the reduced-cues perspective on communication." *Journal of Computer-Mediated Communication* 10 (3).

Beyers, Hans. 2004. "Interactivity and online newspapers: A case study on discussion boards." *Convergence* 10 (4): 11–20.

Bhatia, Vijay K. 2001. "Applied genre analysis: A multi-perspective model." *Ibérica* 3: 3–17.

Biesenbach-Lucas, Sigrun. 2007. "Students writing emails to faculty: An examination of e-politeness among native and non-native speakers of English." *Language Learning & Technology* 11 (2): 59–81.

Biesenbach-Lucas, Sigrun and Weasenforth, Donald. 1998. "The appropriateness of e-mail in composition instruction: A linguistic and rhetorical analysis." In *Les Nouvelles Technologies Educatives dans L'apprentissage des Langues Vivantes: Réflexion Théorique et Applications Practiques*. Centre de Ressources en Langues.

Birnholtz, Jeremy. 2010. "Adopt, adapt, abandon: Understanding why some young adults start, and then stop, using instant messaging." *Computers in Human Behavior* 26: 1427–1433.

Birnie, Sarah A. and Horvath, Peter. 2002. "Psychological predictors of Internet social communication." *Journal of Computer-Mediated Communication* 7 (4).

Bjorge, Anne K. 2007. "Power distance in English lingua franca email communication." *International Journal of Applied Linguistics* 17 (1): 60–80.

Blakemore, Diane. 1987. *Semantic Constraints on Relevance*. Oxford: Blackwell.

Blakemore, Diane. 1992. *Understanding Utterances*. Oxford: Blackwell.

Blanco Rodriguez, María J. 2002. "El chat: La conversación escrita." *Estudios de Lingüística* 16: 43–87.

Blincoe, Robert. 2009. "Luv is all around MySpace, lol," *The Guardian*, May 7, 2009, Technology, p. 3.

Blood, Rebecca. 2003. "Weblogs and journalism: Do they connect?" *Niema Reports* 57 (3): 61–63.

Blum-Kulka, Shoshana. 1990. "You don't touch lettuce with your fingers: Parental politeness in family discourse." *Journal of Pragmatics* 14: 259–288.

Blum-Kulka, Soshana, Danet, Brenda and Gerson, Rimona. 1985. "The language of requesting in Israeli society." In *Language and Social Situations*, Joseph Forgas (ed.), 113–141. New York: Springer.

Bodoff, David. 2006. "Relevance for browsing, relevance for searching." *Journal of the American Society for Information Science and Technology* 57 (1): 69–86.

Boellstorff, Tom. 2008. *Coming of Age in Second Life*. Princeton: Princeton University Press.

Boneva, Bonka S., Quinn, Amy, Kraut, Robert E., Kiesler, Sara and Shklovski, Irina. 2006. "Teenage communication in the instant messaging era." In *Computers, Phones, and the Internet: Domesticating Information Technology*, Robert Kraut, Malcolm Brynin and Sara Kiesler (eds), 201–218. New York: Oxford University Press.

Book, Betsy. 2004. "Moving beyond the game: Social virtual worlds." Paper presented at *The State of Play Conference II*, New York, October.

Borlund, Pia. 2003. "The concept of relevance in IR." *Journal of the American Society for Information Science and Technology* 54 (10): 913–925.

Bortree, Denise S. 2005. "Presentation of self on the web: An ethnographic study of teenage girls' weblogs." *Education, Communication & Information* 5 (1): 25–39.

Bou-Franch, Patricia. 2006. "Solidarity and deference in Spanish computer-mediated communication: A discourse-pragmatic analysis of students' emails to lecturers. In *Ways into Discourse*, Patricia Bou Franch (ed.), 61–79. Granada (Spain): Comares.

Bou-Franch, Patricia. 2011. "Openings and closings in Spanish email conversations." *Journal of Pragmatics* 43: 1772–1785.

Bou-Franch, Patricia and Lorenzo-Dus, Nuria. 2008. "Natural versus elicited data in cross-cultural speech act realisation: The case of requests in Peninsular Spanish and British English." *Spanish in Context* 5 (2): 246–277.

Boudourides, Moses A. 1997. "Accounts of sociality in the information society." Paper presented at *Electronic Commerce*. Metsovo (Greece).

Boxer, Diana and Cortés-Conde, Florencia. 1997. "From bonding to biting: Conversational joking and identity display." *Journal of Pragmatics* 27: 275–294.

boyd, danah. 2004a. "Broken metaphors: Blogging as liminal practice." Paper presented at *Media Ecology Conference 2005*, New York, USA.

boyd, danah. 2004b. "Friendster and publicly articulated social networking." Paper presented at the *Conference on Human Factors and Computing Systems* (CHI 2004), Vienna, ACM.

boyd, danah. 2005. "Blogging out loud: Shifts in public voice." Paper presented at *Lita Conference*, San José, USA.

boyd, danah. 2006a. "A blogger's blog: Exploring the definition of a medium." *Reconstruction* 6 (4).

boyd, danah. 2006b. "Friends, Friendsters, and Top 8: Writing community into being on social network sites." *First Monday* 11 (12).

boyd, danah. 2007a. "Social network sites: Public, private, or what?" *The Knowledge Tree* 13 (May).

boyd, danah. 2007b. "Why youth (heart) social network sites: The role of networked publics in teenage social life." In *Youth, Identity, and Digital Media*. David Buckingham (ed.), 119–142. Cambridge, MA: The MIT Press.

boyd, danah. 2008. *Taken out of Context. American Teen Sociality in Networked Publics*. Unpublished PhD thesis. University of California.

boyd, danah. 2011. "Social network sites as networked publics. Affordances, dynamics, and implications." In *A Networked Self. Identity, Community, and Culture on Social Network Sites*, Zizi Papacharissi (ed.), 39–58. London: Routledge.
boyd, danah and Ellison, Nicole B. 2007. "Social network sites: Definition, history, and scholarship." *Journal of Computer-Mediated Communication* 13 (1).
boyd, danah and Heer, Jeffrey. 2006. "Profiles as conversation: Networked identity performance on *Friendster*." In *Proceedings of the Hawai'i International Conference on System Sciences* (HICSS-39). Kauai, HI: IEEE Computer Society.
boyd, danah, Golder, Scott and Lotan, Gilad. 2010. "Tweet, tweet, retweet: Conversational aspects of retweeting on Twitter." In *Proceedings of HICSS-43*, IEEE: Kauai, January.
Brandtzæg, Peter B. and Heim, Jan. 2009. "Why people use social networking sites." In *Online Communities*, LNCS 5621. A. Ant Ozok and Panayiotis Zaphiris (eds), 143–152. Berlin: Springer-Verlag.
Brody, Herb. 1996. "Session with the cybershrink: An interview with Sherry Turkle." *Technology Review*, February/March 1996.
Brookey, Robert A. and Cannon, Kristopher L. 2009. "Sex lives in second life." *Critical Studies in Media Communication* 26 (2): 145–164.
Brouard, Christophe and Nie, Jian-Yun. 2004. "Relevance as resonance: A new theoretical perspective and a practical utilization in information filtering." *Information Processing and Management* 40 (1): 1–19.
Brown, Penelope and Levinson, Stephen C. 1987. *Politeness: Some Universals in Language Use*. Cambridge: Cambridge University Press.
Bruckman, Amy. 1996. "Finding one's own in cyberspace." *Technology Review*, January 1996.
Bruner, Gordon C. and Kumar, Anand. 2000. "Web commercials and advertising hierarchy-of-effects." *Journal of Advertising Research* 40 (1/2): 35–42.
Bruns, Axel. 1998a. "The n-dimensional village: Coming to terms with cyberspatial topography." *M/C* 4.
Bruns, Axel. 1998b. "The future is user-led: The path towards widespread produsage." *Fibreculture* 11.
Bruns, Axel. 2006. "Towards produsage: Futures for user-led content production." In *Cultural Attitudes towards Communication and Technology*, Fray Sudweeks, Herbert Hrachovec and Charles Ess (eds), 275–284. Perth: Murdoch University.
Bryant, J. Alison, Sanders-Jackson, Ashley and Smallwood, Amber M. K. 2006. "IMing, text messaging, and adolescent social networks." *Journal of Computer-Mediated Communication* 11 (2): 577–592.
Bucher, Hans-Jrgen and Schumacher, Peter. 2006. "The relevance of attention for selecting news content. An eye-tracking study on attention patterns in the reception of print and online media." *Communications* 31: 347–368.
Bulkley, Kate. 2008. "Beyond the banner ad," *The Guardian*, November 3, 2008, Internet Advertising, p. 1.
Bunting, Madeleine. 1999. "On the Internet, words dominate, but touching and feeling lose out," *The Guardian*, January 15, 1999, p. 18.
Burbules, Nicholas C. 2002. "The web as a rhetorical place." In *Silicon Literacies*, Ilana Snyder (ed.), 75–84. London: Routledge.
Burkeman, Oliver. 2001. "Post modern." *The Guardian*, June 20, 2001, G2.
Burkeman, Oliver. 2007. "Brilliantly boring," *The Guardian*, April 3, 2007, G2, pp. 4–7.

Burton, Meredith and Maitland, Peter. 1995. "The effect of electronic mail on traditional postal services." *Working Papers in Communication, Technology and Culture* 4.

Byrne, Elizabeth. 1994. *The Formation of Relationships on Internet Relay Chat*. Sydney: University of Western Sydney, BA Thesis.

Byron, Kristin. 2008. "Carrying too heavy a load? The communication and miscommunication of emotion by email." *Academy of Management Review* 33 (2): 309–327.

Byron, Kristin and Baldridge, David C. 2007. "E-mail recipients' impressions of senders' likability. The interactive effect of nonverbal cues and recipients' personality." *Journal of Business Communication* 44: 137–160.

Caballero, Rosario. 2008. "Theorizing about genre and cybergenre." *CORELL* 2: 14–27.

Cabrera González, M. Ángeles. 2001. "Convivencia de la prensa escrita y la prensa on line en su transición hacia el modelo de comunicación multimedia." *Estudios sobre el mensaje periodístico* 7.

Caffi, Claudia and Janney, Richard W. 1994. "Toward a pragmatics of emotive communication." *Journal of Pragmatics* 22: 325–373.

Caldwell, Barrett S. and Taha, Lilas H. 1993. "Starving at the banquet: Social isolation in electronic communication media." *Interpersonal Computing and Technology* 1 (1).

Calisir, Fethi and Karaali, Demet. 2008. "The impacts of banner location, banner content and navigation style on banner recognition." *Computers in Human Behavior* 24: 535–543.

Cambra González, Toni. 2009. "*Facebook* y los trastornos de la media(ción) tecnológica en la socialización." Paper presented at *IV Congreso Online de la Cibersociedad*.

Cameron, Ann F. and Webster, Jane. 2005. "Unintended consequences of emerging communication technologies: Instant messaging in the workplace." *Computers in Human Behavior* 21: 85–103.

Campbell, Cynthia and Wixkman, Scott A. 2000. "Familiars in a strange land: A case study of friends chatting online." *M/C* 3 (4).

Campbell, Denis. 2007. "Email stress – The new office workers' plague," *The Observer*, August 12, 2007, p. 20.

Carr, Nicholas. 2010. *The Shallows. What the Internet is Doing to Our Brains*. New York: Norton & Company.

Carston, Robyn. 2002. *Thoughts and Utterances*. Oxford: Blackwell.

Carston, R. and Powell, George. 2005. "Relevance theory – New directions and developments." *UCL Working Papers in Linguistics* 17: 279–299.

Carter, Meg. 2004. "Drivetime on the web?," *The Guardian*, March 24, 2004, *Media*, pp. 38–39.

Casaló, Luis, Flavian, Carlos and Guinaliu, Miguel. 2008. "The role of perceived usability, reputation, satisfaction and consumer familiarity on the website loyalty formation process." *Computers in Human Behavior* 24: 325–345.

Castells, Manuel. 1997. *The Information Age: Economy, Society and Culture*, vol. 2: *The Power of Identity*. Oxford: Blackwell.

Cauberghe, Verolien and De Pelsmacker, Patrick. 2010. "The impact of brand prominence and game repetition on brand responses." *Journal of Advertising* 39 (1): 5–18.

Chandler, Daniel. 1997. "Writing oneself in cyberspace." University of Wales, Internet document. <http://www.aber.ac.uk/media/Documents/short/homepgid.html>.

Chang, Hsin H. and Chen, Sue W. 2008. "The impact of customer interface quality, satisfaction and switching costs on e-loyalty: Internet experience as a moderator." *Computers in Human Behavior* 24: 2927–2944.

Charman-Anderson, Suw. 2008. "Breaking the email compulsion," *The Guardian*, August 28, 2008, Technology, pp. 1–2.

Chen, Gina M. 2011. "Tweet this: A uses and gratifications perspective on how active Twitter use gratifies a need to connect with others." *Computers in Human Behavior* 27: 755–762.
Cherny, Lynn. 1995a. "Objectifying the body in the discourse of an object-oriented MUD." *Works & Days* 25/26.
Cherny, Lynn. 1995b. "The modal complexity of speech events in a social MUD." *Electronic Journal of Communication* 5 (4).
Cherny, Lynn. 1999. *Conversation and Community. Chat in a Virtual World.* Stanford: CSLI.
Chesher, Chris. 2005. "Blogs and the crisis of authorship," Paper presented at *Blogtalk Downunder*, Sydney, Australia.
Cheung, Christy M. K., Shen, Aaron X. L., Lee, Matthew K. O. and Wang, WeiPing. 2007. "Let's work together! We-intention to use instant messaging for e-collaboration." In *Proceedings of the 15th European Conference on Information Systems,* Hubert Osterle, Joachim Schelp and Robert Winter (eds), 407–418. St. Gallen, Switzeland.
Cheung, Christy M. K. and Lee, Matthew K. O. 2010. "A theoretical model of intentional social action in online social networks." *Decision Support Systems* 49: 24–30.
Chevalier, Aline and Bonnardel, Nathalie. 2007. "Articulation of web site design constraints: Effects of the task and designers' expertise." *Computers in Human Behavior* 23: 2455–2472.
Chiang, I.-Ping, Huang, Chun-Yao and Huang, Chien-Wen. 2009. "Characterizing web users' degree of web 2.0–ness." *Journal of the American Society for Information Science and Technology* 60 (7): 1349–1357.
Cho, Thomas, 2010. "Linguistic features of electronic mail in the workplace: A comparison with memoranda." *Language@Internet* 7.
Choi, Junho H. and Danowski, James. 2002. "Making a global community on the net – global village or global metropolis?: A network analysis of Usenet newsgroups." *Journal of Computer-Mediated Communication* 7 (3).
Choo, Chun W., Detlor, Brian and Turnbull, Don. 2000. "Information seeking on the web: An integrated model of browsing and searching." *First Monday* 5 (2).
Chung, Deborah S. 2008. "Interactive features of online newspapers: Identifying patterns and predicting use of engaged readers." *Journal of Computer-Mediated Communication* 13: 658–679.
Chung, Donghun and Nam, Chang S. 2007. "An analysis of the variables predicting instant messenger use." *New Media & Society* 9 (2): 212–234.
Chung, Hwiman and Zhao, Xinshu. 2004. "Effects of perceived interactivity on web site preference and memory: Role of personal motivation." *Journal of Computer-Mediated Communication* 10 (1).
Cicognani, Anna. 1998. "On the linguistic nature of cyberspace and virtual communities." *Virtual Reality* 3: 16–24.
Clark, Herbert H. and Brennan, Susan E. 1991. "Grounding in communication." In *Perspectives on Socially Shared Cognition*, Lauren B. Resnik, John H. Levine and Stephanie D. Teasley (eds), 127–149. American Psychological Association.
Cohen, Kris R. 2005. "What does the photoblog want?" *Media, Culture & Society* 27 (6): 883–901.
Collins, Mauri. 1992. "Flaming: The relationship between social context cues and uninhibited verbal behavior in computer-mediated communication." *Emoderators* (Internet Resources).
Collins, Mauri P. and Berge, Zane L. 1997. "Moderating online electronic discussion groups." Paper presented at the *American Educational Research Association Conference*, Chicago (USA).

Condon, Sherry and Cech, Claude G. 1995. "Discourse strategies management in face-to-face vs. computer-mediated decision making interactions." Paper presented at *Computer-Mediated Discourse Analysis* (*GURT'95*), Washington, Georgetown University.
Cooper, William S. 1971. "A definition of relevance for information retrieval." *Information Storage and Retrieval* 7 (1): 19–37.
Cormode, Graham and Krishnamurthy, Balachander. 2008. "Key differences between Web 1.0 and Web 2.0." *First Monday* 13 (6).
Cornelius, Caroline and Boos, Margarete. 2003. "Enhancing mutual understanding in synchronous computer-mediated communication by training. Trade-offs in judgmental tasks." *Communication Research* 30 (2): 147–177.
Cosijn, Enca and Ingwersen, Peter. 2000. "Dimensions of relevance." *Information Processing and Management* 36: 533–550.
Coupland, Nikolas and Jaworski, Adam. 1997. "Relevance, accommodation and conversation: Modeling the social dimension of communication." *Multilingua* 16 (2/3): 187–207.
Croon, Anna. 1997. "R U out there? On personal communities in cyberspace." In *Proceedings of the 20th IRIS Conference. "Social Informatics"*, K Braa and E. Monteiro (eds), 591–604. Oslo: Department of Information and Computer Science, University of Oslo.
Crowston, Kevin and Williams, Marie. 2000. "Reproduced and emergent genres of communication on the World Wide Web." *The Information Society* 16: 201–215.
Crystal, David. 1987. *The Cambridge Encyclopedia of Language*. Cambridge: Cambridge University Press.
Cutler, Richard H. 1995. "Distributed presence and community in cyberspace." *Interpersonal Computing and Technology* 3 (2): 12–32.
Daft, Richard L. and Lengel, Robert H. 1984. "Information richness: A new approach to managerial behavior and organizational design." *Research in Organizational Behavior* 6: 191–233.
Danet, Brenda. 1996a. "Hmmm.. Where's that smoke coming from? Writing, play and performance on Internet Relay Chat." *Journal of Computer-Mediated Communication* 2 (4).
Danet, Brenda. 1996b. "'Talk to you soon': Literacy, letter-writing and the language of electronic mail." Paper presented at *Attending to Technology: Implications for Teaching and Research in the Humanities*, University of Maryland, College Park, November.
Danet, Brenda. 1998. "Computer-mediated communication." In *The Garland Encyclopedia of Semiotics*, Paul Bouissac (ed.). Oxford: Oxford University Press.
Danet, Brenda. 2002. "The language of email." Paper presented at *European Union Summer School*, University of Rome (Italy).
Danet, Brenda, Ruedenberg, Lucia and Rosenbaum-Tamari, Yehudit. 1998. "'Smoking dope' at a virtual party: Writing, play and performance on Internet relay chat." In *Network and Netplay: Virtual Groups on the Internet*, Sheizaf Rafaeli, Fay Sudweeks and Margaret McLaughlin (eds), 41–76. Cambridge, Mass.: MIT Press.
Danielson, David R. 2002. *Transitional Volatility in Web Navigation. Usability Metrics and User Behavior*. Unpublished MA thesis, Stanford University (USA).
Darics, Erika. 2010. "Politeness in computer-mediated discourse of a virtual team." *Journal of Politeness Research* 6: 129–150.
Daugherty, Terry, Eastin, Matthew S. and Bright, Laura. 2008. "Exploring consumer motivations for creating user-generated content." *Journal of Interactive Advertising* 8 (2).
Davies, Bronwyn. and Harre, Rom. 1990. "Positioning: The discursive production of selves." *Journal of the Theory of Social Behavior* 20: 43–63.
Davies, Julia. 2007. "Display, identity and the everyday: Self-presentation through online image sharing." *Discourse* 28 (4): 549–564.

Davis, Joseph E. 1999. "An interview with Sherry Turkle." *The Hedgehog Review*, Autumn, 71–84.
Davis, Mike. 1997. "Fragmented by technologies: A community in cyberspace." *Electronic Computing and Technology* 5 (1–2): 7–18.
de Cindio, Fiorella and Ripamonti, Laura A. 2010. "Nature and roles for community networks in the information society." *AI & Society* 25: 265–278.
de Klerk, Vivian and Bosch, Barbara. 1999. "Nicknames as evidence of verbal playfulness." *Multilingua* 18 (1): 1–16.
de la Fuente, Inmaculada. 1997. "No llegará carta ni del banco," *El País*, March 30, 1997, Domingo, p. 13.
de Moor, Aldo. 2010. Conversations in context: A Twitter case for social media systems design." Paper presented at *I-SEMANTICS 2010*, Graz, Austria, September.
de Moor, Aldo and Efimova, Lilia. 2004. "An argumentation analysis of weblog conversations." In *Proceedings of the 9th International Working Conference on the Language-Action Perspective on Communication Modelling (LAP 2004).*
de Salvador Agra, Saleta. 2009. "Descorporización o de los límites del cibercuerpo." Paper presented at *I Congreso Internacional de Cultura y Género*, Elche (Spain).
de Souza, Zaineb and Dick, Geoffrey N. 2007. "What explains the MySpace phenomenon? Extending the technology acceptance model to explain the use of social networking by schoolchildren." In *Proceedings of the 22nd AIS SIG-ED IAIM Conference*, Montreal, Canada.
DeAndrea, David C., Shaw, Allison S. and Levine, Timothy R. 2010. "Online language: The role of culture in self-expression and self-construal on Facebook." *Journal of Language and Social Psychology* 29 (4): 425–442.
December, J. 1993. "Characteristics of oral culture in discourse on the net." Paper presented at the *12th Penn State Conference on Rethoric and Composition,* University of Pennsylvania, USA.
Deka, Sanjib K. and Lahkar, Narendra. 2010. "Performance evaluation and comparison of the five most used search engines in retrieving web resources." *Online Information Review* 34 (5): 757–771.
Del'Aquila, Pablo. 1999. "Critical consumption and virtual communities." *Paideusis* 2.
del-Teso-Craviotto, Marisol. 2008. "Gender and sexual identity authentication in language use: The case of chat rooms." *Discourse Processes* 10 (2): 251–270.
Derks, Daantje, Bos, Arjan E. R. and von Grumbkow, Jasper. 2007. "Emoticons and social interaction on the internet: The importance of social context." *Computers in Human Behavior* 23: 842–849.
Derks, Daantje, Bos, Arjan E. R. and von Grumbkow, Jasper. 2008. "Emoticons and online message interpretation." *Social Science Computer Review* 26 (3): 379–388.
Derks, Daantje, Fischer, Agneta H. and Bos, Arjan E. 2008. "The role of emotion in computer-mediated communication: A review." *Computers in Human Behavior* 24: 766–785.
Dertouzos, Michael. 2000. "E-mail: Freedom or jail," *Technology Review*, March-April, 2000.
Detlor, Brian, Sproule, Susan and Gupta, Chris. 2003. "Pre-purchase online information seeking: Search versus browse." *Journal of Electronic Commerce Research* 4 (2).
Devitt, Amy J. 2009. "Re-fusing form in genre study." In *Genres in the Internet*, Janet Giltrow and Dieter Stein (eds), 27–47. Amsterdam/Philadelphia: John Benjamins.
D'Haenens, Leen, Jankowski, Nicholas and Heuvelman, Ard. 2004. "News in online and print newspapers: Differences in reader consumption and recall." *New Media & Society* 6 (3): 363–382.
Diago Marco, Marta. 2002. "La máscara en el chat. Nick y avatar en el mundo virtual." *Revista de Antropología Experimental* 2.

Dietrich, Noel S. 2004. "From face-to-face to cyberspace: Examining the impact of instant-messaging within already-established relationships." Paper presented at *Eastern Communication Association Conference*, Boston (USA).

Dillon, Andrew and Gushrowski, Barbara. 2000. "Genres and the Web: Is the personal home page the first uniquely digital genre?" *Journal of the American Society of Information Science* 51 (2): 202–205.

Di Segni, Yaara. 1997a. *"Can't You See You Were Shouting?" The Experience of Visual Impairment in the Virtual World.* Unpublished MA thesis, University of Sheffield (UK).

Di Segni, Yaara. 1997b. "Flame wars, flooding, kicking and spamming: Expressions of aggression in the virtual community." University of Sheffield (UK), MA in *Disability Studies.*

Dobson, Roger. 1996. "Yours truly logs on and signs off," *The Independent*, December 29, 1996, p. 8.

Donath, Judith S. 1996. *Inhabiting the Virtual City: The Design of Social Environments for Electronic Communities.* Unpublished PhD thesis. Cambridge, Mass.: The MIT.

Donath, Judith S. 1997. "Body language without the body: Situating social cues in the virtual world." Paper presented at *International IFIP-Conference on Women, Work and Computerization (WWC)*, Berlin.

Donath, Judith S. 1999. "Identity and deception in the virtual community." In *Communities in Cyberspace*, Mark Smith and Peter Kollock (eds), 29–59. London: Routledge.

Donath, Judith S. 2001. "Mediated faces." In *Cognitive Technology: Instruments of Mind. Proceedings of the 4th International Conference on Cognitive Technology: Instruments of Mind*, Meurig Beynon, Chrystophe L. Nehaniv and Kerstin Dautenhahn (eds), 373–390. Berlin: Springer.

Donath, Judith S. 2007. "Signals in social supernets." *Journal of Computer-Mediated Communication* 13 (1).

Donath, Judith and boyd, danah. 2004. "Public displays of connection." *BT Technology Journal* 22 (4): 71–82.

Donath, Judith, Karahalios, Karrie and Viegas, Fernanda. 1999. "Visualizing conversation." *Journal of Computer-Mediated Communication* 4 (4).

Dor, Daniel. 2003. "On newspaper headlines as relevance optimizers." *Journal of Pragmatics* 35 (5): 695–721.

Döring, Nicola. 2002. "Personal home pages on the net: A review of research." *Journal of Computer-Mediated Communication* 7 (3).

Downes, William. 2000. "The language of felt experience: Emotional, evaluative and intuitive." *Language and Literature* 9: 99–121.

Dresner, Eli and Herring, Susan C. 2010. "Functions of the nonverbal in CMC: Emoticons and illocutionary force." *Communication Theory* 20: 249–268

Ducheneaut, Nicolas, Wen, Ming-Hui, Yee, Nicholas and Wadley, Greg. 2009. "Body and mind: A study of avatar personalization in three virtual worlds." Paper presented at *CHI 2009*, Boston, Mass. (USA), April.

Duke, Shearlean. 2001. "E-mail: Essential in media relations, but no replacement for face-to-face communication." *Public Relations Quarterly*, Winter, pp. 19–22.

Duran, Robert L., Kelly, Lynne and Keaten, James A. 2005. "College faculty use and perceptions of electronic mail to communicate with students." *Communication Quarterly* 43 (2): 159–176.

Duthler, Kirk W. 2006. "The politeness of requests via email and voicemail: Support for the hyperpersonal model." *Journal of Computer-Mediated Communication* 11 (2): 500–521.

Echeverría, Javier. 1994. *Telépolis*. Barcelona: Destino.

Echeverría, Javier. 1999. *Los señores del aire*. Barcelona: Destino.

Edemariam, Aida. 2009. "What are you doing?," *The Guardian*, January 6, 2009, G2, pp. 12–13.

Edlington, Sara. 1995. "Obscene letters in the e-mail," *The Independent*, January 23, 1995, p. 21.

Efimova, Lilia. 2003. "Blogs: the stickiness factor." Paper presented at *BlogTalk: A European Conference on Weblogs*, Vienna (Austria).

Efimova, Lilia and Hendrick, Stephanie. 2005. "In search for a virtual settlement: An exploration of weblog community boundaries." Paper presented at *Communities & Technologies '05*, Milan (Italy).

Efimova, Lilia, Hendrick, Stephanie and Anjewierden, Anjo. 2005. "Finding 'the life between buildings': An approach for defining a weblog community." Paper presented at *Internet Research 6.0: Internet Generations*, Chicago, USA.

Efimova, Lilia and Lassoued, Andrea B. 2008. "On how relationships develop via weblogs: A co-constructed narrative." In *Remote Relationships in a Small World*, Samantha Holland (ed.), 137–154. Berlin: Peter Lang.

Ekman, Paul and Friesen, Wallace. 1969. "The repertoire of nonverbal behavior: Categories, origins, usage and coding." In *Nonverbal Communication, Interaction and Gesture*. Adam Kendon (ed.), 57–105. The Hague: Mouton.

Ellison, Nicole, Heino, Rebecca and Gibbs, Jennifer. 2006. "Managing impressions online: Self-presentation processes in the online dating environment." *Journal of Computer-Mediated Communication* 11 (2): 415–441.

Ellison, Nicole B., Steinfield, Charles and Lampe, Cliff. 2007. "The benefits of Facebook 'friends': Social capital and college students' use of online social network sites." *Journal of Computer-Mediated Communication* 12 (4): 1143–1168.

Ellison, Nicole B., Lampe, Cliff, Steinfield, Charles and Vitak, Jessica. 2011. "With a little help from my friends. How social network sites affect social capital processes." In *A Networked Self. Identity, Community, and Culture on Social Network Sites*, Zizi Papacharissi (ed.), 124–145. London: Routledge.

Elmer-Dewitt, Philip. 1994. "Bards of the Internet," *Time Magazine*, July 4, 1994, pp. 66–67.

Engebretsen, Martin. 2000. "Hypernews and coherence." *Journal of Digital Information* 1 (7).

English, Joel Alexander. 1999. "Language play: The functionality of x-pressive text." Chapter 4 in *Assessing the Synchronous Online Classroom*. Unpublished PhD thesis. Ball State University (USA).

Epperson, Kraettli L. 1995. *Patterns of Social Behavior in Computer-Mediated Communication*. Honors Thesis. Rice University, Department of Sociology.

Erickson, Thomas. 1996a. "Social interaction on the Net: Virtual community as participatory genre." In *Proceedings of the 13th Hawaii International Conference on System Sciences*. Manui (Hawaii): IEEE.

Erickson, Thomas. 1996b. "The world wide web as social hypertext." *Viewpoints*, Communications of the ACM.

Escandell Vidal, Victoria.1996a. *Introducción a la Pragmática*. Barcelona: Ariel.

Escandell Vidal, Victoria. 1996b. "Towards a cognitive approach to politeness." *Language Sciences* 18 (3–4): 629–650.

Escandell Vidal, Victoria. 1998. "Politeness: A relevant issue for relevance theory." *Revista Alicantina de Estudios Ingleses* 11: 45–57.

Escandell-Vidal, Victoria. 2004. "Norms and principles. Putting social and cognitive pragmatics together." In *Current Trends in the Pragmatics of Spanish*, Rosina Márquez-Reiter and María Elena Placencia (eds.), 347–371. Amsterdam/Philadelphia: John Benjamins.

Escribano Otero, Juan J. 1998. "Tratado de las ciberbuenas maneras." Paper presented at *Las Libertades de Acceso y Difusión* (II Seminario de Telecomunicaciones e Información). Madrid: Complutense University.

Escribano Otero, Juan J. and Peña, Rosalía. 1999. "Cibermaneras." *Revista Vivat Academia* 4.

Espinosa Villareal, Mónica. 1999. "Estrategias de moderación como mecanismo de participación y construcción de conocimientos en grupos de discusión electrónicos." *Hiper-Textos* 1 (2).

Estalella, Adolfo. 2006. "La construccion de la blogosfera: Yo soy mi blog (y sus conexiones)." In *La Blogosfera Hispana: Pioneros de la Cultura Digital*, José M. Cerezo (ed.), 20–37. Fundación France Telecom.

Éthier, Jean, Hadaya, Pierre, Talbot, Jean and Cadieux, Jean. 2008. "Interface design and emotions experienced on B2C Web sites: Empirical testing of a research model." *Computers in Human Behavior* 24: 2771–2791.

Etzioni, Amitai. 2000. "E-Communities build new ties, but ties that bind." *The New York Times*, February 10, 2000, Technology.

Fabri, Marc, Moore, David and Hobbs, Dave. 2005. "Empathy and enjoyment in instant messaging." In *Proceedings of 19th British HCI Group Annual Conference* (HCI2005), L. McKinnon, O. Bertelsen and N. Bryan-Kinns (eds), 4–9. Edinburgh (UK).

Farnham, Shelly D. and Churchill, Elizabeth F. 2011. "Faceted identity, faceted lives: Social and technical issues with being yourself online." Paper presented at *CSCW 2011*, Hangzhou, China, March.

Fayard, Anne-Laure. 2006. "Interacting on a video-mediated stage. The collaborative construction of an interactional video setting." *Information Technology & People* 19 (2): 152–169.

Feenberg, Andrew. 1989. "The written world: On the theory and practice of computer conferencing." In *Mindweave: Communication, Computers and Distance Education*, Robin Mason and Antohny Kaye (eds), 22–39. Oxford: Pergamon.

Feld, Scott L. 1981. "The focused organisation of social ties." *American Journal of Sociology* 86 (5): 1015–1035.

Ferebee, Susan S. and Davis, James W. 2009. "Factors that persuade continued use of Facebook among new members." In *ACM International Conference Proceeding Series*; Vol. 350. *Proceedings of the 4th International Conference on Persuasive Technology*. Claremont, California, USA.

Fernback, Jan. 1997. "The individual within the collective: Virtual ideology and the realization of collective principles." In *Virtual Culture. Identity & Communication in Cybersociety*, Steven G. Jones (ed.), 36–54. London: Sage.

Ferrara, Kathleen, Brunner, Hans and Whitemore, Greg. 1991. "Interactive written discourse as an emergent register." *Written Communication* 8 (1): 8–34.

Fetscherin, Marc and Lattemann, Christoph. 2007. "User acceptance of virtual worlds: An explorative study about Second Life." Rollins College and Postdam University. *Second Life Research Team.*

Finquelievich, Susana. 2000. "Informática comunitaria: La solidaria red de redes." Paper presented at *Internet: Herramienta para el desarrollo social*, Buenos Aires.

Fisher, Danyel, Smith, Marc and Welser, Howard T. 2006. "You are who you talk to: Detecting roles in Usenet Newsgroups." In *Proceedings of the 39th Annual Hawaii International Conference on System Sciences (HICSS'06)*, Hawaii, January.

Fitzpatrick, Michael. 2007. "Digital generation dismisses email as 'for old people'," *The Guardian*, December 6, 2007, Technology, p. 6.

Flanagin, Andrew J. 2005. "IM online: Instant messaging use among college students." *Communication Research Reports* 22 (3): 175–187.

Flavián-Blanco, Carlos, Gurrea-Sarasa, Raquel and Orús-Sanclemente, Carlos. 2011. "Analyzing the emotional outcomes of the online search behaviour with search engines." *Computers in Human Behavior* 27: 540–551.

Fleck, Matthes, Kirchhoff, Lars, Meckel, Miriam and Stanoevska-Slabeva, Katarina. 2007. "Applications of blogs in corporate communication." *Studies in Communication Sciences* 7 (2): 227–245.

Fong, John and Burton, Suzan. 2006. "Electronic word-of-mouth: A comparison of stated and revealed behavior on electronic discussion boards." *Journal of Interactive Advertising* 6 (2).

Fono, David and Raynes-Goldie, Kate. 2006. "Hyperfriendship and beyond: Friends and social norms on Livejournal." In *Internet Research Annual: Selected Papers from the Association of Internet Researchers Conference*. Volume 4, Mia Consalvo and Caroline Haythornthwaite (eds), 91–103. New York: Peter Lang.

Foster, Derek. 1997. "Community and identity in the electronic village." In *Internet Culture*, David Porter (ed.), 23–37. London: Blackwell.

Fouser, Robert J., Inque, Narahiko and Lee, Chungmin. 2000. "The pragmatics of orality in English, Japanese and Korean computer-mediated communication." In *Words on the Web*, Lyn Pemberton and Simon Shurville (eds), 52–60. Exeter, UK: Intellect Books.

Fox, Jesse and Bailenson, Jeremy N. 2009. "Virtual virgins and vamps: The effects of exposure to female characters' sexualized appearance and gaze in an immersive virtual environment." *Sex Roles* 61: 147–157.

Fraca de Barrera, Lucía. 2007. "El chateo desde la ciberlingüística." *Argos* 24 (46): 25–37.

Fraser, Bruce. 1990. "Perspectives on politeness." *Journal of Pragmatics* 14: 219–236.

Fraser, Bruce and Nolen, William. 1981. "The association of deference with linguistic form." *International Journal of the Sociology of Language* 27: 93–109.

Fraser, Matthew and Dutta, Soumitra. 2008. *Throwing Sheep in the Boardroom. How Online Social Networking Will Transform your Life, Work and World*. Chichester: Wiley.

Freeman, John. 2009. *The Tyranny of E-mail*. New York: Scribner.

Freiermuth, Mark R. 2011. "Debating in an online world: A comparative analysis of speaking, writing, and online chat." *Text & Talk* 31 (2): 127–151.

Frobenius, Maximiliane. 2011. "Beginning a monologue: The opening sequence of video blogs." *Journal of Pragmatics* 43: 814–827.

Fuccella, Jeanette and Pizzolato, Jack. 1998. "Creating web site designs based on user expectations and feedback." *ITG Newsletter* 1 (1).

Fuller, Rodney. 1994. "Human-computer-human interaction: How computers affect interpersonal communication." *The Arachnet Electronic Journal on Virtual Culture* 2 (2).

Furukawa, Tadanobu, Matsuo, Yutaka, Ohmukai, Ikki, Uchiyama, Koki and Ishizuka, Mitsuru. 2007. "Social networks and reading behavior in the blogosphere." Paper presented at *International Conference on Weblogs and Social Media* (ICWSM), Boulder, Colorado (USA).

Gains, Jonathan. 1999. "Electronic mail – A new style of communication or just a new medium?: An investigation into the text features of e-mail." *English for Specific Purposes* 18 (1): 81–101.

Galindo Cáceres, Jesús. 1998. "Cibercultura, ciberciudad, cibersociedad. Hacia la construcción de mundos posibles en nuevas metáforas conceptuales." *Razón y Palabra* 10 (3).

Galindo Cáceres, Jesús. 2010. "La internet y sus redes sociales. Comunicología e ingeniería en comunicacion social de un fenómeno aún emergente." *Razón y Palabra* 71.

Gangadharbatla, Harsha. 2008. "*Facebook* me: Collective self-esteem, need to belong, and internet self-efficacy as predictors of the iGeneration's attitudes toward social networking sites." *Journal of Interactive Advertising* 8 (2).

Garau, Maia, Slater, Mel, Bee, Simon and Sasse, Martina A. 2001. "The impact of eye gaze on communication using humanoid avatars." In *Proceedings of the SIG-CHI Conference on Human factors in Computing Systems*, Seattle (USA).

Garcés-Conejos Blitvich, Pilar. 2010. The *YouTubification* of politics, impoliteness and polarization. In *Handbook of Research on Discourse Behavior and Digital Communication: Language Structures and Social Interaction*, Rotimi Taiwo (ed.), 540–563. Hershey, USA: IGI Global.

Garcés-Conejos Blitvich, Pilar. forthcoming. "Politics, "lies" and YouTube: A genre approach to assessments of im/politeness on Obama's 9/9/2009 presidential address. In *New Perspectives on (Im)politeness and Interpersonal Communication*, Lucía Fernandez, Mariola Herández, Manuel Padilla and Reyes Gómez (eds.). Newcastle, UK: Cambridge Scholars Publishing.

Garcés-Conejos Blitvich, Pilar and Bou-Franch, Patricia. 2008. "Cortesía en las páginas web interactivas. El comercio electrónico." In *Cortesía y Conversación. De lo escrito a lo oral. III Coloquio Internacional del Programa Edice*, Antonio Briz, Antonio Hidalgo, Marta Albelda, Josefa Contreras and Nieves Hernández (eds), 468–488. Valencia (Spain): Universitat de Valencia, Departamento de Filología Española.

Garcés-Conejos Blitvich, Pilar and Lorenzo-Dus, Nuria. 2010. "On-line community organizing: rapport and directness. A case study from the Obama presidential campaign." *Issues in Political Discourse Analysis* 3 (1): 1–24.

García Gómez, Antonio. 2010. "Disembodiment and cyberspace: Gendered discourses in female teenagers' personal information disclosure." *Discourse & Society* 21 (2): 135–160.

Garfield, S. 2006. "How to make 80 million friends and influence people," *The Observer*, June 18, 2006, Review, pp. 6–9.

Garrett, R. Kelly and Danziger, James N. 2007. "IM = Interruption management? Instant messaging and disruption in the workplace." *Journal of Computer-Mediated Communication* 13 (1).

Garton, Laura, Haythornthwaite, Caroline and Wellman, Barry. 1997. "Studying online social networks." *Journal of Computer-Mediated Communication* 3 (1).

Gasca, Luis and Gubern, Román. 1988. *El Discurso del Comic*. Madrid: Cátedra.

Georgalou, Mariza. 2010. "'Pathfinding' discourses of self in social network sites." In *Handbook of Research on Discourse Behavior and Digital Communication: Language Structures and Social Interaction*, Rotimi Taiwo (ed.), 39–65. Hershey: IGI Global.

Gheorghiu, Alina. 2008. "Building one's self in the context of online interactions." In *Teenagers' Actions and Interactions Online in Central and Eastern Europe. Potentials and Empowerment, Risks and Victimization*, Monica Barbovschi and Maria Diaconescu (eds), 55–72. Cluj-Napoka (Romania): Cluj University Press.

Gibbs, Raymond W. 1999. *Intentions in the Experience of Meaning*. Cambridge: Cambridge University Press.

Giese, Mark. 1998. "Self without body: Textual self-representation in an electronic community." *First Monday* 3 (4).

Giltrow, Janet and Stein, Dieter. 2009. "Genres in the Internet: Innovation, evolution, and genre theory." In *Genres in the Internet*, Janet Giltrow and Dieter Stein (eds), i–x. Amsterdam/Philadelphia: John Benjamins.

Gimenez, Julio C. 2000. "Business e-mail communication: Some emerging tendencies in register." *English for Specific Purposes* 19: 237–251.

Giménez, Martha. 1997. "The dialectics between the real and the virtual." In *Mapping Cyberspace: Social Research in the Electronic Frontier*, J. E. Behar (ed.). Nashville: Dowling Press.

Gladwell, Malcolm. 2002. *The Tipping Point*. Boston: Back Bay Books.

Goffman, Erving. 1974. *Frame Analysis*. New York: Harper & Row.

Goffman, Erving. 1983. "The interaction order." In *Los momentos y sus hombres*, 169–205. Barcelona: Paidós.

Goffman, Erving. 1987. *La presentación de la persona en la vida cotidiana*. Buenos Aires: Amorrortu (first edited 1959).

Golbeck, Jennifer. 2007. "The dynamics of web-based social networks: Membership, relationships, and change." *First Monday* 12 (11).

Goleman, Daniel. 2007. "E-mail is easy to write (and to misread)," *The New York Times*, October 7, 2007.

Gómez, Edgar. 1998. "La CMC: Comunicación mediada por computadora. Caso BBS." *Razón y Palabra* 10 (3).

Gómez, Rosario G. 2010. "El (incierto) papel de la prensa," *El País*, May 31, 2010, pp. 34–35.

González-Bueno, Manuela. 1998. "The effects of electronic mail on Spanish L2 discourse." *Language Learning and Technology* 1 (2): 55–70.

Gosálvez, Patricia. 2010. "¿Me 'ajuntas'?, versión 2.0," *El País*, August 1, 2010, pp. 28–29.

Gousseva, Julia. 1998. "An experience in cyberspace communication: Listserv interaction in a freshman composition class." *Linguistik Online* 1.

Gouti, Gérard. 2003. "Intercambios en los foros de debate: Algunos elementos de reflexión para un acercamiento lingüístico." In *Nuevos Géneros Discursivos: Los Textos Electrónicos*, Covadonga López Alonso and Arlette Séré (eds), 63–75. Madrid: Biblioteca Nueva.

Goutsos, Dionysis. 2005. "The interaction of generic structure and interpersonal relations in two-party e-chat discourse." *Language@Internet* 3.

Govella, Austin. 2008. "Some thoughts about Twitter." *Thinking & Making* (Blog), April 16, 2008.

Graham, Sage L. 2007. "Disagreeing to agree: Conflict, (im)politeness and identity in a computer-mediated community." *Journal of Pragmatics* 39: 742–759.

Grau, Abel. 2008. "Internet cambia la forma de leer… ¿y de pensar?," *El País*, October 10, 2008, pp. 30–31.

Grau, Abel. 2010. "La Red desafía el reinado de la televisión," *El País*, January 2, 2010, pp. 30–31.

Greenhill, Anita and Fletcher, Gordon. 1996. "The social construction of electronic space." *Social Semiotics* 6 (2): 179–198.

Greisdorf, Howard. 2000. "Relevance: An interdisciplinary and information science perspective." *Informing Science* 3 (2): 67–71

Greisdorf, Howard. 2003. "Relevance thresholds: A multi-stage predictive model of how users evaluate information." *Information Processing and Management* 39: 403–423.

Greisdorf, Howard and Spink, Amanda. 2000. "Recent relevance research: Implications for information professionals." *Online Information Review* 24 (5): 389–395.

Greller, Leonore M. and Barnes, Sue. 1993. "Groupware and interpersonal text: The computer as a medium of communication." *Interpersonal Computing and Technology* 1 (2).

Grice, H. Paul. 1975. "Logic and conversation." In *Speech Acts (Syntax and Semantics 3)*, Peter Cole and Jerry Morgan (eds), 41–58. New York: Academic Press.

Grinter, Rebecca E. and Palen, Leysia. 2002. "Instant messaging in teen life." In *Proceedings of the 2002 ACM Conference on Computer Supported Cooperative Work* (CSCW '02), New Orleans (USA).

Gross, Ralph and Acquisti, Alessandro. 2005. "Information revelation and privacy in online social networks." In *Proceedings of the 2005 ACM Workshop on Privacy in the Electronic Society*, 71–80. New York: ACM.

Grossman, Lev. 2009. "Quitting Twitter," *Time*, March 16, 2009, p. 38.

Gruber, Helmut. 1998. "Computer-mediated communication and scholarly discourse: Forms of topic initiation and thematic development." *Pragmatics* 8 (1): 21–45.

Gruber, Helmut. 2000a. "Scholarly email discussion list postings: A single new genre of academic communication?" In *Words on the Web*, Lyn Pemberton and Simon Shurville (eds), 36–43. Exeter, UK: Intellect Books.

Gruber, Helmut. 2000b. "Theme and intertextuality in scholarly e-mail messages." *Functions of Language* 7 (1): 79–116.

Guadagno, Rosanna E., Muscanell, Nicole L., Okdie, Bradley M., Burk, Nanci M. and Ward, Thomas B. 2011. "Even in virtual environments women shop and men build: A social role perspective on Second Life." *Computers in Human Behavior* 27: 304–308.

Gubern, Román. 1992. *La Mirada Opulenta. Exploración de la Iconosfera Contemporánea.* Barcelona: GG.

Gelléri, Péter. 1998. *The IRC Vernacular. A Linguistic Analysis of Internet Relay Chat.* Unpublished MA thesis. Témavezetõ.

Gumperz, John J. 1971. *Language in Social Groups.* Stanford: Stanford University Press.

Gumperz, John J. 1977. "Sociocultural knowledge in conversational inference." In *Georgetown University Round Table of Languages and Linguistics*, Muriel Saville-Troike (ed.), 191–211. Washington: Georgetown University Press.

Gumperz, John J. 1989. "Contextualization cues and metapragmatics: The retrieval of cultural knowledge." In *CLS 25, Part II*, Caroline Wiltshire, Randolph Graczyk and Bradley Music (eds), 77–88. Chicago: Chicago Linguistic Society.

Gwynne, S. C. and Dickerson, John F. 1997. "Lost in the e-mail." *Time* 149 (16), April 21, 1997.

Gygax, Pascal, Oakhill, Jane and Garnham, Alan. 2003. "The representation of characters' emotional responses: Do readers infer specific emotions?" *Cognition and Emotion* 17: 413–428.

Halliday, Josh. 2011. "Don't think of Foursquare as a game," *The Guardian*, April 4, 2011, Media, p. 5.

Halliday, M. A. K. 1995. *An Introduction to Functional Grammar.* London: Arnold.

Hamman, Robin B. 1999. "Computer networks linking network communities: A study of the effects of computer network use upon pre-existing communities." In *Virtuelle Gruppen. Charakteristika und Problemdimensionen*, Udo Thiedeke (ed.). Westdeutscher Verlag.

Hampton, Keith N., Livio, Oren and Goulet, Lauren S. 2010. "The social life of wireless urban spaces: Internet use, social networks, and the public realm." *Journal of Communication* 60: 701–722.

Hanley, Lynsey. 2005. "Geek translations," *The Observer*, November 13, 2005, Review, p. 8.

Hard af Segerstad, Ylva and Ljungstrand, Peter. 2002. "Instant messaging and awareness of presence in *Webwho*." *International Journal of Human-Computer Studies* 56: 147–171.

Harmon, Amy. 1999. "E-mail spawns debate on changing language," *The New York Times*, February 22, 1999.

Harmon, Amy. 2001. "'You've got mail', more and more, and mostly, it is junk," *The New York Times*, December 24, 2001, Technology.

Harris, Helen, Bailenson, Jeremy N., Nielsen, Alexia and Yee, Nick. 2009. "The evolution of social behavior over time in Second Life." *Presence* 18 (6): 434–448.

Harrison, Sandra. 1998a. "E-mail discussions as conversation: Moves and acts in a sample from a listserv discussion." *Linguistik Online* 1.

Harrison, Sandra. 1998b. "E-mail at work." Institute for Professional Communication, Archive of Papers/Articles.

Harrison, Sandra. 2000. "Maintaining the virtual community: Use of politeness strategies in an email discussion forum." In *Words on the Web*, Lyn Pemberton and Simon Shurville (eds), 69–78. Exeter (UK): Intellect Books.

Harter, Stephen P. 1992. "Psychological relevance and information science." *Journal of the American Society for Information Science* 43 (9): 602–615.

Hasan, Bassam and Ahmed, Mesbah U. 2007. "Effects of interface style on user perceptions and behavioral intention to use computer systems." *Computers in Human Behavior* 23: 3025–3037.

Hassell, Lewis. 1998. *Media, Speech Act Theory and Computer Supported Cooperative Work.* Unpublished PhD thesis. Philadelphia: Drexel University.

Hastrdlová, Šárka. 2009. *Language of Internet Relay Chat (Reconsidering (Im)Politeness in the Context of IRC).* Unpublished PhD thesis. Brno: Masarykova Univerzita.

Hauben, Michael and Hauben, Ronda. 1998. "Netizens: On the history and impact of Usenet and the Internet." *First Monday* 3 (7).

Haugh, Michael. 2003. "Anticipated versus inferred politeness." *Multilingua* 22: 397–413.

Haugh, Michael. 2010. "When is an email really offensive?: Argumentativity and variability in evaluations of impoliteness." *Journal of Politeness Research* 6: 7–31.

Haverkate, Henk. 1988. "Politeness strategies in verbal interaction: An analysis of directness and indirectness in speech acts." *Semiotica* 71: 59–71.

Haverkate, Henk. 1994. *La Cortesía Verbal. Estudio Pragmalingüístico.* Madrid: Gredos.

Haythornthwaite, Caroline. 2009. "Social networks and Internet connectivity effects." *Information, Communication & Society* 8 (2): 125–147.

Headlam, Bruce. 2001. "How to e-mail like a C.E.O.," *The New York Times*, April 8, 2001, Magazine.

Heike, Masayuki, Fujita, Kinya and Tanaka, Takahiro. 2009. "Personalized and deformed avatars for realizing avatarmediated real person-to-person communication." In *Online Communities*, A. Ant Ozok and Panayotis, Zaphiris (eds), 207–215. Berlin: Springer-Verlag.

Hemp, Paul. 2009. "Is that message really necessary?," *The Guardian*, September 24, 2009, G2, pp. 10–13.

Hermida, Alfred. 2010. "From TV to Twitter: How ambient news became ambient journalism." *M/C Journal* 13 (2).

Herring, Susan C. 1994. "Politeness in computer culture: Why women thank and men flame." In *Cultural Performances: Proceedings of the Third Berkeley Women and Language Conference*, Mary Bucholtz, A. C. Liang, Laurel Sutton and Caitlin Hines (eds), 278–294. Berkeley: Berkeley Women and Language Group.

Herring, Susan C. 1996a. "Two variants of an electronic message schema." In *Computer-Mediated Communication. Linguistic, Social and Cross-Cultural Perspectives*, Susan Herring (ed.), 81–106. Amsterdam/Philadelphia: John Benjamins.

Herring, Susan C. 1996b. "Posting in a different voice: Gender and ethics in CMC." In *Philosophical Perspectives on Computer-Mediated Communication*, Charles Ess (ed.), 115–146. New York: Suny.

Herring, Susan C. 1999. "Interactional coherence in CMC." *Journal of Computer-Mediated Communication* 4 (4).

Herring, Susan C. 2001. "Computer-mediated discourse." In *The Handbook of Discourse Analysis*, Deborah Tannen, Deborah Schiffrin and Heidi Hamilton (eds), 612–634. Oxford: Blackwell.

Herring, Susan C. 2003. "The past, present and future of weblogs." Paper presented at *School of Library and Information Science Colloquium*, Indiana University, USA.

Herring, Susan C. forthcoming. Grammar and electronic communication." In *The Encyclopedia of Applied Linguistics*, Carol A. Chapelle (ed.). Oxford: Wiley-Blackwell.

Herring, Susan C., Scheidt, Lois A., Bonus, Sabrina and Wright, Elijah. 2004. "Bridging the gap: A genre analysis of weblogs." In *Proceedings of the 37th Hawaii International Conference on System Sciences (HICSS-37)*. Los Alamitos, CA: IEEE.

Herring, Susan C., Scheidt, Lois A., Wright, Elijah and Bonus, Sabrina. 2005. "Weblogs as a bridging genre." *Information Technology & People* 18 (2): 142–171.

Higashimori, Isao and Wilson, Deirdre. 1996. "Questions on relevance." *UCL Working Papers in Linguistics* 8: 1–14.

Himelboim, Itai. 2008. "Reply distribution in online discussions: A comparative network analysis of political and health newsgroups." *Journal of Computer-Mediated Communication* 14 (1): 156–177.

Hinduja, Sameer and Patchin, Justion W. 2008. "Personal information of adolescents on the Internet: A quantitative content analysis of *MySpace*." *Journal of Adolescence* 31: 125–146.

Holbrook, Daniel L. 2006. *Theorizing the Diary Weblog*. Unpublished MA thesis. University of Chicago (USA).

Holland, Samantha and Harpin, Julie. 2008. "'It's only MySpace': Teenagers and social networking online." In *Remote Relationships in a Small World*, Samantha Holland (eds), 117–134. Berlin: Peter Lang.

Hollenbaugh, Erin E. 2010. "Personal journal bloggers: Profiles of disclosiveness." *Computers in Human Behavior* 26: 1657–1666.

Holmes, Janet. 1984. "Women's language: A functional approach." *General Linguistics* 24: 149–178.

Honan, Matthew. 2001. "Will microads save online content?" *Online Journalism Review*, December 11, 2001.

Honeycutt, Courtenay and Herring, Susan C. 2009. "Beyond microblogging: Conversation and collaboration via Twitter." In *Proceedings of the Forty-Second Hawai'i International Conference on System Sciences* (HICSS-42). Los Alamitos, USA: IEEE Press.

Horrigan, John. 2009. *The Mobile Difference*. Pew Internet & American Life Project.

Horrigan, John B. and Rainie, Lee. 2002. "Emails that matter: Changing patterns of Internet use over a year's time." *IT&Society* 1 (1): 135–150.

Hourihan, Meg. 2002. "What we're doing when we blog." *O'Reilly Network*, June 13, 2002.

Hsieh, Yu-Chen and Chen, Kuo-Hsiang. 2011. "How different information types affect viewer's attention on internet advertising." *Computers in Human Behavior* 27: 935–945.

Hua, Guangying and Haughton, Dominique. 2009. "Virtual worlds adoption: A research framework and empirical study." *Online Information Review* 33 (5): 889–1000.

Huang, Albert H. 2002. "E-mail communication and supervisor-subordinate exchange quality: An empirical study." *Human Systems Management* 21: 193–204.

Hughes, Sarah. 2009. "Twitter proves to be a page-turner," *The Guardian*, June 1, 2009, Media, p. 7.

Humphrey, Caroline. 2009. "The mask and the face: Imagination and social life in Russian chat rooms and beyond." *Ethnos* 74 (1): 31–50.

Hussain, Malik, Nakamura, Brandon and Marino, John. 2011. "Avatar appearance & information credibility in Second Life." Paper presented at *iConference'11*, Seattle, USA, February.

Ihlström, Carina and Henfridsson, Ola. 2005. "Online newspapers in Scandinavia. A longitudinal study of genre change and interdependency." *Information Technology & People* 18 (2): 172–192.

Im, Hyun-Gyung, Yates, JoAnne and Orlikowski, Wanda. 2009. "Temporal coordination through communication: Using genres in a virtual start-up organization." *Information Technology & People* 18 (2): 89–119.

Islas, Octavio. 2008. "La sociedad de la ubicuidad, los prosumidores y un modelo de comunicación para comprender la complejidad de las comunicaciones digitales." *Razón y Palabra* 65.

Jackson, Michelle H. 1997. "Assessing the structure of communication on the World Wide Web." *Journal of Computer-Mediated Communication* 3 (1).

Jaffe, J. Michael, Lee, Young-Eum, Huang, Li-Ning and Oshagan, Hayg. 1995. "Gender, pseudonyms and CMC: Masking identities and baring souls." Paper presented at the *45th Annual Conference of the International Communication Association*, Alburquerque (USA).

Jamali, Hamid R. and Asadi, Jaeid. 2010. "Google and the scholar: The role of Google in scientists' information-seeking behaviour." *Online Information Review* 34 (2): 282–294.

Janoschka, Anja. 2004. *Web Advertising*. Amsterdam/Philadelphia: John Benjamins.

Jarrett, Kylie. 2008. "Interactivity is evil! A critical investigation of Web 2.0." *First Monday* 13 (3).

Jary, Mark. 1998a. "Is relevance theory asocial?" *Revista Alicantina de Estudios Ingleses* 11: 157–169.

Jary, Mark. 1998b. "Relevance theory and the communication of politeness." *Journal of Pragmatics* 30 (1): 1–19.

Java, Akahay, Song, Xiaodan, Finin, Tim and Tseng, Belle. 2007. "Why we Twitter: Understanding microblogging and community." Paper presented at *9th WEBKDD and 1st SNA-KDD Workshop '07*, San Jose, California (USA).

Jenkins, Henry, Clinton, Katie, Purushotma, Ravie, Robinson, Alice J. and Weigel, Margaret. 2006. "Confronting the challenges of participatory culture: Media education for the 21st century." *MacArthur Foundation.*

Jiménez Gómez, Juan J. and Vela Delfa, Cristina. 2004. "Los malentendidos en la comunicación electrónica." *Interlingüística* 15: 753–769.

Johnova, Markéta. 2004. "The language of chat." *Philologica.net*, January 13, 2004.

Johnson, Steven. 2009. "How Twitter will change the way we live," *Time*, June 5, 2009.

Joinson, Adam N. 2008. "'Looking at', 'looking up' or 'keeping up with' people? Motives and Uses of Facebook." In *Proceedings of the 26th Annual SIGCHI Conference on Human Factors in Computing Systems*, 1027–1036. Florence (Italy), April.

Joly, Adrien, Maret, Pierre and Daigremont, Johann. 2009. "Context-awareness, the missing block of social networking." *International Journal of Computer Science and Applications* 6 (2): 50–65.

Jones, Donald E. 2005. "I, avatar: Constructions of self and place in Second Life and the technological imagination." *Gnovis* IV.

Jones, Linda. 2007. "Member of the Twittering classes," *The Guardian*, May 17, 2007, Technology, p. 1.

Jones, Ninrod. 1997. *nEt.SPeAk. Tomorrow's World?* Unpublished MA thesis. Woodlane: University College Falmouth, Darlington College of Arts.

Jones, Quentin. 1997. "Virtual-communities, virtual settlements & cyber-archaeology: A theoretical outline." *Journal of Computer-Mediated Communication* 3 (3).

Jones, Steven G (ed.). 1995a. *Cybersociety*. London: Sage.
Jones, Steven G. 1995b. "Understanding community in the information age." In *Cybersociety*, Steven G. Jones (ed.), 10–35. London: Sage.
Jones, Steven G. 1997a. "The Internet and its social landscape." In *Virtual Culture. Identity & Communication in Cybersociety*, Steven G. Jones (ed.), 7–35. London: Sage.
Jones, Steven G. 1997b. "Using the news: An examination of the value and use of news sources in CMC." *Journal of Computer-Mediated Communication* 2 (4).
Jones, Steven G. (ed.). 1998a. *Cybersociety 2.0*. London: Sage.
Jones, Steven G. 1998b. "Information, Internet and community: Notes toward an understanding of community in the information age." In *Cybersociety 2.0*, Steven G. Jones (ed.), 1–34. London: Sage.
Jones, Steve, Millermaier, Sarah, Goya-Martínez, Mariana and Schuler, Jessica. 2008. "Whose space is MySpace? A content analysis of MySpace profiles." *First Monday* 13 (9).
Jonsson, Ewa. 1998. *Electronic Discourse. On Speech and Writing on the Internet*. Lulea University of Technology, Department of Communicación and Languages, *D-Course on English*.
Jucker, Andreas. H. 1988. "The relevance of politeness." *Multilingua* 7 (4): 375–384.
Kalbach, James. 2006. "'I'm feeling lucky': The role of emotions in seeking information on the web." *Journal of the American Society for Information Science and Technology* 57 (6): 813–818.
Kalman, Yoram M. and Rafaeli, Sheizaf. 2005. "Email chronemics: Unobtrusive profiling of response times." In *Proceedings of the 38th Hawaii International Conference on System Sciences*, Hawaii, January.
Kalyanaraman, Sriram and Sundar, S. Shyam. 2006. "The psychological appeal of personalized content in web portals: Does customization affect attitudes and behavior?" *Journal of Communication* 56: 110–132.
Kalyanaraman, Sriram and Sundar, S. Shyam. 2008. "Impression formation effects in online mediated communication." In *Mediated Interpersonal Communication*, Elly A. Konijn, Sonja Utz, Martin Tanis and Susan B. Barnes (eds), 217–233. London: Routledge.
Kang, Hye-Seung and Yang, Hee-Dong. 2006. "The visual characteristics of avatars in computer-mediated communication: Comparison of Internet Relay Chat and Instant Messenger as of 2003." *International Journal of Human-Computer Studies* 64: 1173–1183.
Karlgren, Jussi. 2010. "Conventions and mutual expectations. Understanding sources for web genres." In *Genres on the Web: Computational Models and Empirical Studies*, Alexander Mehler, Serge Sharoff and Marina Santini (eds), 33–46. Berlin: Springer.
Karlsson, Lena. 2006. "Acts of reading diary weblogs." *Human IT* 8 (2): 1–59.
Kasper, Gabrielle. 1990. "Linguistic politeness. Current research issues." *Journal of Pragmatics* 14: 193–218.
Kataoka, Kuniyoshi. 2003. "Emotion and youth identities in personal letter writing. An analysis of pictorial signs and unconventional punctuation." In *Discourse Constructions of Youth Identities*, Jannis K. Androutsopoulos and Alexandra Georgakopoulou (eds), 121–149. Amsterdam/Philadelphia: John Benjamins.
Kato, Yuuki, Kato, Shogo and Akahori, Kanji. 2007. "Effects of emotional cues transmitted in e-mail communication on the emotions experienced by senders and receivers." *Computers in Human Behavior* 23: 1894–1905.
Kavada, Anastasia. 2010. "Email lists and participatory democracy in the European Social Forum." *Media, Culture & Society* 32 (3): 355–372.

Kavanaugh, Andrea, Carroll, John M., Rosson, Mary B., Zin, Than T. and Reese, Debbie D. 2005. "Community networks: Where offline communities meet online," *Journal of Computer-Mediated Communication* 10 (4).

Keegan, Victor. 2007. "Screen grabbers – crime hits the digital frontier," *The Guardian*, November 17, 2007, pp. 16–17.

Keegan, Victor. 2008. "Art, music, gossip – it's (virtually) all there in my parallel universe," *The Observer*, November 17, 2008, p. 28.

Keenan, Andrew and Shiri, Ali. 2009. "Sociability and social interaction on social networking websites." *Library Review* 58 (6): 438–450.

Kendall, Lori. 2007. "'Shout into the wind and it shouts back'. Identity and interactional tensions on *LiveJournal*." *First Monday* 12 (9).

Kenney, Keith, Gorelik, Alexander and Mwangi, Sam. 2000. "Interactive features of online papers." *First Monday* 5 (1).

Kervin, Lisa, Mantei, Jessica and Herrington, Anthony. 2010. "Blogs as a social networking tool to build community. " In *Handbook of Research on Social Interaction Technologies and Collaboration Software: Concepts and Trends*, Tatyana Dumova and Richard Fiordo (eds), 685–700. Hershey: IGI Global.

Kibby, Marjorie D. 2005. "Email forwardables: Folklore in the age of the Internet." *New Media & Society* 7 (6): 770–790.

Kidron, Yael and Kuzar, Ron. 2002. "My face is paling against my will. Emotion and control in English and Hebrew." *Pragmatics & Cognition* 10: 129–157.

Kiesler, Sara, Siegel, Jane and McGuire, Timothy W. 1984. "Social psychological aspects of computer-mediated communication." *American Psychologist* 39: 1123–1134.

Kim, Jinsuk and Dindia, Kathryn. 2011. "Online self-disclosure: A review of research." In *Computer-Mediated Communication in Personal Relationships*, Kevin B. Wright and Lynne M. Webb (eds), 156–180. Berlin: Peter Lang.

Kim, Yoojung, Sohn, Dongyoung and Choi, Sejung M. 2011. "Cultural difference in motivations for using social network sites: A comparative study of American and Korean college students." *Computers in Human Behavior* 27: 365–372.

Kiousis, Spiro. 2002. "Interactivity: A concept explication." *New Media & Society* 4: 355–384.

Kirkgöz, Yasemin. 2010. "Analyzing the discourse of e-mail communication." In *Handbook of Research on Discourse Behavior and Digital Communication: Language Structures and Social Interaction*, Rotimi Taiwo (ed.), 335–348. Hershey, USA: IGI Global.

Klamma, Ralf, Cao, Yiwei and Spaniol, Marc. 2007. "Watching the blogosphere: Knowledge sharing in the Web 2.0." Paper presented at *International Conference on Weblogs and Social Media*, Boulder, USA.

Kling, Rob. 1996a. "Social relationships in electronic forums: Hangouts, salons, workplaces, and communities." *Computer-Mediated Communication Magazine*, Spring 1996.

Kling, Rob. 1996b. "Synergies and competition between life in cyberspace and face-to-face communitities." *Social Science Computer Review* 14 (1): 50–54.

Kneepkens, E. W. E. M. and Zwaan, Rolf A. 1994. "Emotions in literary text comprehension." *Poetics* 23: 125–138.

Knupsky, Aimee C. and Nagy-Bell, Natalie M. 2011. "Dear professor: The influence of recipient sex and status on personalization and politeness in e-mail." *Journal of Language and Social Psychology* 30 (1): 103–113.

Ko, Hsiu-Chia, Yin, Chun-Po and Kuo, Feng-Yang. 2008. "Exploring individual communication power in the blogosphere." *Internet Research* 18 (5): 541–561.

Kochman, Thomas. 1984. "The politics of politeness: Social warrants in mainstream American public etiquette." In *Meaning, Form and Use in Context: Linguistic Applications (GURT'84)*. Deborah Schiffrin (ed.), 200–209. Washington: Georgetown University Press.

Koda, Tomoko, Ishida, Toru, Rehm, Matthias adn André, Elisabeth. 2009. "Avatar culture: Cross-cultural evaluations of avatar facial expressions." *AI & Society* 24: 237–250.

Kolko, Beth and Reid, Elizabeth. 1998. "Dissolution and fragmentation: Problems in on-line communities." In *Cybersociety 2.0*, Steven G. Jones (ed.), 212–229. London: Sage.

Kollock, Peter and Smith, Mark. 1996. "Managing the virtual commons: Cooperation and conflict in computer communities." *Computer-Mediated Communication. Linguistic, Social and Cross-Cultural Perspectives*, Susan Herring (ed.), 109–128. Amsterdam/Philadelphia: John Benjamins.

Kollock, Peter and Smith, Mark. 1999. "Communities in cyberspace." In *Communities in Cyberspace*, Mark Smith and Peter Kollock (eds.), 3–25. London: Routledge.

Kuhn, Thomas. 1975. *La Estructura De Las Revoluciones Científicas*. Mexico: Fondo de Cultura Económica.

Kuiper, Koenraad. 1997. "The relevance of face calibration." In *Proceedings of the University of Hertfordshire Relevance Theory Workshop*, Marjolein Groefsema (ed.), 105–111. Chelmsford (UK): Peter Thomas.

Kumar, Nanda and Benbasat, Izak. 2001. "Shopping as experience and web site as a social actor: Web interface design and para-social presence." Paper presented at *Twenty-Second International Conference on Information Systems*, New Orleans (USA).

Kurlander, David, Skelly, Tim and Salesin, David. 1996. "Comic Chat." In *Proceedings of the 6th Conference SIGGRAPH*, 225–236. New Orleans (USA).

Kwasnik, Barbara H. and Crowston, Kevin. 2005. "Introduction to the special issue. Genres of digital documents." *Information Technology & People* 18 (2): 76–88.

Laborda Gil, Xabier. 2005. "Tecnologías, redes y comunicación interpersonal. Efectos en las formas de la comunicación digital." *Anales de Documentación* 8: 101–116.

Lakoff, Robin. 1973. "The logic of politeness; or minding your p's and q's." In *Papers from the Ninth Regional Meeting*, 292–305. Chicago (USA): Chicago Linguistic Society.

Lakoff, Robin. 1977. "What you can do with words: Politeness, pragmatics and performatives." In *Proceedings of the Texas Conference on Performatives, Presuppositions and Implicatures*, Andy Rogers, Bob Wall and John P. Murphy (eds), 79–106. Arlington (USA): Center for Applied Linguistics.

Lakoff, Robin. 1989. "The limits of politeness: Therapeutic and courtroom discourse." *Multilingua* 8: 101–129.

Lamb, Roberta and Davidson, Elizabeth. 2002. "Social scientists: Managing identity in socio-technical networks." Hawaii: *Hawaii International Conference on Systems Sciences* 35.

Lameiro, Máximo and Sánchez, Roberto. 1998. "Vínculos e Internet. Investigación cualitativa acerca de nuevas formas de vincularse." *Boletín de la Sociedad Española de Psicoterapia y Técnicas de Grupo*, Época IV, 14: 45–66.

Lampe, Cliff, Ellison, Nicolle and Steinfield, Charles. 2006. "A Face(book) in the crowd: Social searching vs. social browsing." In *Proceedings of ACM Special Interest Group on Computer-Supported Cooperative Work*, ACM Press, 167–170.

Lampe, Cliff, Ellison, Nicole and Steinfeld, Charles. 2007. "A familiar Face(book): Profile elements as signals in an online social network." In *Proceedings of Conference on Human Factors in Computing Systems*, 435–444. New York: ACM Press.

Lan, Li. 2000. "Email: A challenge to standard English?" *English Today* 64 (16, 4): 23–29/55.

Lang, Candace. 1995. "Body language: The resurrection of the corpus in text-based VR." *Works and Days* 26.

Lankshear, Colin and Knobel, Michele. 2003. "Do-it-yourself broadcasting: Writing weblogs in a knowledge society." Paper presented at *American Education Research Association Annual Meeting Symposia*, Chicago (USA).

Launspach, Sonja. 2000. "Literal or loose talk: The negotiation of meaning on an Internet discussion list." In *Words on the Web*, Lyn Pemberton and Simon Shurville (eds), 87–95. Exeter, UK: Intellect Books.

Lawley, Liz. 2004. "Blog research issues." *Many-to-Many*, June 24, 2004.

Le Page, Robert B. 1986. "Acts of identity." *English Today* 8: 21–24.

Lee, Gia B. 1996. "Addressing anonymous messages in cyberspace." *Journal of Computer-Mediated Communication* 2 (1).

Lee, Matthew W. 2001. *Chatscape: A Behavior-Enhanced Graphical Chat Built on A Versatile Client-Server Architecture*." Unpublished MA thesis. Massachusetts: M.I.T.

Leech, Geoffrey. 1983. *Principles of Pragmatics*. London: Longman.

Lenarcic, John and Sarkar, Pradip. 2008. "The long tail of blogging: A nurturing mechanism for sustainable online communities with niche interests." Paper presented at *21st Bled eConference eCollaboration: Overcoming Boundaries through Multi-Channel Interaction*, Bled, Slovenia.

Lenhart, Amanda B. 2005. *Unstable Texts: An Ethnographic Look at How Bloggers and their Audience Negotiate Self-presentation, Authenticity and Norm Formation*. Unpublished MA thesis. Washington: Georgetown University.

Lenhart, Amanda B. 2009. "Twitter and status updating." *Pew Internet Project Data Memo*.

Lenhart, Amanda and Madden, Mary. 2007. "Social networking sites and teens: An overview." *Pew Internet Project*.

Leung, Louis. 2003. "Impacts of net-generation attributes, seductive properties of the Internet, and gratifications obtained on Internet use." *Telematics and Informatics* 20: 107–129.

Levinson, Stephen C. 1983. *Pragmatics*. Cambridge: Cambridge University Press.

Lewis, Cynthia and Fabos, Bettina. 2005. "Instant messaging, literacies, and social identities." *Reading Research Quarterly* 40 (4): 470–501.

Li, Dan. 2005. *Why Do You Blog: A Uses-and-Gratifications Inquiry Into Bloggers' Motivations*. Unpublished MA thesis. Milwaukee: Marquette University.

Li, Xigen. 2011. "Factors influencing the willingness to contribute information to online communities." *New Media & Society* 13 (2): 279–296.

Lim, Sun. 2002. "The self-confrontation interview: Towards an enhanced understanding of human factors in web-based interaction for improved website usability." *Journal of Electronic Commerce Research* 3 (3): 162–173.

Lin, Kuan-Yu and Lu, Hsi-Peng. 2011. "Why people use social networking sites: An empirical study integrating network externalities and motivation theory." *Computers in Human Behavior* 27: 1152–1161.

Lin, Yu-Ru, Sundaram, Hari, Chi, Yun, Tatemura, Jun and Tseng, Belle. 2006. "Discovery of blog communities based on mutual awareness." Paper presented at *WWW2006 Workshop on the Weblogging Ecosystem*, Edinburgh (England).

Lindsay, Peter H. and Norman, Donald A. 1983. *Introducción a la Psicología Cognitiva*. Madrid: Tecnos.

Ling, Rich and Baron, Naomi S. 2007. "Text messaging and IM: A linguistic comparison of American college data." *Journal of Language and Social Psychology* 26 (3): 291–298.

Liu, Geoffrey Z. 1999. "Virtual community presence in Internet Relay chatting." *Journal of Computer-Mediated Communication* 5 (1).
Livingood, Jeb. 1995. "Revenge of the introverts." *Computer-Mediated Communication* 2 (4).
Livingstone, Sonia. 2007. "The challenge of engaging youth online: Contrasting producers' and teenagers' interpretations of websites." *European Journal of Communication* 22 (2): 165–184.
Livingstone, Sonia. 2010. "Taking risky opportunities in youthful content creation. Teenagers' use of social networking sites for intimacy, privacy and self-expression." In *The New Media and Cybercultures Anthology*, Pramod K. Nayar (ed.), 468–482. Oxford: Wiley-Blackwell.
Lombard, Matthew and Ditton, Teresa. 1997. "At the heart of it all: The concept of telepresence." *Journal of Computer-Mediated Communication* 3 (2).
Lomborg, Stine. 2009. "Navigating the blogosphere: Towards a genre-based typology of weblogs." *First Monday* 14 (5).
London, Scott. 1997. "Civic networks: Building community on the net." Kettering Foundation.
López Alonso, C. 2003. "El correo electrónico." In *Nuevos Géneros Discursivos: Los Textos Electrónicos*, Covadonga López Alonso and Arlette Séré (eds), 21–44. Madrid: Biblioteca Nueva.
López Quero, Salvador. 2010. "Marcas gramaticales de oralidad en los chats y foros de debate: Incorporación de marcadores discursivos del español hablado." *Oralia* 13: 173–195.
López Rúa, Paula. 2005. "Written conversations or spoken genres? A multilingual approach to the linguistic and pragmatic features of e-mails." In *Actas del V Congreso Internacional de AELFE*. Zaragoza (Spain).
Lorenzo-Dus, Nuria, Garcés-Conejos Blitvich, Pilar and Bou-Franch, Patricia. 2011. "On-line polylogues and impoliteness: The case of postings sent in response to the Obama Reggaeton YouTube video." *Journal of Pragmatics* 43 (10): 2578–2593.
Lu, Hsi-Peng and Hsiao, Kuo-Lun. 2007. "Understanding intention to continuously share information on weblogs." *Internet Research* 17: 345–361.
Lu, Hsi-Peng and Lee, Ming-Ren. 2010. "Demographic differences and the antecedents of blog stickiness." *Online Information Review* 34 (1): 21–38.
Lüders, Marika, Proitz, Lin and Rasmussen, Terje. 2010. "Emerging personal media genres." *New Media & Society* 12 (6): 947–963.
Luzón, María J. 2009. "Scholarly hyperwriting: The function of links in academic weblogs." *Journal of the American Society for Information Science and Technology* 60 (1): 75–89.
Mabrito, Mark. 1991. "Electronic mail as a vehicle for peer response." *Written Communication* 8 (4): 509–532.
Mabry, Edward A. 1997. "Framing flames: The structure of argumentative messages on the net." *Journal of Computer-Mediated Communication* 2 (4).
Macdougall, Alan. 1999. "'And the word was made flesh and dwelt among us...': Towards pseudonymous life on the Internet." *M/C* 2 (3).
Maglaughlin, Kelly L. and Sonnenwald, Diane H. 2002. "User perspectives on relevance criteria: A comparison among relevant, partially relevant, and not-relevant judgments." *Journal of the American Society for Information Science and Technology* 53 (5): 327–342.
Maldonado, Tomás. 1998. *Telemática y Nuevos Escenarios Urbanos*. Barcelona: Paidós.
Marcoccia, Michel. 2004. "On-line polylogues: Conversation structure and participation framework in Internet newsgroups." *Journal of Pragmatics* 36: 115–145.
Marcoccia, Michel, Atifi, Hassan and Gauducheau, Nadia. 2008. "Text-centered versus multimodal analysis of instant messaging conversation." *Language@Internet* 5.

Margolis, Michael and Resnick, David. 1999. "Third Voice: Vox populi or vox dei?" *First Monday* 4 (10).

Maricic, Ibolya. 2000. "Cyberpoliteness: Requesting strategies on *The Linguist List*." Paper presented at the *Seventh Conference of the International Pragmatics Association*, Budapest (Hungary), July.

Mariottini, Laura. 2004. *Linguistica e Nuovi Media. Fenomeni di Variazione della Lingua del le Chat in Italiano e in Spagnolo*. Unpublished PhD thesis. Rome: Universitá Degli Studi Roma Tre.

Mariottini, Laura. 2006. "El uso de los diminutivos y su relación con la cortesía lingüística en los chats. Análisis contrastivo de comunidades virtuales españolas e italianas." *Cultura, Lenguaje y Representación* 3: 103–131.

Mariottini, Laura. 2008. "Cibercortesía: Estrategias de toma del turno en los chats." In *Cortesía y Conversación. De lo Escrito a lo Oral. III Coloquio Internacional del Programa Edice*, Antonio Briz, Antonio Hidalgo, Marta Albelda, Josefa Contreras and Nieves Hernández (eds), 489–502. Valencia (Spain): Universitat de Valencia, Departamento de Filología Española.

Mariottini, Laura. 2011a. "Ciberpragmática: Hacia la fijación de nuevos géneros telemáticos." Paper presented at the Seminar *Meaning in Interaction*, Seville (Spain), University of Seville, May.

Mariottini, Laura. 2011b. "Lenguaje y nuevas tecnologías: Análisis de los blogs de abogados." In *Multiple Voices in Academic and Professional Discourse: Current Issues in Specialized Language Research, Teaching and New Technologies*, Sergio Maruenda-Bataller and Begoña Clavel-Arroitia (eds), 187–198. Newcastle: Cambridge Scholars Publishing.

Mariottini, Laura. forthcoming. "Las páginas web de turismo. ¿Género móvil o modelo mental?" In *Il Discorso del Turismo. Aspetti Linguistici e Varietà Textuali*, Elena Liberani and Jordi Canals (eds). Trieste (Italy): Tangram.

Marmaridou, A. Sophia. 1989. "Proper names in communication." *Journal of Linguistics* 25: 355–372.

Martín, Javier. 2008. "Angustiados por el 'e-mail'," *El País*, September 16, 2008, pp. 34–35.

Martín, Javier. 2009. "Internet no nos cambia tanto," *El País*, March 26, pp. 28–29.

Marwick, Alice. 2005. "'I'm a lot more interesting than a *Friendster* profile': Identity presentation, authenticity and power in social networking services." Paper presented at *The Conference of the Association of Internet Researchers* 6, Chicago (USA), October.

Marwick, Alice and boyd, danah. 2010. "I tweet honestly, I tweet passionately: Twitter users, context collapse, and the imagined audience." *New Media & Society* 13 (1): 114–133.

Masterson, John T. 1996. *Nonverbal Communication in Text Based Virtual Realities*. Unpublished MA thesis. University of Miami (USA).

Mateo, José and Yus, Francisco. 2000. "Insults: A relevance-theoretic taxonomical approach to their translation." *International Journal of Translation* 12 (1): 97–130.

Mateo, José and Yus, Francisco. 2009. "Business language from a cognitive perspective." In *Intercultural Business Communication and Simulation and Gaming Methodology*, Victoria Guillén-Nieto, Carmen Marimón-Llorca and Chelo Vargas-Sierra (eds), 97–115. Berlin: Peter Lang.

Mateo, José and Yus, Francisco. 2010. "Towards an intercultural pragmatic taxonomy of insults." Paper presented at the *4th International Conference on Intercultural Pragmatics*, Madrid, October.

Matthews, Heather. 2000. "Effects of group identity on discussions in public on-line fora." In *Words on the Web*, Lyn Pemberton and Simon Shurville (eds), 79–86. Exeter: Intellect Books.

Matzat, Uwe. 1998. "Informal academic communication and scientific usage of Internet discussion groups." Paper presented at *Iriss'98*, Bristol (UK).
Mayans, Joan. 2000a. "MUDs: Rol en línea," *iWorld* 26, April, 2000.
Mayans, Joan. 2000b. "Chats. Un nuevo lenguaje," *iWorld*, 29, July–August 2000, pp. 42–50.
Mayans, Joan. 2002a. *Género Chat. O Cómo la Etnografía Puso un Pie en el Ciberespacio*. Barcelona: Gedisa.
Mayans, Joan. 2002b. "De la incorrección normativa en los chats." *Revista de Investigación Lingüística* V (2): 101–116.
McElhearn, Kirk. 1996. *Writing Conversation: An Analysis of Speech Events in E-mail Mailing Lists*. Unpublished MA thesis. Aston University, Language Studies Unit.
McIlvenny, Paul. 1999. "Avatars R Us? Discourses of community and embodiment in intercultural cyberspace." *Intercultural Communication* 1.
McInnes, Alice. 1997. "The agency of the infozone: Exploring the effects of a community network." *First Monday* 2 (2).
McKee, Heidi. 2002. "'YOUR VIEWS SHOWED TRUE IGNORANCE!!!': (Mis)communication in an online interracial discussion forum." *Computers and Composition* 19: 411–434.
McKenna, Katelyn Y. A., Green, Amie S. and Glenson, Marci E. J. 2002. "Relationship formation on the Internet: What's the big attraction?" *Journal of Social Issues* 58 (1): 9–31.
McMillan, Sally J. 2002. "Exploring models of interactivity from multiple research traditions: Users, documents, and systems." In *Handbook of New Media*, Leah A. Lievrouw and Sonia Livingston (eds), 162–182. London: Sage.
Meier, Ardith J. 1992. "Brown and Levinson's legacy of politeness." *Views* 1 (1): 15–35.
Meltzer, Tom. 2010. "Failure to connect," *The Guardian*, August 7, 2010, Weekend, pp. 24–29.
Menges, Joyce. 1996. "Feeling between the lines." *Computer-Mediated Communication* 3 (10).
Merchant, Guy. 2001. "Teenagers in cyberspace: An investigation of language use and language change in internet chatrooms." *Journal of Research in Reading* 24 (3): 293–306.
Mesch, Gustavo S. and Beker, Guy. 2010. Are norms of disclosure of online and offline personal information associated with the disclosure of personal information online?" *Human Communication Research* 36: 570–592.
Metz, J. Michel. 1994. "Computer-mediated communication: Literature review of a new context." *Interpersonal Computing and Technology* 2 (2): 31–49.
Meyrowitz, Joshua. 1985. *No Sense of Place: The Impact of Electronic Media on Social Behavior*. New York: Oxford University Press.
Mika, Karin. 1998. "Information vs. commercialization: The Internet and unsolicited electronic mail." *The Richmond Journal of Law and Technology* IV (3).
Millard, William B. 1995. "'A great flame follows a little spark': Metaflaming, conditions of closure, and functions of the 'dis' in the rhetoric of a discussion list." *Works & Days* 25/26, vol. 13 (1–2).
Miller, Hugh. 1995. "The presentation of self in electronic life: Goffman on the Internet." Paper presented at *Embodied Knowledge and Virtual Space*, University of London.
Miller, Vincent. 2008. "New media, networking and phatic culture." *Convergence* 14 (4): 387–400.
Miller, Hugh and Mather, Russell. 1998. "The presentation of self in WWW home pages." Paper presented at *Iriss'98*, Bristol (UK).
Miller, Carolyn R. and Shepherd, Dawn. 2004. "Blogging as social action: A genre analysis of the weblog." In *Into the Blogosphere: Rhetoric, Community, and Culture of Weblogs*, Laura J. Gurak, Smiljana Antonijevic, Laurie Johnson, Clancy Ratliff and Jessica Reyman (eds). Los Alamitos: IEEE Press.
Milroy, James. 1978. *Language and Social Networks*. Oxford: Blackwell.

Milroy, James. 1992. *Linguistic Variation and Change*. Oxford: Blackwell.
Milroy, James and Milroy, Lesley. 1992. "Social networks and social class: Toward an integrated sociolinguistic model." *Language in Society* 21 (1): 1–26.
Minsky, Mervin. 1975. "Frame-system theory." In *Thinking. Readings in Cognitive Science*, Philip N. Johnson-Laird and Peter Wason (eds), 355–376. Cambridge: Cambridge University Press.
Mischaud, Edward. 2007. *Twitter: Expressions of the Whole Self. An Investigation into User Appropriation of a Web-Based Communications Platform*. Unpublished MA thesis. London School of Economics, England.
Mishne, Gilad and Glance, Natalie. 2006. "Leave a reply: An analysis of weblog comments." Paper presented at *Weblogging Ecosystem: Aggregation, Analysis and Dynamics*, Edinburgh.
Misoch, Sabina. 2008. "Avatars as user visualization and their social function in virtual environments." *Studies in Communication Sciences* 8 (1): 51–65.
Mitchell, William J. 1995. *City of Bits*. Cambridge, Mass.: MIT e-book.
Mitra, Ananda. 1997. "Virtual commonality: Looking for India on the Internet." In *Virtual Culture. Identity & Communication in Cybersociety*, Steven G. Jones (ed.), 55–79. London: Sage.
Mitra, Ananda. 2010. *Digital Communications. From E-Mail to the Cyber Community*. New York: Chelsea House.
Mitra, Ananda and Cohen, Elisia. 1999. "Analysing the Web." In *Doing Internet Research. Critical Issues and Methods for Examining the Net*, Steve Jones (ed.), 179–202. London: Sage.
Mitra, Ananda, Willyard, Jennifer, Platt, Carrie A. and Parsons, Michael. 2005. "Exploring web usage and selection criteria among male and female students." *Journal of Computer-Mediated Communication* 10 (3).
Miura, Asako and Shinohara, Kazumitsu. 2005. "Social intelligence design in online chat communication: A psychological study on the effects of 'congestion.'" *AI & Society* 19: 93–109.
Mizzaro, Stefano. 1997. "Relevance: The whole history." *Journal of the American Society for Information Science* 48 (9): 810–832.
Mobrand, Kathryn and Spyridakis, Jan H. 2007. "Explicitness of local navigational links: Comprehension, perceptions of use, and browsing behavior." *Journal of Information Science* 33: 41–61.
Molist, Mercé. 2009. "Los grupos de noticias tienden a su extinción en todo el mundo," *El País*, June 25, 2009, Ciberpaís, p. 6.
Montañés Brunet, Elivra M., Bosch Abarca, Elena and Giménez Moreno, Rosa. 2005. "Protocolos comunicativos en la promoción de servicios turísticos." In *Las Lenguas De Especialidad: Nuevas Perspectivas de Investigación (Quaderns de Filologia. Estudis Lingüístics X)*, María A. Olivares Pardo and Francisca Suau Jiménez (eds), 205–223. Valencia (Spain): University of Valencia.
Montero-Fleta, Begoña, Montesinos-López, Anna, Pérez-Sabater, Carmen and Turney, Ed. 2009. "Computer mediated communication and informalization of discourse: The influence of culture and subject matter." *Journal of Pragmatics* 41: 770–779.
Moor, Peter J., Heuvelman, Ard and Verleur, Ria. 2010. "Flaming on YouTube." *Computers in Human Behavior* 26: 1536–1546.
Moran, Charles and Hawisher, Gail. 1998. "The rhetorics and languages of electronic mail." In *Page to Screen*, Ilana Snyder (ed.), 80–101. London: Routledge.
Morand, Davd A. and Ocker, Rosalie J. 2002. "Politeness theory and computer-mediated communication: A sociolinguistic approach to analyzing relational messages." In *Proceedings of the 36th Hawaii International Conference on System Sciences (HICSS'03)*, Hawaii (USA).

Morie, Jacquelyn F. 2008. "The performance of the self and its effect on presence in virtual worlds." In *Proceedings of the 11th Annual International Workshop on Presence*, Anna Spagnolli and Luciano Gamberini (eds), 265–269. Padova (Italy): CLEUP.

Morrás, A. 2009. "El reconocimiento emocional y gesticular del avatar llega a Second Life" (blog entry), *El Metaverso*, March 29, 2009.

Morrisett, Lloyd N. 1996. "Habits of mind and a new technology of freedom." *First Monday* 1 (3).

Mortensen, Torill and Walker, Jill. 2002. "Blogging thoughts: Personal publication as an online research tool." In *Researching ICTs in Context*, Andrew Morrison (ed.), 249–279. Oslo: Intermedia.

Mouzourakis, Panayotis. 1996. "Videoconferencing: techniques and challenges." *Interpreting* 1 (1): 21–38.

Muñoz, Ramón. 2009. "El móvil omnipotente," *El País*, December 29, 2009, pp. 24–25.

Muñoz, Ramón and Riveiro, Aitor. 2009. "Que Obama use Twitter le humaniza," *El País*, March 26, 2009, p. 53.

Murphy, Jamie, Hofacker, Charles and Mizerski, Richard. 2006. "Primacy and recency effects on clicking behaviour." *Journal of Computer-Mediated Communication* 11: 522–535.

Myers, Greg. 2010. *The Discourse of Blogs and Wikis*. London: Continuum.

Nardi, Bonnie A., Whittaker, Steve and Bradner, Erin. 2000. "Interaction and outeraction: Instant messaging in action." In *Proceedings of Conference on Computer Supported Cooperative Work*, 79–88. New York: ACM Press.

Nardi, Bonnie A., Schiano, Diane J. and Gumbrecht, Michelle. 2004. "Blogging as a social activity, or would you let 900 million people read your diary?" In *Proceedings of the 2004 ACM Conference on Computer Supported Cooperative Work*. Chicago (USA).

Nardi, Bonnie A., Schiano, Diane J., Gumbrecht, Michelle and Swartz, Luke. 2005. "Why we blog." *Communications of the ACM* 47 (12): 41–46.

Nastri, Jacqueline, Peña, Jorge and Hancock, Jeffrey T. 2006. "The construction of away messages: A speech act analysis." *Journal of Computer-Mediated Communication* 11 (4): 1025–1045.

Newitz, Annalee. 1995. "Surplus identity on-line." *Bad Subjects* 18.

Ngwenyama, Ojelanki K. and Lee, Allen S. 1997. "Communication richness in electronic mail: Critical social theory and the contextuality of meaning." *MS Quarterly* 21 (2): 145–167.

Nicolle, Steve. 2000. "Communicated and non-communicated acts in relevance theory." *Pragmatics* 10 (2): 233–245.

Niekamp, Raymond A. 2003. *Audience Activity Among Users of the World Wide Web*. Unpublished PhD thesis, Penn State University (USA).

Nielsen, Anne E. 2002. "The emergence of web-mediated genres." Paper presented at the *International Conference on Advances in Infrastructure for: e-Business, e-Education, e-Science, e-Medicine on the Internet*, L'Aquila, Italy, February.

Nielsen, Jakob. 2008. "Bridging the designer-user gap." *Useit.com*, March 17, 2008.

Nilsen, Ake. 1999. "Simulation and the intersubjective creation of meaning." *Paideusis* 2.

Nilsen, Mona and Mäkitalo, Åsa. 2010. "Towards a conversational culture? How participants establish strategies for co-ordinating chat postings in the context of in-service training." *Discourse Studies* 12 (1): 90–105.

Nowak, Kristine L. 2004. "The influence of anthromorphism and agency on social judgement in virtual environments." *Journal of Computer-Mediated Communication* 9 (2).

Nuyts, Jan. 1987. "Pragmatics and cognition: On explaining language." In *The Pragmatic Perspective*, Jef Verschueren and Marcella Bertuccelli-Papi (eds), 715–732. Amsterdam/Philadelphia: John Benjamins.
Ochs, Elinor. 1979. "Planned and unplanned discourse." In *Syntax and Semantics*, vol. 12: *Discourse and Syntax*, Talmy Givon (ed.), 51–80. London: Academic Press.
Ogiermann, Eva. 2009. "Politeness and in-directness across cultures: A comparison of English, German, Polish and Russian requests." *Journal of Politeness Research* 5: 189–216.
O'Neill, Jacki and Martin, David. 2003. "Text chat in action." In *Proceedings of Group 2003*, 40–49. Florida (USA): ACM.
O'Reilly, Tim. 2007. "What is Web 2.0: Design patterns and business models for the next generation of software." *International Journal of Digital Economics* 65: 17–37.
Origgi, Gloria. 2002. "Authors and authority (an interview with Umberto Eco)." *Text-e*, February 28, 2002.
Orihuela, José L. 2000. "Las nuevas tecnologías de la información: Claves para el debate." *Nueva Revista de Política, Cultura y Arte* 70 (July–August): 44–50.
Orihuela, José L. 2005. "Weblogs y blogosfera: El medio y la comunidad." In *Blogs. La conversación en Internet que está revolucionando medios, empresas y a ciudadanos*, 13–48. Madrid: ESIC.
Orihuela, José L. 2008. "Internet: la hora de las redes sociales." *Nueva Revista* 119: 57–62.
Orlikowski, Wanda and Yates, JoAnne. 1994. "Genre repertoire: The structuring of communicative practices in organizations." *Administrative Science Quarterly* 39: 541–574.
Orlikowski, Wanda and Yates, JoAnne. 2002. "Genre systems: Structuring interaction through communicative norms." *Journal of Business Communication* 39 (1): 13–35.
O'Sullivan, Patrick B. and Flanagin, Andrew J. 2000. "An interactional reconceptualization of 'flaming' and other problematic messages." Paper presented at the *International Communication Association's Annual Conference*, Acapulco (México).
Outing, Steve. 2000. "Why is my mail talking to me?," *The Guardian*, April 17, 2000, New Media.
Pagendarm, Magnus and Schaumburg, Heike. 2001. "Why are users banner-blind? The impact of navigation style on the perception of web banners." *Journal of Digital Information* 2 (1).
Pajares Tosca, Susana. 2000. "A pragmatics of links." *Journal of Digital Information* 1 (6).
Palme, Jacob. 1999. "Support for decisions by e-mail." Paper presented at *Terena Conference*, Lund.
Pano, Ana. 2008a. *Dialogar en la Red. La Lengua Española en Chats, E-Mails, Foros y Blogs*. Berlin: Peter Lang.
Pano, Ana. 2008b. "Estrategias comunicativas y lingüísticas en listas y foros de debate en Internet." *CÍRCULO de Lingüística Aplicada a la Comunicación* 33: 28–44.
Panyametheekul, Siriporn and Herring, Susan C. 2003. "Gender and turn allocation in a Thai chat room." *Journal of Computer-Mediated Communication* 9 (1).
Paolillo, John. 1999. "The virtual speech community: Social network and language variation on IRC." *Journal of Computer-Mediated Communication* 4 (4).
Paolillo, John. 2001. "Language variation on Internet Relay Chat: A social network approach." *Journal of Sociolinguistics* 5 (2): 180–213.
Paolillo, John, Warren, Jonathan and Kunz, Breanne. 2010. "Genre emergence in amateur flash." In *Genres on the Web: Computational Models and Empirical Studies*, Alexander Mehler, Serge Sharoff and Marina Santini (eds), 277–302. Berlin: Springer.
Papacharissi, Zizi. 2009. "The virtual geographies of social networks: A comparative analysis of *Facebook, Linkedin* and *ASmallWorld*." *New Media & Society* 11 (1–2): 199–120.

Park, Jung-ran. 2008a. "Linguistic politeness and face-work in computer-mediated communication, part 1: A theoretical Framework." *Journal of the American Society for Information Science and Technology* 59 (13): 2051–2059.

Park, Jung-ran. 2008b. "Linguistic politeness and face-work in computer mediated communication, part 2: An application of the theoretical framework." *Journal of the American Society for Information Science and Technology* 59 (14): 2199–2209.

Park, Taemin K. 1994. "Toward a theory of user-based relevance: A call for a new paradigm of inquiry." *Journal of the American Society for Information Science* 45 (3): 135–141.

Patel, Nayna, Coughlan, Jane and Macredie, Robert D. 2006. "A method for identifying communication breakdowns in user-interfaces." *Studies in Communication Sciences* 6 (1): 27–46.

Patterson, Holly. 1996. *Computer-Mediated Groups. A Study of a Culture in Usenet*. Unpublished PhD thesis. University of Texas A&M.

Peccei, Jean Stilwell. 1999. *Pragmatics*. London: Routledge.

Pérez-Lanzac, Carmen and Rincón. 2009. "Tu 'extimidad' contra mi intimidad." *El País* 24-03-2009, pp. 32–33.

Pérez Sabater, Carmen. 2007. *Los Elementos Conversacionales en la Comunicación Escrita via Internet en Lengua Inglesa*. Unpublished PhD thesis. Castelló (Spain): Jaume I University.

Pérez Sabater, Carmen, Turney, Ed and Montero Fleta, Begoña. 2008. "Orality and literacy, formality and informality in email communication." *Ibérica* 15: 71–88.

Peris, Rosana, Gimeno, Miguel A., Pinazo, Daniel, Ortet, Generós, Carrero, Virginia, Sanchiz, María L. and Ibáñez, Ignacio. 2002. "Online chat rooms: Virtual spaces of interaction for socially oriented people." *CyberPsychology & Behavior* 5 (1): 43–51.

Peter, Jochen, Valkenburg, Patti M. and Schouten, Alexander P. 2007. "Precursors of adolescents' use of visual and audio devices during online communication." *Computers in Human Behavior* 23: 2473–2487.

Peterson, Katherine and Siek, Katie A. 2009. "Analysis of information disclosure on a social networking site." In *Online Communities*, LNCS 5621. A. Ant Ozok and Panayiotis Zaphiris (eds), 256–264. Berlin: Springer-Verlag.

Philips, Gerald M. and Barnes, Susan B. 1995. "Is your e-pal an ax-murderer?" *Interpersonal Computing and Technology* 3 (4).

Pilkington, Adrian. 2000. *Poetic Effects. A Relevance Theory Perspective*. Amsterdam/Philadelphia: John Benjamins.

Pisani, Francis. 2007. "Web 2.0: Twitter o la tendencia haiku," *El País*, April 19, 2007, Ciberpaís, p. 2.

Pitts, Mary J. 2003. "Television web sites and changes in the nature of storytelling." *Simile* 3 (3): 1–8.

Porter, David (ed.). 1997. *Internet Culture*. London: Blackwell.

Porter, Lance and Golan, Guy J. 2006. "From subservient chickens to brawny men: A comparison of viral advertising to television advertising." *Journal of Interactive Advertising* 6 (2).

Portillo, Eloy and Hartza, Juan. 1995. "Los sujetos ante el mundo digital." *Archipiélago* 23: 21–26.

Posteguillo, Santiago. 1997. "Acronyms in computer science: A genre indicator of audience." In *Estudios de lingüística aplicada*, José L. Otal, Inmaculada Fortanet and Victoria Codina (eds), 283–289. Castelló (Spain), Jaume I University, Servicio de Publicaciones.

Posteguillo, Santiago. 2003. *Netlinguistics. Language, Discourse and Ideology in Internet*. Castelló (Spain), Jaume I University, Servicio de Publicaciones.

Poster, Mark. 1995. *The Second Media Age*. Oxford: Blackwell.

Poyatos, Fernando. 1975. "Cross-cultural analysis of paralinguistic 'alternants' in face-to-face interaction." In *Organization of Behavior in Face-to-Face Interaction*, Adam Kendon, Richard Harris and Mary Ritchie Key (eds), 285–314. The Hague: Mouton.

Poyatos, Fernando. 2002. *Nonverbal Communication Across Disciplines*. Volumen II: *Paralanguage, Kinesics, Silence, Personal and Environmental Interaction*. Amsterdam/Philadelphia: John Benjamins.

Preisler, Brent. 1986. *Linguistic Sex Roles in Conversation: Social Variation in the Expression of Tentativeness in English*. Berlin: Mouton de Gruyter.

Provine, Robert R., Spencer, Robert J. and Mandell, Darcy L. 2007. "Emotional expression online. Emoticons punctuate website text messages." *Journal of Language and Social Psychology* 26 (3): 299–307.

Pujolà, Joan-Tomas and Palomeque, Cristina. 2010. "Developing a multimodal transcription to account for interaction in 3D virtual worlds: The 3M method." In *Analizar Datos > Describir Variación*, Jorge L. Bueno Alonso et al. (eds), 134–145. Vigo (Spain): University of Vigo, Servizo de Publicacións.

Quan-Haase, Anabel. 2008. "Instant messaging on campus: Use and integration in university students' everyday communication." *The Information Society* 24: 105–115.

Quan-Haase, Anabel and Collins, Jessica L. 2008. "'I'm there, but I might not want to talk to you.'" *Information, Communication & Society* 11 (4): 526–543.

Quinion, Michael B. 1996. "Only joking? Should the smiley be outlawed?," *World Wide Words*, August 8, 1996.

Rafaeli, Sheizaf and Sudweeks, Fay. 1997. "Networked interactivity." *Journal of Computer-Mediated Communication* 2 (4).

Rains, Stephen A. and Young, Anna M. 2006. "A sign of the times: An analysis of organizational members' email signatures." *Journal of Computer-Mediated Communication* 11 (4).

Rains, Stephen A., Tumlin, Geoffrey R. and Knapp, Mark L. 2009. "Electronic bumper stickers: The content and interpersonal functions of messages attached to e-mail signatures." *Discourse Studies* 11 (1): 105–120.

Rajani, Rakhi and Rosenberg, Duska. 2000. "Usable?.. or not?.. Factors affecting the usability of web pages." *CMC Magazine* 6 (1).

Raudaskoski, Pirkko. 2000. "The use of communicative resources in internet video conferencing." In *Words on the Web*, Lyn Pemberton and Simon Shurville (eds), 52–60. Exeter, UK: Intellect Books, 44–51.

Recanati, Francois. 1993. *Direct Reference*. Oxford: Blackwell.

Reicher, Steve, Spears, Russell and Postmes, Tom. 1995. "A social identity model of deindividuation phenomena." In *European Review of Social Psychology*, vol. 6, W. Stroebe and M. Hewstone (eds), 161–198. Chichester: Wiley.

Reid, Elizabeth M. 1991. *Electropolis: Communication and Community on Internet Relay Chat*. Unpublished PhD thesis. Melbourne: University of Melbourne.

Reid, Elizabeth M. 1994. *Cultural Formations in Text-Based Virtual Realities*. Unpublished MA thesis. University of Melbourne (Australia).

Reitbauer, Margit. 2006. "Hypertextual information structures and their influence on reading comprehension: An empirical study." *Miscelánea* 33: 65–87.

Reventós, Laura. 2008. "Mister Twitter," *El País*, January 17, 2008, Ciberpaís, pp. 1/6.

Rheingold, Howard. 1993. *The Virtual Community*. Reading, Mass.: Addison-Wesley.

Ribes, Xabier. 2007. "La Web 2.0. El valor de los metadatos y de la inteligencia colectiva." *Telos* 73: 36–43.

Rice, Ronald E. 1992. "Task analyzability, use of new medium and effectiveness: A multi-site exploration of media richness." *Organization Science* 3 (4): 475–500.

Richmond, Shane. 2009. "Twitter looks like a publishing platform but it isn't one," *The Telegraph*, January 5, 2009.

Richtel, Matt. 2000. "BACKLASH!; Now that I have your attention." *The New York Times*, October 8, 2000, Technology.

Rintel, E. Sean and Pittam, Jeffery. 1997. "Strangers in a strange land: Interaction management on Internet Relay Chat." *Human Communication Research* 23 (4): 507–534.

Riordan, Monica A. and Kreuz, Roger J. 2010. "Emotion encoding and interpretation in computer-mediated communication: Reasons for use." *Computers in Human Behavior* 26: 1667–1673.

Rivas Monroy, María U. 1996. "Nombres propios: Realidad y mundos posibles." In *Mundos de ficción*, vol. II, José M. Pozuelo Yvancos and Francisco Vicente Gómez (eds), 1319–1326. Murcia (Spain): University of Murcia and AES.

Ruane, Michael E. 1999. "An instant language, packed with meaning: Online chatters invent shorthand for fast talk," *Washington Post*, December 14, 1999, A1.

Ruedenberg, Lucia, Danet, Brenda and Rosenbaum-Tamari, Yehudit. 1994. "Virtual virtuosos: Play and performance at the computer keyboard." *Electronic Journal of Communication* 5 (4).

Ryu, Eun J. 2008. "3D avatars in instant messenger." Paper presented at *Computer-Mediated Communication '08*, Pittsburgh (USA), Carnegie Mellon University.

Sachdev, V., Nerur, S. and Teng, J. T. C. 2010. "Interactivity redefined for the social web." In *Handbook of Research on Social Interaction Technologies and Collaboration Software: Concepts and Trends*, Tatyana Dumova and Richard Fiordo (eds), 586–600. Hershey: IGI Global.

Salmerón, Ladislao, Kintsch, Walter and Kintsch, Eileen. 2010. "Self-regulation and link selection strategies in hypertext." *Discourse Processes* 47: 175–211.

Salvucci, Dario D. and Taatgen, Niels A. 2011. *The Multitasking Mind*. Oxford: Oxford University Press.

Sandbothe, Mike. 1998. "Media temporalities in the Internet." *Journal of Computer-Mediated Communication* 4 (2).

Santini, Marina, Mehler, Alexander and Sharoff, Serge. 2010. "Riding the rough waves of genre on the web. Concepts and research questions." In *Genres on the Web: Computational Models and Empirical Studies*, Alexander Mehler, Serge Sharoff and Marina Santini (eds), 3–30. Berlin: Springer.

Sanz Álava, Inmaculada. 2006. "El correo electrónico profesional: ¿género, sub-género o cibergénero?" Paper presented at the *5th International AELFE Conference*, Zaragoza (Spain), September.

Saracevic, Tefko. 1996. "Relevance reconsidered." In *Information Science: Integration in Perspectives. Proceedings of the Second Conference on Conceptions of Library and Information Science* (CoLIS 2), P. Ingwersen & N. O. Pors (eds), 201–218. Copenhaguen (Denmark): Royal School of Librarianship.

Saracevic, Tefko. 2006. "Relevance: A review of the literature and a framework for thinking on the notion in information science. Part II." *Advances in Librarianship* 30: 3–77.

Saracevic, Tefko. 2007. "Relevance: A review of the literature and a framework for thinking on the notion in information science. Part III: Behavior and effects of relevance." *Journal of the American Society for Information Science and Technology* 58 (13): 2126–2144.

Savas, Perihan. 2011. "A case study of contextual and individual factors that shape linguistic variation in synchronous text-based computer-mediated communication." *Journal of Pragmatics* 43: 298–313.

Saville-Troike, Muriel. 1989. *The Ethnography of Speaking*. Oxford: Blackwell.

Schamber, Linda, Eisenberg, Michael S. and Nilan, Michael S. 1990. "A re-examination of relevance: Toward a dynamic, situational definition." *Information Processing & Management* 26: 755–775.

Scheidt, Lois A. and Wright, Elijah. 2004. "Common visual design elements of weblogs." In *Into the Blogosphere: Rhetoric, Community, and Culture of Weblogs*. Laura J. Gurak, Smiljana Antonijevic, Laurie Johnson, Clancy Ratliff, and Jessica Reyman (eds). Los Alamitos: IEEE Press.

Schmidt, Jan. 2007. "Blogging practices: An analytical framework." *Journal of Computer-Mediated Communication* 12 (4).

Schneider, Steven M. and Foot, Kirsten A. 2004. "The web as an object of study." *New Media & Society* 6 (1): 114–122.

Schulze, Markus. 1999. "Substitution of paraverbal and nonverbal cues in the written medium of IRC." In *Dialogue Analysis and the Mass Media*, Bernd Naumann (ed.), 65–82. Niemeyer: Tübingen.

Schwarz, Ori. 2010. "On Friendship, boobs and the logic of the catalogue. Online self-portraits as a means for the exchange of capital." *Convergence* 16 (2): 163–183.

Schwarz, Ori. 2011. "Who moved my conversation? Instant messaging, intertextuality and new regimes of intimacy and truth." *Media, Culture & Society* 33 (1): 71–87.

Sell, Roger D. 1991. "The politeness of literary texts." In *Literary Pragmatics*, Roger D. Sell (ed.), 208–224. London: Routledge.

Serpentelli, Jill. 1992. "Conversational structure and personality correlates of electronic communication." University of Cologne (Germany), Haverford College, Technical report.

Severinson Eklundh, Kerstin. 2010. "To quote or not to quote: Setting the context for computer-mediated dialogues." *Language@Internet* 7.

Sevillano, Elena. 2010. "Internet cambia las reglas," *El País*, January 30, 2010, Extra Publicidad, pp. 12–14.

Shank, Gary. 1993. "Abductive multiloguing. The semiotic dynamics of navigating the net." *Arachnet Electronic Journal on Virtual Culture* 1 (1).

Shao, Guosong. 2009. "Understanding the appeal of user-generated media: A uses and gratification perspective." *Internet Research* 19 (1): 7–25.

Shapiro, Norman Z. and Anderson, Robert H. 1985. "Towards an ethics and etiquette for electronic mail." Santa Mónica, CA. (USA): The Rand Corporation, Report R-3283–NSF/RC.

Shea, Virginia. 1994. *Netiquette*. San Rafael, CA.: Albion.

Shepherd, Michael and Watters, Carolyn R. 1998. "The evolution of cybergenres." In *Proceedings of the Thirty-First Annual Hawaii International Conference on System Sciences (HICSS '98)*, vol. II, 97–109. Hawaii.

Simich-Dudgeon, Carmen. 1999. "Becoming an online community: The role of involvement strategies in the textual messages of graduate students." Paper presented at the *International Online Conference on Teaching Online in Higher Education*, Bloomington (USA): Indiana University.

Simmons, Thomas L. 1999. "Face threats in a faceless medium: Negotiating ideological parameters in computer-mediated communication." In *Language and Ideology* (selected papers from the 6th International Pragmatics Conference), Jef Verschueren (ed.), 514–543. Antwerp: IPrA.

Simpson, J. 2005. 'Conversational floors in synchronous text-based CMC discourse." *Discourse Studies* 7 (3): 337–361.
Singh, Surendra N., Dalal, Nikunj, Mishra, Sanjay and Patil, Vivek H. 2005. "A communications view of web page perception." *Journal of Current Issues and Research in Advertising* 27 (1): 31–52.
Skovholt, Karianne. 2009. *Email Literacy in the Workplace*. Unpublished PhD thesis. University of Oslo (Norway).
Slembrouck, Stef. 2000. What is meant by discourse analysis?" University of Gent, Internet document.
Smith, Christine B., McLaughlin, Margaret L. and Osborne, Kerry K. 1997. "Conduct control on Usenet." *Journal of Computer-Mediated Communication* 2 (4).
Smith, David. 2006. "How to get the life you really want." *The Observer*, July 9, 2006, p. 13.
Smith, Marc A., Farnham, Shelly A. and Drucker, Steven M. 2000. "The social life of small graphical chat spaces." In *Proceedings of the SIGCHI Conference on Human Factors in Computing Systems*, 462–469. The Hague (Netherlands): ACM.
Smith, Marc, Cadiz, J. J. and Burkhalter, Byron. 2000. "Conversation trees and threaded chats." In *Proceedings of the 2000 ACM Conference on Computer Supported Cooperative Work*.
Smith, Mark and Kollock, Peter (eds). 1999. *Communities in Cyberspace*. London: Routledge.
Smith, Neil and Wilson, Deirdre. 1983. *La Lingüística Moderna. Los Resultados de la Revolución de Chomsky*. Barcelona: Anagrama.
Smith, P. A., Newman, I. A. and Parks, L. M. 1997. "Virtual hierarchies and virtual networks: Some lessons from hypermedia usability research applied to the World Wide Web." *International Journal of Human-Computer Studies* 47: 67–95.
Solanilla, Laura. 2002. "¿Qué queremos decir cuando hablamos de interactividad? El caso de los webs de los museos de historia y arqueología." *Digit-Hum* 4.
Sotillo, Susana M. 2000. "Discourse functions and syntactic complexity in synchronous and asynchronous communication." *Language Learning and Technology* 4 (1): 82–119.
Soucek, Roman and Moser, Klaus. 2010. "Coping with information overload in email communication: Evaluation of a training intervention." *Computers in Human Behavior* 26: 1458–1466.
Spears, Russell and Lea, Martin. 1992. "Social influence and the influence of the 'social' in computer-mediated communication." In *Contexts of Computer-Mediated Communication*, Martin Lea (ed.), 30–65. New York: Harvester Wheatsheaf.
Spears, Russell, Lea, Martin and Lee, Stephen. 1990. "De-individuation and group polarization in computer-mediated communication." *British Journal of Social Psychology* 29: 121–134.
Sperber, Dan. 1994. "Understanding verbal understanding." In *What is Intelligence*, J. Khalfa (ed.), 179–198. Cambridge: Cambridge University Press.
Sperber, Dan. 1996. *Explaining Culture. A Naturalistic Approach*. Oxford: Blackwell.
Sperber, Dan. 2000. "Metarepresentations in an evolutionary perspective." In *Metarepresentations*, Dan Sperber (ed.), 117–137. Oxford: Oxford University Press.
Sperber, Dan and Wilson, Deirdre. 1986. *Relevance. Communication and Cognition*. Oxford: Blackwell.
Sperber, Dan and Wilson, Deirdre. 1987. "Précis of *Relevance. Communication and Cognition*." *Behavioral and Brain Sciences* 10: 697–754.
Sperber, Dan and Wilson, Deirdre. 1995. *Relevance. Communication and Cognition* (2nd edition). Oxford: Blackwell.

Sperber, Dan and Wilson, Deirdre. 1997. "Remarks on relevance theory and the social sciences." *Multilingua* 16 (2/3): 145–151.
Sperber, Dan and Wilson, Deirdre. 2002. "Pragmatics, modularity and mind-reading." *Mind & Language* 17 (1–2): 3–23.
Spiegel, Dana S. 2001. "Coterie: A visualization of the conversational dynamics within IRC." Massachusetts: M.I.T. Master in Media Arts and Sciences.
Spink, Amanda and Greisdorf, Howard. 2000. "Regions of relevance study: Part I. Measuring and mapping users' relevance judgments." *Journal of the American Society for Information Science* 51 (13).
Spink, Amanda and Greisdorf, Howard. 2001. "Regions and levels: Measuring and mapping users' relevance judgments." *Journal of the American Society for Information Science and Technology* 52 (2): 161–173.
Spink, Amanda, Greisdorf, Howard and Bateman, Judy. 1998. "From highly relevant to nonrelevant: Examining different regions of relevance." *Information Processing and Management* 34 (5): 599–622.
Sproull, Lee and Kiesler, Sara. 1986. "Reducing context cues: Electronic mail in organizational communication." *Management Science* 32: 1492–1512.
Squires, Lauren. 2010. "Enregistering internet language." *Language in Society* 39: 457–492.
Stallabrass, J. 1998. "Formas de identidad en el ciberespacio." *Revista de Occidente* 206: 77–97.
Stanford, Peter. 1999. "Getting the message," *The Sunday Times*, July 4, 1999, Culture.
Stefanone, Michael A., Kwon, Kyounghee and Lackaff, Derek. 2011. The value of online friends: Networked resources via social network sites." *First Monday* 16 (2).
Stein, Dieter. 2006. "E-mail, Internet, chatroom talk: Pragmatics." In *Encyclopedia of Language and Linguistics* (2nd edition). Keith Brown (ed.), 116–124. Amsterdam: Elsevier.
Stenström, Anna-Brita. 1994. *An Introduction to Spoken Interaction*. London: Longman.
Stephens, Keri K., Cowan, Renee L. and Houser, Marian L. 2011. "Organizational norm congruency and interpersonal familiarity in e-mail: Examining messages from two different status perspectives." *Journal of Computer-Mediated Communication* 16: 228–249.
Stille, Alexander. 2000. "Marshall McLuhan is back from the dustbin of history," *The New York Times*, October 14, 2000, Technology.
Stone, Roland L. 1995. "Patterns of politeness in a listserv discussion." Paper presented at *Computer-Mediated Discourse Analysis*(*GURT'95*), Washington, Georgetown University.
Stromer-Galley, Jennifer and Martey, Rosa M. 2009. "Visual spaces, norm governed places: The influence of spatial context online." *New Media & Society* 11 (6): 1041–1060.
Stuart, Dong. 1999. "Digital domains: Nation, state and virtual territory." *Paideusis* 2.
Subrahmanyam, Kaveri and Šmahel, David. 2011. *Digital Youth: The Role of Media in Development*. Berlin: Springer.
Suler, John. 1997a. "Psychological dynamics of online synchronous conversations in text-driven chat environments." In *The Psychology of Cyberspace*. Internet document.
Suler, John. 1997b. "Communicative subtlety in multimedia chat." In *The Psychology of Cyberspace*. Internet document.
Suler, John. 1997c. "The final showdown between in-person and cyberspace relationships." In *The Psychology of Cyberspace*. Internet document.
Suler, John. 1998. "E-mail communication and relationships." In *The Psychology of Cyberspace*. Internet document.

Suler, John. 1999. "The psychology of avatars and graphical space in multimedia chat communities or how I learned to stop worrying and love My Palace Props." In *The Psychology of Cyberspace*. Internet document.

Suler, John. 2000. "Hypotheses about online text relationships." In *The Psychology of Cyberspace*. Internet document.

Suler, John. 2007. "Second life, second chance." In *The Psychology of Cyberspace*. Internet document.

Sun, Jun and Poole, Marshall S. 2010. "Beyond connection: Situated wireless communities." *Communications of the ACM* 53 (6): 121–125.

Sundar, S. Shyam. 2008. "Self as source: agency and customization in interactive media." In *Mediated Interpersonal Communication*, Elly A. Konijn, Sonja Utz, Martin Tanis and Susan B. Barnes (eds), 58–74. London: Routledge.

Swales, John M. 1990. *Genre Analysis*. Cambridge: Cambridge University Press.

Swanson, Don R. 1986. "Subjective versus objective relevance in bibliographic retrieval systems." *The Library Quarterly* 56: 389–398.

Sweeney, Camille. 1999. "In a chat room you can be N E 1," *The New York Times*, October 17, 1999, Magazine.

Szóstek, Agnieszka Matysiak. 2011. "'Dealing with My Emails': Latent user needs in email management." *Computers in Human Behavior* 27: 723–729.

Tanaka, Keiko. 1994. *Advertising Language. A Pragmatic Approach to Advertisements in Britain and Japan*. London: Routledge.

Tanaka, Jennifer. 2000. "Crammed with spam," *Newsweek*, April 10, 2000.

Tannen, Deborah. 1979. "What's in a frame? Surface evidence for underlying expectations." In *New Directions in Discourse Processing*, Roy O. Freedle (ed.), 137–181. Norwood (NJ.): Ablex.

Taylor, Jacqui. 2000. "Electronic mail, communication and social identity: A social psychological analysis of computer-mediated group interaction." In *Words on the Web*, Lyn Pemberton and Simon Shurville (eds), 96–105. Exeter, UK: Intellect Books.

Taylor, T. L. 2002. "Living digitally. Embodiment in virtual worlds." In *The Social Life of Avatars. Presence and Interaction in Shared Virtual Environments*, Ralph Schroeder (ed.), 40–62. London: Springer.

Tella, Seppo. 1992. "The adoption of international commnication networks and electronic mail into foreign language education." *Scandinavian Journal of Educational Research* 36 (4): 305–312.

Ten Have, Paul. 2000. "Computer-mediated chat: Ways of finding chat partners." *M/C* 3 (4).

Tewksbury, David. 2003. "What do Americans really want to know? Tracking the behavior of news readers on the Internet." *Journal of Communication* 53 (4): 694–710.

Thelwall, Mike. 2009a. "MySpace comments." *Online Information Review* 33 (1): 58–76.

Thelwall, Mike. 2009b. "Homophily in MySpace." *Journal of the American Society for Information Science and Technology* 60 (2): 219–231.

Thelwall, Mike and Stuart, David. 2010. Social network sites. An exploration of features and diversity." In *Social Computing and Virtual Communities*, Panayiotis Zaphiris and Chee Siang Ang (eds), 263–282. London: CRC Press.

Thompson, Clive. 2008. "Brave new world of digital intimacy." *The New York Times*, September 5, 2008, Magazine.

Thoms, Lesley and Thelwall, Mike. 2005. "Academic home pages: Reconstruction of the self." *First Monday* 10 (12).

Thomsen, Steven R., Straubhaar, Joseph D. and Bolyard, Drew M. 1998. "Ethnomethodology and the study of online communities: Exploring the cyber streets." Paper presented at *Iriss'98*, Bristol (UK).
Tidwell, Lisa C. and Walther, Joseph B. 2002. "Computer-mediated communication effects on disclosure, impressions, and interpersonal evaluations: Getting to know one another a bit at a time." *Human Communication Research* 28 (3): 317–348.
Toma, Catalina L. 2010. "Affirming the self through online profiles: Beneficial effects of social networking sites." In *Proceedings of the 28th International Conference on Human Factors In Computing Systems*, 1749–1752. New York: ACM.
Trammell, Kaye D., Williams, Andrew P., Postelnicu, Monica and Landreville, Kristen D. 2006. "Evolution of online campaigning: Increasing interactivity in candidate web sites and blogs through text and technical features." *Mass Communication & Society* 9 (1): 21–44.
Trott, Patricia. 1996. "Body-surfing the net: Questions of corporeality and identity on the Internet." Carleton University, *Working Papers in Communication, Technology, and Culture* 5.
Turkle, Sherry. 1994. "Constructions and reconstructions of self in virtual reality: Playing in the MUDs." *Mind, Culture, and Activity* 1 (3).
Turkle, Sherry. 1996a. "Virtuality and its discontents: Searching for community in cyberspace." *The American Prospect* 24, Winter 1996, 50–57.
Turkle, Sherry. 1996b. "Parallel lives: Working on identity in virtual space." In *Constructing the Self in a Mediated World*, Debra Grodin and Thomas R. Lindlof (eds), 156–176. London: Sage.
Turkle, Sherry. 1996c. "Who am we?" *Wired* 4 (1). January 1996.
Turkle, Sherry. 1997. *La Vida en la Pantalla*. Barcelona: Paidós.
Turkle, Sherry. 1998. "Repensar la identidad de la comunidad virtual." *El Paseante* 27/28: 48–51.
Turkle, Sherry. 1999. "What are we thinking about when we are thinking about computers?" In *The Science Studies Reader*, Mario Biagioli (ed.), 543–552. London: Routledge.
Turkle, Sherry. 2011. *Alone Together: Why We Expect More from Technology and Less from Each Other*. New York: Basic Books.
Turnage, Anna K. 2007. "Email flaming behaviors and organizational conflict." *Journal of Computer-Mediated Communication* 13 (1).
Valenzuela, Sebastián, Park, Namsu and Kee, Kerk F. 2009. "Is there social capital in a social network site?: Facebook use and college students' life satisfaction, trust, and participation": *Journal of Computer-Mediated Communication* 14: 875–901.
Valtersson, Maria. 1996. *Virtual Communities*. Unpublished MA thesis. University of Umea.
van Alstyne, Marshall and Brynjolfsson, Erik. 1997. *Electronic Communities: Global Village or Cyberbalkans?* Cambridge, Mass.: MIT Sloan School.
van Dijck, José. 2004. "Composing the self: Of diaries and lifelogs." *Fibreculture* 3.
van Dijck, José. 2009. "Users like you? Theorizing agency in user-generated content." *Media, Culture & Society* 31 (1): 41–58.
van Doorn, Niels. 2010. "The ties that bind: The networked performance of gender, sexuality and friendship on MySpace." *New Media & Society* 12 (4): 583–602.
van Schaik, Paul and Ling, Jonathan. 2006. "The effects of graphical display and screen ratio on information retrieval in web pages." *Computers in Human Behavior* 22: 870–884.
Varnhagen, Connie K., McFall, G. Peggy, Pugh, Nicole, Routledge, Lisa, Sumida-MacDonald, Heather and Kwong, Trudy E. 2009. "Lol: New language and spelling in instant messaging." *Reading and Writing* 23 (6): 719–733.

Vela Delfa, Cristina. 2002. "Un nuevo prototipo textual: El correo electrónico." *Interlingüística* 13: 447–465.
Vela Delfa, Cristina. 2005. *El Correo Electrónico: El Nacimiento de un Nuevo Género*. Unpublished PhD thesis. Madrid: Complutense University.
Vicdan, Handan and Ulusoy, Ebru. 2008. "Symbolic and experiential consumption of body in virtual worlds: From (dis)embodiment to symembodiment." *Journal of Virtual Worlds Research* 1 (2).
Vicente, Begoña. 1999. "La teoría de la pertinencia." In *Filosofía del Lenguaje II: Pragmática*, Marcelo Dascal (ed.), 115–136. Madrid: Trotta.
Vidal Jiménez, Rafael. 2000. "La red y la destrucción de la identidad." *TTC* III, 10–3–2000.
Viegas, Fernanda B. and Donath, Judith S. 1999. "Chat Circles." In *Proceedings of CHI'99*. Pittsburgh (USA). ACM.
Vilhjálmsson, Hannes H. and Cassell, Justine. 1998. "*BodyChat*: Autonomous communicative behaviors in avatars." In *AGENTS '98. Proceedings of the Second International Conference on Autonomous Agents*, 269–276. Minneapolis (USA).
Voida, Amy, Newstetter, Wendy C. and Mynatt, Elizabeth D. 2002. "When conventions collide: The tensions of instant messaging attributed." Paper presented at *CHI 2002*, Minneapolis (USA).
Voiskounsky, Alexander E. 1997. "Telelogue conversations." *Journal of Computer-Mediated Communication* 2 (4).
von der Pütten, Astrid M., Krämer, Nicole C., Gratch, Jonathan and Kang, Sin-Hwa. 2010. "'It doesn't matter what you are!' Explaining social effects of agents and avatars." *Computers in Human Behavior* 26: 1641–1650.
Vuorinen, Kimmo. 2005. *Using Weblogs for Discussion*. Unpublished MA thesis. University of Tampere.
Wagg, Stephen. 1998. "Punching your weight. Conversations with Jo Brand." In *Because I Tell a Joke or Two*, Stephen Wagg (ed.), 111–136. London: Routledge.
Wainer, Jaclyn, Dabbish, Laura and Kraut, Robert. 2011. "Should I open this email?: Inbox-level cues, curiosity and attention to email." Paper presented at *CHI 2011*, Vancouver, Canada, May.
Walker, Rob and Bakopoulos, Babis. 2005. "Conversations in the dark: How young people manage chatroom relationships." *First Monday* 10 (4).
Wallace, Derek. 2000. "E-mail and the problems of communication." *M/C* 3 (4).
Walther, Joseph B. 1992. "Interpersonal effects in computer-mediated interaction. A relational perspective." *Communication Research* 19 (1): 52–90.
Walther, Joseph B. 1996. "Computer-mediated communication: Impersonal, interpersonal, and hyperpersonal interaction." *Communication Research* 23 (1): 3–43.
Walther, Joseph B. 2007. "Selective self-presentation in computer-mediated communication: Hyperpersonal dimensions of technology, language, and cognition." *Computers in Human Behavior* 23: 2538–2557.
Walther, Joseph B. and D'Addario, Kyle P. 2001. "The impacts of emoticons on message interpretation in computer-mediated communication." *Social Science Computer Review* 19 (3): 324–347.
Walther, Joseph B., van der Heide, Brandon, Kim, Sang-Yeon, Westerman, David and Tong, Stephanie T. 2008. "The role of friends' appearance and behavior on evaluations of individuals on Facebook: Are we known by the company we keep?" *Human Communication Research* 34: 28–49.

Walther, Joseph B., Carr, Caleb T., Choi, Scott S., DeAndrea, David C., Kim, Jinsuk, Tong, Stephanie T. and Van Der Heide, Brandon. 2011. "Interaction of interpersonal, peer, and media influence sources online. A research agenda for technology convergence." In *A Networked Self. Identity, Community, and Culture on Social Network Sites*, Zizi Papacharissi (ed.), 17–38. London: Routledge.

Wang, Shaojung S., Moon, Shin-Il, Kwon, Kyounghee H., Evans, Carolyn A. and Stefanone, Michael A. 2010. "Face off: Implications of visual cues on initiating friendship on Facebook." *Computers in Human Behavior* 26: 226–234.

Warschauer, Mark. 2000. "Language, identity, and the Internet." In *Race in Cyberspace*, Beth Kolko, Lisa Nakamura and Gilbert Rodman (eds), 151–170. London: Routledge.

Watson, AnnMarie. 1996. "Formality of computer-mediated communication between faculty and students." *CTA Senior Thesis Papers*, Spring 1996.

Watson, N. 1997. "Why we argue about virtual community: A case study of the *phish.net* fan community." In *Virtual Culture. Identity & Communication in Cybersociety*, Steven G. Jones (ed.), 102–132. London: Sage.

Watts, Richard J. 2003. *Politeness*. Cambridge: Cambridge University Press.

Weasenforth, Donald and Lucas, Sigrun. 1997. "On-line and off-line texts of non-native speakers: Distinguishable text types?" Paper presented at *E-mail, The Web, and MOOs*, Washington, University of Washington.

Weinreich, Frank. 1997. "Establishing a point of view toward virtual communities." *Computer-Mediated Communication* 3 (2).

Welford, Heather. 1999. "Net-aholics anonymous." *The Guardian*, September 30, 1999, Online.

Wellman, Barry and Gulia, Milena. 1999. "Virtual communities as communities: Net surfers don't ride alone." In *Communities in Cyberspace*, Mark Smith and Peter Kollock (eds), 167–194. London: Routledge.

Welsh, Edward. 1997. "'Information fatigue' saps the e-mail set." *The Sunday Times*, April 20, 1997, Britain.

Werry, Christopher. 1996. "Linguistic and interactional features of Internet Relay Chat." In *Computer-Mediated Communication. Linguistic, Social and Cross-Cultural Perspectives*, Susan Herring (ed.), 47–64. Amsterdam/Philadelphia: John Benjamins.

Weston, Jay. 1994. "Old freedoms and new technologies: The evolution of community networking." Paper presented at *Free Speech and Privacy in the Information Age*, University of Waterloo, Canada.

Wharton, Tim. 2000. "Interjections, language and the 'showing'/'saying' continuum." *UCL Working Papers in Linguistics* 12: 173–213.

Wharton, Tim. 2009. *Pragmatics and Non-Verbal Communication*. Cambridge: Cambridge University Press.

White, Howard D. 2007a. "Combining bibliometrics, information retrieval, and relevance theory, Part 1: First examples of a synthesis." *Journal of the American Society for Information Science and Technology* 58 (4): 536–559.

White, Howard D. 2007b. "Combining bibliometrics, information retrieval, and relevance theory, Part 2: Some implications for information science." *Journal of the American Society for Information Science and Technology* 58 (4): 583–605.

Wilkins, Harriet. 1991. "Computer talk. Long-distance conversations by computer." *Written Communication* 8 (1): 56–78.

Wilkinson, Lamar and Buboltz, Walter C. 1998. "E-mail: Communication of the future?" In *AACE, Proceedings of SITE '98*, Washington.

Williams, Eric. 1999. "E-mail and the effects of future developments." *First Monday* 4 (8).
Williams, Kipling D., Cheung, Christopher K. T. and Choi, Wilma. 2000. "Cyberostracism: Effects of being ignored over the Internet." *Journal of Personality and Social Psychology* 79 (5): 748–762.
Willis, Darnéy. 1999. "Effects of using enhancing visual elements in web site design." *ACJ* 3 (1).
Willson, Michele. 2010. "The possibilities of network sociality." In *International Handbook of Internet Research*, Jeremy Hunsinger, Lisbeth Klastrup and Matthew Allen (eds), 493–505. Berlin: Springer.
Wilson, Andrew. 1993. "A pragmatic device in electronic communication." *Journal of Pragmatics* 19: 389–398.
Wilson, Deirdre. 1994. "Relevance and understanding." In *Language and Understanding*, Gillian Brown, Kirsten Malmkjær, Alastair Pollitt and John Williams (eds), 35–58. Oxford: Oxford University Press.
Wilson, Deirdre. 1999. "Metarepresentation in linguistic communication." *UCL Working Papers in Linguistics* 11: 127–162.
Wilson, Deirdre. 2000. "Metarepresentation in linguistic communication." In *Metarepresentations*, Dan Sperber (ed.), 411–448. Oxford: Oxford University Press.
Wilson, Deirdre. 2002. "Relevance: The cognitive principle." Pragmatic Theory (Lecture notes). University College London.
Wilson, Deirdre and Sperber, Dan. 1986. "Sobre la definición de relevancia." In *La Búsqueda del Significado*, L. M. Valdés Villanueva (ed.), 583–598. Madrid: Tecnos.
Wilson, Deirdre and Sperber, Dan. 1993. "Linguistic form and relevance." *Lingua* 90: 1–15.
Wilson, Deirdre and Sperber, Dan. 2002a. "Relevance Theory." *UCL Working Papers in Linguistics* 14: 249–290.
Wilson, Deirdre and Sperber, Dan. 2002b. "Relevance theory: A tutorial." In *Proceedings of the Third Tokyo Conference on Psycholinguistics*, Y. Otsu (ed.), 45–70. Tokyo: Hituzi Syobo Publishing Company.
Wilson, Deirdre and Sperber, Dan. 2002c. "Truthfulness and relevance." *Mind* 111 (443): 583–632.
Wilson, Deirdre and Sperber, Dan. 2004. "Relevance theory." In *Handbook of Pragmatics*, Gregory Ward and Laurence Horn (eds), 607–632. Oxford: Blackwell.
Wilson, Patrick. 1973. "Situational relevance." *Information Storage and Retrieval* 9 (8): 457–471.
Wirth, Uwe. 2002. "As we may surf: The relevance of abductive inference for surfing through the internet." *Semiotica* 141 (1/4): 159–168.
Witmer, Diane F. and Katzman, Sandra L. 1997. "On-line smiles: Does gender make a difference in the use of graphic accents?" *Journal of Computer-Mediated Communication* 2 (4).
Witmer, Diane F. and Taweesuk, Chutatip. 1998. "Why business people use the world wide web." In *Proceedings of Cultural Attitudes Towards Communication and Technology '98*, Charles Ess and Fay Sudweeks (eds), 173–184. Sydney, Australia: University of Sydney.
Wortzel, Adrianne. 1998. "Screening our lives. An interview with Sherry Turkle." *Intelligent Agent* 2 (2).
Wu, Ling L., Chuang, Ya-Lan and Joung, Yuh-Jzer. 2008. "Contextual multi-dimensional browsing." *Computers in Human Behavior* 24: 2873–2888.
Wynn, Eleanor and Katz, James E. 1997. "Hyperbole over cyberspace: Self-presentation & social boundaries in Internet home pages and discourse." *The Information Society* 13 (4): 297–328.

Xu, Lingling, Yi, Cheng and Xu, Yunjie. 2007. "Emotional expression online: The impact of task, relationship and personality perception on emoticon usage in Instant Messenger." Paper presented at *11 Pacific Asia Conference on Information Systems*, Auckland, New Zealand.

Yates, Simeon J. 1996. "Oral and written linguistic aspects of computer conferencing." In *Computer-Mediated Communication. Linguistic, Social and Cross-Cultural Perspectives.* Susan Herring (ed.), 29–46. Amsterdam/Philadelphia: John Benjamins.

Yates, Simeon J. and Ornikowski, Wanda J. 1993. "Genres of organizational communication: A structurational approach to studying communication and media." *Academy of Management Review* 17 (2): 299–326.

Yee, Nick, Bailenson, Jeremy N. and Ducheneaut, Nicolas. 2009. "The Proteus effect. Implications of transformed digital self-representation on online and offline behavior." *Communication Research* 36 (2): 285–312.

Yee, Nick, Ellis, Jason and Ducheneaut, Nicolas. 2009. "The tyranny of embodiment." *Artifact* 9: 1–6.

Yongyan, Li. 2000. "Surfing e-mails." *English Today* 64 (14, 4): 30–34/55.

Yoo, Chan Y. 2011. "Modeling audience interactivity as the gratification-seeking process in online newspapers." *Communication Theory* 21: 67–89.

Young, Jeffrey R. 1994. "Textuality in cyberspace: MUDs and written experience." *Eserver*, Collections in the Arts and Humanities.

Yule, George. 1996. *Pragmatics.* Oxford: Oxford University Press.

Yus, Francisco. 1995. *Conversational Cooperation in Alternative Comics.* Alicante: University of Alicante, Department of English Studies (Working Papers, 4).

Yus, Francisco. 1996a. *Pragmática y relevancia. Un Modelo Escripto-Icónico Aplicado al Cómic Inglés.* Alicante: University of Alicante, Servicio de Publicaciones (microfiche).

Yus, Francisco. 1996b. "Iconicidad y doble articulación: Dos proyectos infructuosos de la semiótica respecto al arte." In *La interdisciplinariedad en el discurso artístico: ¿Realidad o utopía?* Vol. II, José L. Caramés Lage, Carmen Escobedo de Tapia and Jorge L. Bueno Alonso (eds), 435–458. Oviedo (Spain): University of Oviedo.

Yus, Francisco. 1997a. *Cooperación y Relevancia. Dos Aproximaciones Pragmáticas a la Interpretación.* Alicante: University of Alicante, Servicio de Publicaciones.

Yus, Francisco. 1997b. *La Interpretación y la Imagen de Masas. Un Modelo Pragmático Aplicado al Discurso del Cómic Inglés.* Alicante: Instituto Juan Gil-Albert.

Yus, Francisco. 1997c. "La teoría de la relevancia y la estrategia humorística de la incongruencia-resolución." *Pragmalingüística* 3–4: 497–508.

Yus, Francisco. 1998a. "A decade of relevance theory." *Journal of Pragmatics* 30: 305–345.

Yus, Francisco. 1998b. "Relevance: A thematic bibliographical list." *Revista Alicantina de Estudios Ingleses* 11 (special issue devoted to relevance theory): 261–285.

Yus, Francisco. 1998c. "Pragmática del malentendido." In *Perspectivas Pragmáticas en Lingüística Aplicada*, Ignacio Guillén and Ignacio Vázquez (eds), 105–111. Zaragoza: Anúbar.

Yus, Francisco. 1998d. "The *what-do-you-mean syndrome.* A taxonomy of misunderstandings in Harold Pinter's plays." *Estudios Ingleses de la Universidad Complutense* 6: 81–100.

Yus, Francisco. 1998e. "Relevance theory and media discourse: A verbal-visual model of communication." *Poetics* 25: 293–309.

Yus, Francisco. 1998f. "Irony: Context accessibility and processing effort." *Pragmalingüística* 5-6: 391–411.

Yus, Francisco. 1998g. *La Preeminencia de la Voz.* Alicante (Spain): University of Alicante, Servicio de Publicaciones.

Yus, Francisco. 1999a. "Towards a pragmatic taxonomy of misunderstandings." *Revista Canaria de Estudios Ingleses* 38: 217–239.
Yus, Francisco. 1999b. "Misunderstandings and explicit/implicit communication." *Pragmatics* 9 (4): 487–517.
Yus, Francisco. 2000a. "Relevance theory online bibliographic service." University of Alicante (Spain), Department of English Studies. [http://www.ua.es/personal/francisco.yus/rt.html] (thematic sections). [http://www.ua.es/personal/francisco.yus/rt2.html] (alphabetical order by author).
Yus, Francisco. 2000b. "On reaching the intended ironic interpretation." *International Journal of Communication* 10 (1–2): 27–78.
Yus, Francisco. 2000c. "Literal/nonliteral and the processing of verbal irony." *Pragmalingüística* 8–9: 349–374.
Yus, Francisco. 2001a. *Ciberpragmática. El Uso del Lenguaje en Internet*. Barcelona: Ariel.
Yus, Francisco. 2001b. "Ciberpragmática: Entre la compensación y el desconcierto." Paper presented at *Segundo Congreso de la Lengua Española*, Valladolid (Spain).
Yus, Francisco. 2001c. *El Discurso Femenino en el Cómic Alternativo Inglés*. Alicante: University of Alicante, Servicio de Publicaciones.
Yus, Francisco. 2002a. "Discourse and identity." In *International Encyclopedia of the Social and Behavioral Sciences*, N. J. Smelser and P. B. Baltes (eds), 3728–3732. Amsterdam: Elsevier.
Yus, Francisco. 2002b. "Review of A. Pilkington's *Poetic Effects*." *Journal of Pragmatics* 34: 619–628.
Yus, Francisco. 2002c. "Stand-up comedy and cultural spread: The case of sex roles." *Babel A.F.I.A.L.*, Special issue on humour, 245–292.
Yus, Francisco. 2003a. *Cooperación y Relevancia. Dos Aproximaciones Pragmáticas a la Interpretación* (2nd revised edition). Alicante (Spain): University of Alicante.
Yus, Francisco. 2003b. "El *chat* como doble filtro comunicativo." *Revista de Investigación Lingüística* V (2): 41–169.
Yus, Francisco. 2003c. "The linguistic-cognitive essence of virtual community." Paper presented at the *First International Conference on Internet and Language*, Castelló (Spain): Jaume I University.
Yus, Francisco. 2003d. "La lengua de los medios en la red y su interpretación." In *Manual de redacción ciberperiodística*, Javier Díaz Noci and Ramón Salaverría (eds), 309–352. Barcelona: Ariel.
Yus, Francisco. 2003e. "Humor and the search for relevance." *Journal of Pragmatics* 35 (9): 1295–1331.
Yus, Francisco. 2003f. "El papel del contexto en la comunicación por Internet." Paper delivered at *Jornades de Foment de la Investigació* 7, Castelló (Spain), Jaume I University.
Yus, Francisco. 2004. "Pragmatics of humorous strategies in *El club de la comedia*." In *Current Trends in the Pragmatics of Spanish*, Rosina Márquez-Reiter and María Elena Placencia (eds), 320–344. Amsterdam/Philadelphia: John Benjamins.
Yus, Francisco. 2005a. "Attitudes and emotions through written text: The case of textual deformation in Internet chat rooms." *Pragmalingüística* 13: 147–174.
Yus, Francisco. 2005b. "The linguistic-cognitive essence of virtual community." *Ibérica* 9: 79–102.
Yus, Francisco. 2005c. "In search of cognitively relevant Internet banners." *Image [&] Narrative* 11.

Yus, Francisco. 2005d. "Dave Allen's stand-up monologues: An epidemiological approach." In *Thistles. A homage to Brian Hughes*. Volume 2: *Essays in Memoriam*, José Mateo and Francisco Yus (eds), 317–344. Alicante: University of Alicante, Department of English Studies.

Yus, Francisco. 2006a. "Relevance theory." In *Encyclopedia of Language and Linguistics* (2nd edition), volume 10, Keith Brown (ed.), 512–519. Amsterdam: Elsevier.

Yus, Francisco. 2006b. "Saturación contextual en la comprensión de la ironía." Paper presented at *VIII Jornadas de Estudios de Lingüística. Análisis del Discurso*, Alicante (Spain), University of Alicante.

Yus, Francisco. 2007a. "Enunciados irónicos: Contexto y relevancia." Paper presented at the University of La Laguna, Tenerife (Spain), March.

Yus, Francisco. 2007b. *Virtualidades Reales. Nuevas Formas de Comunidad en la Era de Internet.* Alicante: University of Alicante, Servicio de Publicaciones.

Yus, Francisco. 2007c. "Estrategias interpretativas en los ciber-medios." Paper presented at *VIII Ciclo de Conferencias "Adquisición y uso de segundas lenguas" (Nuevas tecnologías en la enseñanza / aprendizaje de segundas lenguas)*, Seville (Spain), Pablo de Olavide University.

Yus, Francisco. 2007d. "La comunicación en las lenguas de especialidad." In *Las Lenguas Profesionales y Académicas*. Enrique Alcaraz, José Mateo and Francisco Yus (eds), 81–92. Barcelona: Ariel.

Yus, Francisco. 2008a. "Estrategias lingüísticas de los hablantes en la Red." Paper presented at the Seminar *Lenguas y Culturas Hispánicas en Internet*, Madrid, Residencia de Estudiantes.

Yus, Francisco. 2008b. "Alterations of relevance in cyber-media." *Universitas Psychologica* 7(3): 623–636.

Yus, Francisco. 2008c. "A relevance-theoretic classification of jokes." *Lodz Papers in Pragmatics* 4 (1): 131–157.

Yus, Francisco. 2008d. "Towards a pragmatics of weblogs." In *Pragmática, Discurso y Sociedad* (*Quaderns de Filologia XII*), Patricia Bou Franch, A. Emma Sopeña and Antonio Briz (eds), 15–33. Valencia: University of Valencia.

Yus, Francisco. 2008e. "Weblogs: Web pages in search of a genre?" In *The Texture of Internet*, Santiago Posteguillo, María J. Esteve and María L. Gea-Valor (eds), 118–142. Newcastle: Cambridge Scholars Press.

Yus, Francisco. 2008f. "Inferring from comics: A multi-stage account." In *El discurs del Còmic* (*Quaderns de Filologia, Estudis de Comunicació*, vol. III), Pelegrí Sancho Cremades, Carmen Gregori Signes and Santiago Renard (eds), 223–249. Valencia (Spain): University of Valencia.

Yus, Francisco. 2009a. "Saturación contextual en la comprensión de la ironía." In *Dime Cómo Ironizas y te Diré Quién Eres: Una aproximación pragmática a la ironía*, Leonor Ruiz Gurillo and Xosé Padilla García (eds), 309–331. Frankfurt: Peter Lang.

Yus, Francisco. 2009b. "La interacción comunicativa por Internet." Paper presented at Polytechnic University of Valencia (Spain), May.

Yus, Francisco. 2009c. "Estrategias y efectos en el discurso humorístico." Paper presented at *II Jornadas de Lingüística y Cognición*, Cádiz (Spain), University of Cádiz, March.

Yus, Francisco. 2009d. "Visual metaphor versus verbal metaphor: A unified account." In *Multimodal Metaphor*, Charles Forceville and Eduardo Uriós-Aparisi (eds), 147–172. Berlin: Mouton de Gruyter.

Yus, Francisco. 2009e. "La interacción comunicativa por Internet," Paper presented at Polytechnic University of Valencia, Spain, May.

Yus, Francisco. 2010a. "Relevance theory." In *The Oxford Handbook of Linguistic Analysis*, Bernd Heine and Heiko Narrog (eds), 679–701. Oxford: Oxford University Press.

Yus, Francisco. 2010b. *Ciberpragmática 2.0. Nuevos Usos del Lenguaje en Internet*. Barcelona: Ariel.

Yus, Francisco. 2010c. "Users' relevance on the web." In *Handbook of Research on Discourse Behavior and Digital Communication: Language Structures and Social Interaction*, Rotimi Taiwo (ed.), 411–423. Hershey, USA: IGI Global.

Yus, Francisco. 2010d. "La relevancia como motor de comunicación en Internet." Paper presented at *Semiótica 3.0. Conocimiento, Contemporaneidad, Comunicación*, Bogotá (Colombia), November.

Yus, Francisco. 2010e. "Relevance, humour and translation." Paper presented at the *5th Conference Interpreting for Relevance: Discourse and Translation*, Kazimierz Dolny (Poland), June.

Yus, Francisco. 2011a. "Relevance equations of effective Internet communication." In *Interdisciplinarity and Languages. Current Issues in Research, Teaching, Professional Applications and ICT*, Barry Pennock and Tina Suau (eds), 65–85. Berlin: Peter Lang.

Yus, Francisco. 2011b. "A pragmatic analysis of jokes." Paper presented at the Seminar *Meaning in Interaction*, Seville (Spain): University of Seville, May.

Yus, Francisco. forthcoming a. "Cyberpragmatics." In *The Encyclopedia of Applied Linguistics*, Carol A. Chapelle (ed.). Oxford: Wiley-Blackwell.

Yus, Francisco. forthcoming b. "Strategies and effects in humorous discourse: The case of jokes." *Pragmalingüística*.

Žegarac, Vladimir. 2007. "A cognitive-pragmatic perspective on communication and culture." In *Handbooks of Applied Linguistics. Vol. 7: Intercultural Communication*, Helga Kotthoff and Helen Spencer-Oatey (eds), 31–53. Berlin: Mouton de Gruyter.

Zhao, Shanyang, Grasmuck, Sherri and Martin, Jason. 2008. "Identity construction on Facebook: Digital empowerment in anchored relationships." *Computers in Human Behavior* 24: 1816–1836.

Zhao, Dejin and Rosson, Mary B. 2009. "How and why people Twitter: The role that micro-blogging plays in informal communication at work." In *Proceedings of the ACM 2009 International Conference on Supporting Group Work*. Sanibel Island, Florida (USA).

Zimmerman, Eric and Bar-Ilan, Judit. 2009. "PIM @ academia: How e-mail is used by scholars." *Online Information Review* 33 (1): 22–42.

Zitzen, Michaela and Stein, Dieter. 2004. "Chat and conversation: A case of transmedial stability?" *Linguistics* 42 (5): 983–1021.

Ziv, Oren. 1996. "Writing to work: How using e-mail can reflect technological and organizational change." In *Computer-Mediated Communication. Linguistic, Social and Cross-Cultural Perspectives*, Susan Herring (ed.), 243–264. Amsterdam/Philadelphia: John Benjamins.

Name index

L

M

Subject index